GENETICS - RESEARCH AND ISSUES

EPIGENETICS

MECHANISMS AND CLINICAL PERSPECTIVES

GENETICS - RESEARCH AND ISSUES

Additional books in this series can be found on Nova's website under the Series tab.

Additional e-books in this series can be found on Nova's website under the e-book tab.

GENETICS - RESEARCH AND ISSUES

EPIGENETICS

MECHANISMS AND CLINICAL PERSPECTIVES

YILDIZ DINCER
EDITOR

NOVA BIOMEDICAL
New York

Library of Congress Cataloging-in-Publication Data

Names: Dincer, Yildiz, editor.
Title: Epigenetics : mechanisms and clinical perspectives / editor, Yildiz Dincer, Ph.D. (Istanbul University, Cerrahpasa Medical Faculty, Department of Biochemistry, Turkey).
Description: Hauppauge, New York : Nova Science Publisher's, Inc., [2016] | Series: Genetics - research and issues | Includes index.
Identifiers: LCCN 2015050167 (print) | LCCN 2016000949 (ebook) | ISBN 9781634844901 (hardcover) | ISBN 9781634845199 (ebook) | ISBN 9781634845199 ()
Subjects: LCSH: Epigenetics. | Genetic regulation.
Classification: LCC QH450 .E67 2016 (print) | LCC QH450 (ebook) | DDC 572.8/65--dc23
LC record available at http://lccn.loc.gov/2015050167

Published by Nova Science Publishers, Inc. † New York

CONTENTS

PREFACE

Epigenetics refers to changes in gene expression that occur without alteration in DNA base sequences. Epigenetic modifications activate or deactivate parts of the genome at strategic times and in specific locations. Epigenetic regulations are central to many cellular processes such as imprinting, X chromosome inactivation, DNA damage response, cellular reprogramming and senescence. Although all cells in one individual have the same DNA sequence, epigenetic regulation occurs at the specific gene loci in specific cells to yield specific cellular phenotypes. Epigenetic dysregulations have been linked to pathogenesis of various diseases such as cancer, diabetes, obesity, infertility, cardiovascular diseases, neurodegenerative disorders and chronic inflammatory diseases. Epigenetic modifications are acquired throughout life and influenced by environmental factors. They are reversible and inheritable. The interaction between diet and the epigenome is well documented. Manipulation of epigenetic mechanisms by nutrients may affect the vulnerability of cells to degeneration that is responsible for various diseases. The underlying pathological mechanisms are established early in life in diseases such as cancer and Alzheimer's disease. In this context, a diet containing nutritional components that are essential for proper epigenetic regulations may be protective against these diseases. On the other hand, epigenetic modifications are potential targets for novel therapeutic strategies due to their reversibility. A good understanding of epigenetic dysregulations underlying various pathologies may provide important information to prevent the onset and progression of diseases. This book is intended to provide an overview for the influence of early life nutrition on the epigenome and the role of epigenetic dysregulations in the pathogenesis of Alzheimer's disease, cancer, cardiovascular disorders, inflammation, obesity and infertility. The author hopes it is useful for students and clinicians who are curious about this field and/or intend to find careers in related subjects.

The author would like to thank Prof. İlhan Onaran for his interest, advice and support.

Yıldız Dincer

Istanbul University

Cerrahpasa Medical Faculty

Department of Biochemistry

Email: yldz.dincer@gmail.com

October 2015, Istanbul

In: Epigenetics: Mechanisms and Clinical Perspectives
Editor: Yildiz Dincer
ISBN: 978-1-63484-490-1

Chapter 1

EPIGENETIC MODIFICATIONS AND DEVELOPMENTAL ORIGIN OF HEALTH AND DISEASES (DOHAD)

Jia Zheng and Xinhua Xiao*

Department of Endocrinology, Key Laboratory of Endocrinology, Ministry of Health, Peking Union Medical College Hospital, Diabetes Research Center of Chinese Academy of Medical Sciences and Peking Union Medical College, Beijing, China
Department of Endocrinology, Peking Union Medical College Hospital, Chinese Academy of Medical Sciences and Peking Union Medical College, Beijing, China

ABSTRACT

It is generally believed that genotype and adult lifestyle elements are primary risks of some metabolic diseases such as insulin resistance, obesity and diabetes mellitus in later life. However, increasing evidence demonstrates that early life malnutrition during the period of gestation and/or lactation may increase our susceptibility to such metabolic diseases in later life. The underlying mechanism is still not very clear. Recently, epigenetics is hypothesized to be the important molecular basis of the imbalanced early life nutrition and glucose metabolism disorders, which is known as "Developmental Origin of Health and Diseases" (DOHaD). Currently, there are substantial epidemiological studies and experimental animal models that have demonstrated nutritional disturbances during the critical periods of early life development can significantly impact the predisposition to developing some metabolic diseases in later life. The fundamental mechanism is that early developmental nutrition can regulate epigenetic modifications of some genes associated with development and metabolism. DNA methylation is the first discovered and one important epigenetic modification. MicroRNAs are recognized as an important epigenetic modification and they are a major class of small non-coding RNAs (about 20-22 nucleotides) which can mediate posttranscriptional regulation of target genes with cell differentiation and apoptosis. Recent studies suggest that DNA methylation and microRNAs maybe the crucial modulators of fetal epigenetic programming in nutrition and metabolic disorders. This chapter will focus on how early life nutrition can alter the epigenome, produce different phenotypes and alter disease susceptibilities, especially for impaired glucose metabolism.

* Correspondence author: Xinhua Xiao. Tel: +86-10-69155073, E-mail: xiaoxinhua@ medmail.com.cn.

Keywords: epigenetics, DNA methylation, microRNAs, early life nutrition, glucose metabolism

Abbreviations

AUC	Area Under the ROC Curve;
CVD	Cardiovascular diseases;
DMR	Differentially methylated region;
DOHaD	Developmental Origin of Health and Diseases;
GR	Glucocorticoid receptor;
GDM	Gestational diabetes mellitus;
F1	First-generation;
FGR	Fetal growth restriction;
HMGCR	3-hydroxy-3-methylglutaryl coenzyme A reductase;
HNF4α	Hepatocyte nuclear factor 4 receptor α;
IDF	International Diabetes Federation;
IGF2	Insulin like growth factor 2;
IGT	Impaired glucose tolerance;
Il13ra2	Interleukin 13 receptor alpha 2;
IUGR	Intrauterine growth restriction;
LncRNAs	Long non-coding RNAs;
LP	Low protein;
MeCP2	Methyl CpG-binding protein 2;
MOR	μ-opioid receptor;
PPARα	Peroxisome proliferator-activated receptor alpha;
RXRA	Retinoid X receptor-α;
SAM	S-adenosylmethionine;
T2DM	Type 2 diabetes mellitus.

1. Introduction

Nowadays, the prevalence of diabetes mellitus is increasing very rapidly worldwide, which is now considered a pandemic non-communicable disease. According to the Diabetes Atlas published by the International Diabetes Federation (IDF) on World Diabetes Day in 2017, 387 million people are suffering from diabetes and the number will rise to 592 million by 2035, which implies one twelfth of the world's population will endure diabetes. In particular, more than 21 million infants acquired diabetes from their mother during pregnancy in 2013. It is estimated that diabetes caused 4.9 million deaths in 2014, that is to say, a person died from the disease every seven seconds [1]. In summary, diabetes is becoming a more and more severe problem for our society.

Unfortunately, the pathogenesis of diabetes has not been clearly understood yet. Although it is generally believed that genes together with adult lifestyle factors are major risks of diabetes. There is substantial evidence that the prenatal and early postnatal nutrition play a key role in determining our susceptibility to diabetes in later life [2]. Little is known about the molecular mechanisms underlying the interaction between maternal diet and aging. However, the hypothesis that epigenetic mechanisms may link such imbalanced nutrition with altered disease risk has been widely accepted in recent years. Epigenetics is one of the major mechanisms explaining the theory of "Developmental Origin of Health and Diseases" (DOHaD), which focuses on the association between perinatal nutrition and late-onset disease, such as obesity, insulin resistance, impaired glucose tolerance (IGT) and type 2 diabetes mellitus (T2DM) [3]. Therefore, epigenetics is likely to be an important molecular basis of malnutrition during early life and glucose metabolism disorders in later life.

2. What Is Epigenetics?

Traditionally, epigenetics has been used to explain the phenotypic events which cannot be explained by genetic mechanisms. The term "epigenetics" was first proposed by Waddingtong in 1942 and defined as a branch of biology that studied the causal interactions between genes and their products [4]. In 2006, the definition of epigenetics was updated and considered as a mechanism that could affect gene expression without altering the nucleotide sequence. Most importantly, it could be inherited between generations steadily by mitosis and meiosis through cell differentiation [5]. Epigenetic conditions can illustrate the reason why an organism produces many different cell types during its development, despite the fact that most of the cells in a multicellular organism share the same genetic information. Epigenetic modifications can regulate gene expression reversibly. The three major epigenetic processes are DNA methylation, histone modifications and non-coding RNAs. DNA methylation was the first recognized and the most well-characterized epigenetic modification in the 1970s [6]. Histone modification can influence gene expression by altering chromatin structure [7]. Non-coding RNAs, such as microRNA and long non-coding RNA, are also included in epigenetic modifications [8]. Epigenetic processes play a critical role in normal development and differentiation in mammals [9]. Recently, epigenetics was defined as "molecular factors and processes around DNA were mitotically stable and regulated genome activity which were independent of DNA sequence" by Skinner et al. in 2010 [10]. Although the definition of epigenetics has been updated constantly, the three core features are always reserved, namely: (1)without any alterations in DNA sequence (2) heritability (3) plasticity and reversibility. In this chapter, we will focus on DNA methylation and microRNAs.

2.1. DNA Methylation

DNA methylation is the first discovered epigenetic modification and it is also one of the best-studied epigenetic modifications in the context of altered environment. It is a biochemical process involving the covalent addition of a methyl group at the 5' position of cytosine in DNA. This normally occurs on a cytosine followed by a guanine, known as a CpG dinucleotide (the p denotes the intervening phosphate group). It is catalyzed by DNA methyltransferases, and S-adenosylmethionine (SAM) is the methyl donor [11]. DNA methylation typically occurs in CpG dinucleotide context. CpG dinucleotides are not randomly distributed throughout the genome but they are often grouped in clusters at the 5' regulatory regions of many genes, called CpG islands. Most CpG dinucleotides are methylated, but those located in CpG islands are usually unmodified [12]. However, hypermethylation of these CpG islands has the specific effect on reducing gene expression and repressing transcription, while hypomethylation of CpG islands is related to transcriptional activation [13]. DNA methylation is a common modification in mammalian genomes. It constitutes a stable epigenetic symbol that can stably alter the expression of genes and transmit through DNA replication as cells divide and differentiate from embryonic stem cells into specific tissues [11]. DNA methylation is essential for normal development and is associated with a number of key processes including genomic imprinting, X-chromosome inactivation [14], suppression of repetitive elements and carcinogenesis [15].

2.2. MicroRNAs

Non-coding RNAs are the latest discovered epigenetic modifications including microRNAs and long non-coding RNAs (LncRNAs) [8]. MicroRNAs (miRNAs) are gradually recognized as an important epigenetic modification in recent years. They are a major class of small non-coding RNAs with about 20-22 nucleotides which can mediate posttranscriptional regulation of gene expressions. More precisely, miRNAs can bind to the 3' untranslated regions (3'-UTR) of target genes specifically, resulting in either translation inhibition or mRNA degradation [16]. They are complicated and multi-functional that up to 60% of our genome is regulated by miRNAs and each miRNA can probably regulate several hundred gene expressions. The modulations of miRNAs can facilitate some key developmental processes such as cell proliferation, cell line differentiation and programmed cell apoptosis [8]. Therefore, we will focus on the regulations of microRNAs between fetal epigenetic programming in nutrition and glucose metabolism.

3. Evidence from Human Cohorts

3.1. Fetal Programming Hypothesis

In 1977, the association between the quality of early life environment and future risk of disease in later life was first proposed by Forsdahl, who discovered that infant mortality rate was positively associated with an increased risk of cardiovascular diseases (CVD) in middle age [17]. Subsequently, Barker and his colleagues found an inverse relationship between birth weight and increased CVD mortality in 1989 [18]. The effects of early life nutrition on the increased risk of metabolic diseases were the most clearly shown in studies of the Dutch famine during the winter of 1944. These studies showed that the individuals whose mothers were exposed to famine in the late stage of gestation had lower birth weight and an increased risk of obesity, CVD, insulin resistance and hypertension in later life compared with unexposed individuals [19]. What is more, apart from nutrition deficiency in early life, overnutrition is also associated with an significantly increased susceptibility to metabolic disease such as obesity, hypertension, atherosclerosis, insulin resistance and diabetes mellitus

[20]. It can involve some organs related to glucose metabolism, including liver, pancreas, adipose tissue and skeletal muscle through different mechanisms. The nutrition during early life has long term phenotypic effects. This idea was first proposed by professor Barker and Hales in the 1990s, known as "fetal programming hypothesis" [21]. After more than 20-year long term research, it has now been accepted worldwide.

3.2. DNA Methylation and Fetal Programming Hypothesis

Epigenetic events are crucial for early development. However, it can be influenced by some environmental factors, potentially programming the genome for later adverse health outcomes. Gestation is the critical time window for maternal nutrition to affect the offspring. Early life nutrition can induce persistent DNA methylation. More specifically, both under-nutrition and over-nutrition will bring about epigenetic modification during the early life, including the embryonic development and neonatal period. A number of clinical studies have shown that people born during the famine in the Netherlands still exhibited lower methylation level of insulin like growth factor 2 (*IGF2*) after 60 yeas, compared with those who had no exposure to prenatal famine [22]. It indicates that epigenetic modification of some special genes affected by periconceptional maternal nutrition restriction can be kept for decades. Another study showed that the children whose mother used folic acid had a 4.5% higher methylation in the *IGF2* gene differentially methylated region (DMR) than those whose mother did not take folic acid. It also indicates an inverse independent association between *IGF2* DMR and birth weight, which means that periconceptional folic acid use is associated with epigenetic changes in *IGF2* in the child that may affect intrauterine programming of growth and development with consequences for health and disease throughout life [23]. A recent cohort study demonstrated that retinoid X receptor-α (*RXRA*) methylation had independent association with sex-adjusted childhood fat mass and it could explain more than 25% of the variance in childhood adiposity. Higher methylation of *RXRA* was associated with lower maternal carbohydrate intake in early pregnancy. These researches suggest that prenatal development is a substantial risk of metabolic disease, such as diabetes, obesity and metabolic syndrome [24].

3.3. MicroRNAs and Fetal Programming Hypothesis

It is well known that nutrition during early life development can impact the health of the adult permanently. Fetal growth restriction (FGR), also known as intrauterine growth restriction (IUGR), is a relatively common, pleiotropic complication of pregnancy. It is not only associated with significantly increased perinatal morbidity and mortality, and is also a major determinant of cardiovascular disease and glucose intolerance in adult life. Some clinical studies have indicated that some programmed changes in miRNAs expression may regulate the pathophysiology of intrauterine growth. The placenta is a crucial organ which can regulate the exchange of fetomaternal nutrients for the developing fetus, thereby modulating intrauterine development. More specifically, recent studies have showed that epigenetic modifications of placenta may play an important role on the health of the offspring and maternal nutrition. Some studies showed that several miRNAs in placentas are involved in pre-eclampsia, preterm labor, FGR and newborn neurobehavior, such as miRNA-15b, miRNA-181a, miRNA-210, miR-377, miR-483-5p and miR-493 [25, 26]. Huang et al. indicated that the levels of miRNA-424 were significantly increased in placentae from women with FGR [27]. Tang et al. also observed that aberrant high expression level of miR-141 might play an important role in the pathogenesis of IUGR which can suppress the expressions of target genes, such as, E2F transcription factor 3 and pleiomorphic adenoma gene 1. Furthermore, they also discovered that miR-141 could serve as a potential biomarker to distinguish between FGR and normal controls with the Area Under the ROC Curve (AUC) of 0.839 and the 88.5% sensitivity and 71.7% specificity [28]. Ferland-McCollough et al. showed that miR-483-3p is upregulated in adipose tissue rather than placentas from low birth weight adult humans [29]. In addition to IUGR, miRNAs programming were also existed in fetal macrosomia. A recent research indicated that aberrant expression of miRNA-21 in placenta is associated with macrosomia which may similarly have long-term effects on health and disease in adult life [30, 31]. Thus, all these evidence support the modifications of miRNAs in the fetal programming.

4. Evidence from Animal Experiments

In addition to the evidence from studies of human cohorts, the association between dietary changes during specific windows of development and epigenomic modifications in later life has also been reported in several animal models. In this chapter, we will describe some well-known epigenetic changes that have been identified from animal models of obesity, insulin resistance and diabetes.

4.1. DNA Methylation and Animal Experiments

4.1.1. Maternal Nutrition

Protein restriction is frequently used as a model for maternal malnutrition. Low protein (LP) diet is associated with impaired fetal growth, the development of obesity, insulin resistance and diabetes in the offspring [32]. Many epigenetic modifications have been reported in offspring exposed to a maternal LP diet. One of the first studies showing the link between nutritional imbalances during intrauterine development and epigenetic modifications was that feeding a LP diet (9% casein vs. 18% casein) to rats during pregnancy resulted in global DNA hypermethylation in the liver of the fetuses [33]. Recent studies confirmed that maternal LP feeding during pregnancy might also result in locus-specific changes in DNA methylation. For example, feeding pregnant rats a protein-restricted diet induced hypomethylation of the glucocorticoid receptor (*GR*) and Peroxisome proliferator-activated receptor alpha (*PPARα*) promoters in the livers of juvenile and adult offspring [34]. Also in Wistar rats, *PPARα* gene methylation was 20.6% lower with expression 10.5-fold higher and *GR* gene methylation was 22.8% lower with expression 200% higher in pups of dams that were fed protein restriction diet throughout pregnancy [35]. Hepatocyte nuclear factor 4 receptor α (*HNF4α*) gene has been implicated in the etiology of T2DM. Sandovici observed that maternal LP diet during pregnancy and lactation could lead to progressive epigenetic silencing of the entire *HNF4α* locus in rat pancreatic islets, which weakened the promoter-enhancer interaction and resulted in a permanent reduction in *HNF4α* expression [36]. Maternal protein-undernutrition during pregnancy and lactation in Balb/c mice affected the balance between food intake and energy expenditure in adults. Moreover, this nutritional stress resulted in the removal of methyls at CpGs located in the promoter of leptin, which was associated with DNA hypomethylation of the leptin promoter in adipose tissue [37]. Interestingly, the DNA methylation of LP diet is not restricted to rodents. Male offspring of maternal LP diet of Meishan sows during pregnancy and lactation demonstrated higher *GR* binding to the mitochondrial DNA promoter, which was accompanied by lower cytosine methylation and hydroxymethylation on mitochondrial DNA promoter [38]. Similarly, piglets with maternal LP diet during pregnancy and lactation showed significantly lower body weight and liver weight at weaning. Hepatic activation of 3-hydroxy-3-methylglutaryl coenzyme A reductase (*HMGCR*) gene transcription in LP piglets was associated with promoter hypomethylation [39].

Nowadays, both the incidence of obesity during pregnancy and gestational diabetes mellitus (GDM) are increasing along with the improved living conditions. Therefore, scientists are increasingly concerned about overnutrition on human health. There is increasing evidence that caloric excess in maternal diet is associated with an altered metabolic phenotype. Recently, researches showed that maternal high fat diet might modify DNA methylation and gene expression in the offspring. Maternal high fat diet during gestation was associated with global and gene-specific promoter DNA hypomethylation, like the dopamine reuptake transporter, the μ-opioid receptor *(MOR)* and preproenkephalin in the brain from the first-generation (F1) offspring of C57BL/6J x DBA/2J hybrids. These changes might affect feeding behavior, which can increase obesity and obesity-associated risk for metabolic syndrome [40]. During the postnatal period of the same model, the offspring of maternal high fat diet showed increased binding of Methyl CpG-binding protein 2 (*MeCP2*) to the *MOR* promoter in reward-related regions of the brain, which can repress the transcription of *MOR* gene [41]. A recent study showed that the offspring whose mothers fed on high fat diet during late gestation had increased weight accrual and food intake, exhibiting insulin resistance and hyperlipidemia. Furthermore, the offspring mice emerged with increased methylation of adiponectin and leptin receptor, and decreased methylation of leptin genes [42].

4.1.2. Paternal Nutrition

Apart from the detrimental impacts of maternal malnutrition on abnormal glucose metabolism in offspring, recent studies showed that parental malnourished exposures could also affect the phenotype of the offspring. Paternal lifestyle and particular nutrition factors can affect spermatogenesis at the level of germ and sertoli cells [43] and the composition of seminal fluid [44]. Similarly, alterations in paternal diet are also associated with altered DNA methylation in the offspring. As shown by Ng et al. a paternal high fat diet consumption of SD rats exposure could induce intergenerational transmission of impaired glucose tolerance and insulin homeostasis in their female offspring. The interleukin 13 receptor alpha 2 (*Il13ra2*) promoter was hypomethylated in female offspring after high fat feeding of their fathers [45]. Offspring of fathers fed on a low protein diet in C57BL/6J mice exhibited elevated hepatic expression of many genes involved in glucose metabolism and lipid biosynthesis. Epigenomic bisulfite sequence analysis on DNA from the liver of offspring revealed numerous modest (about 20%) changes in cytosine methylation depending on paternal diet, including a substantial increase in methylation at an intergenic CpG island 50 kb upstream of the *PPARα* gene [46]. A recent study showed that paternal prediabetes increased the susceptibility of the offspring to diabetes through gametic epigenetic alterations [47]. All these results indicate that parental diet can also induce impaired glucose metabolism in offspring through the epigenetic modification mechanisms, especially, DNA methylation.

4.2. MicroRNAs and Animal Experiments

4.2.1. Maternal Nutrition

Not only the evidence can be observed in humans, the association between maternal nutrition and microRNAs modulations has also been reported in several animal experiments. One recent study suggested that maternal consumption of a high fat diet can induce glucose intolerant and insulin intolerant, obesity and the abnormal lipid metabolism in the early life of the mice offspring by downregulating the expression of miRNA-122 and upregulating miRNA-370, and then they can modulate the expression of target hepatic β-oxidation-related genes, hence, contributing to metabolic disturbances in adult life [48]. Maternal low protein diet which is an animal model of intrauterine growth restriction also can lead to abnormal glucose metabolism by modulating the expression of some miRNAs, for example, one recent research showed that gestational protein restriction in rats could lead to pancreatic failure in offspring which may result from the transgenerational transmission of miRNA-375 misexpression in β-cells [49]. Lie et al. suggested that maternal undernutrition around the time of conception in sheep could induce changes in the expression of about 22 miRNAs, which may have an influence on altering the abundance of the critical insulin-signaling molecules in skeletal muscle and in the association between maternal undernutrition in the periconceptional period and insulin resistance in adult life [50]. The same animal model also showed that maternal undernutrition around conception can impact the insulin signaling and gluconeogenic factors which can be explained by alterations in the expression of a number of specific candidate microRNAs in the liver of fetal sheep [51].

4.2.2. Paternal Nutrition

Not only maternal malnutrition can have a detrimental effect on glucose metabolism in the offspring, animal studies have also shown the potential importance of paternal diets in future metabolic diseases risk. Paternal lifestyle, particularly, nutrition condition can impact spermatogenesis at the level of germ and sertoli cells [43] and the composition of seminal fluid [44]. Similarly, miRNAs are also an important modulators in paternal diet and glucose metabolism in the offspring. One study indicated that offspring of whose father were fed a low protein diet from weaning until sexual maturity, showed decreased levels of cholesterol esters and elevated hepatic expression of many genes involved in lipid and cholesterol biosynthesis. Furthermore, it was exhibited that a number of miRNAs were also altered in the liver tissues of the offspring persistently by microarray hybridization, namely, miRNA-98, miRNA -21, let-7 and miRNA-199 were upregulated and miRNA-210 was downregulated, many of which are associated with cellular proliferation and target lipid biosynthesis genes [46]. Therefore, all these studies suggested that microRNAs could be the crucial modulators of fetal epigenetic programming in nutrition and glucose metabolism.

To conclude, the studies of human cohorts and animal models demonstrate that developmental programming of adult disease occurs at each nutrition spectrum during the early life, due to maternal caloric deprivation, maternal nutritional excess and paternal malnutrion. Although the specific mechanisms leading to aberrant glucose metabolism in each situation are different, the current evidence supports that the epigenetic modification, especially, DNA methylation and microRNAs maybe an important molecular link of them, and this epigenome can be inherited from one generation to another.

5. Developmental Plasticity and Reversibility

Generally speaking, plasticity is viewed as the ability that one genotype can produce different phenotypes in response to different environments. The time window of maximal plasticity appears to be during development, which is termed as "Developmental plasticity". It has evolved to provide the best chances of survival and reproductive success to the organism. Further studies illustrated that developmental plasticity could be regulated by epigenetic modifications. It was traditionally considered that epigenetic modifications were static in the control of gene expression. However, it is now being altered by the idea that these marks are dynamic induced by some factors. Below are examples supporting the idea, for example, intervention with drugs in early life can bring long-term benefits to health. Offspring born to dams with calorie restriction during gestation and lactation are more liable to develop obesity and insulin resistance [52]. However, when leptin is given to neonatal mice born to dams suffered from calorie restriction, the mice show a phenotype close to healthy controls and do not develop obesity in later life even fed with an energy dense high fat diet [53]. It suggests that developmental metabolic programming is potentially reversible by an intervention late in the phase of developmental plasticity. The long-lasting effects of early leptin intervention has been shown to be related with consistent up-regulation of some critical glucose metabolic genes such as transcription factor, *PPARα*, whereas sustained up-regulation of *PPARα* is regulated by reduced DNA methylation in the promoter of *PPARα* [54]. Stress responses in the adult rat are programmed early in life by maternal care, and is associated with altered transcription factor binding to the hippocampal *GR* promoter [55]. Interestingly, central infusion of these adult offspring with L-methionine, a precursor to S-adenosyl-methionine that serves as the donor of methyl groups for DNA methylation, with altered DNA methylation due to maternal behavior programming during gestation can be reversed, together with altered nerve growth factor-inducible protein-A binding to the *GR* promoter and hypothalamic-pituitary-adrenal and behavioral responses to stress [56, 57]. These results demonstrate that despite the inherent stability of the epigenomic marks established early in life through behavioral programming, they are potentially reversible in the adult brain. Therefore, in view of the reversibility of epigenetic mechanisms, intervention with drugs during early life may ameliorate the glucose metabolism disorders in later life if the early disrupting environment has resulted in modified epigenetics, which can generates long-lasting effects.

CONCLUSION AND FURTHER PERSPECTIVE

Today, apart from genotype and adult lifestyle, early life abnormal nutrition including both undernutrition and overnutrition are closely related to aberrant glucose metabolism in adult such as impaired glucose tolerance, gestational diabetes mellitus and type 2 diabetes. It is suggested that DNA methylation and microRNAs may be the crucial epigenetic mechanisms accounting for the relationship between the early life nutrition and glucose metabolism in later life. More importantly, DNA methylation and microRNAs can be utilized as novel biomarkers and targets for the diagnosis, prevention and intervention of diabetes in the near future.

REFERENCES

[1] *IDF Diabetes Atlas, 6th edn. Available online*: http://www.idf.org/diabetesatlas.

[2] Slatkin, M. Epigenetic inheritance and the missing heritability problem. *Genetics.* 2009, 182, 845-850.

[3] Gluckman, P.D.; Hanson, M.A.; Buklijas, T.; Low, F.M.; Beedle, A.S. Epigenetic mechanisms that underpin metabolic and cardiovascular diseases. Nature reviews. *Endocrinology.* 2009, 5, 401-408.

[4] Waddington, C.H. The epigenotype. 1942. *International journal of epidemiology.* 2012, 41, 10-13.

[5] Holliday, R. *Epigenetics A Historical Overview. Epigenetics-Us.* 2006, 1, 76-80.

[6] Holliday, R.; Pugh, J.E. DNA Modification Mechanisms and Gene Activity during Development. *Science.* 1975, 187, 226-232.

[7] Turner, B.M. Histone acetylation as an epigenetic determinant of long-term transcriptional competence. *Cell Mol Life Sci.* 1998, 54, 21-31.

[8] Aguilera, O.; Fernandez, A.F.; Munoz, A.; Fraga, M.F. Epigenetics and environment: a complex relationship. *J Appl Physiol.* 2010, 109, 243-251.

[9] Fraga, M.F. Genetic and epigenetic regulation of aging. *Curr Opin Immunol.* 2009, 21, 446-453.

[10] Skinner, M.K.; Manikkam, M.; Guerrero-Bosagna, C. Epigenetic transgenerational actions of environmental factors in disease etiology. Trends in endocrinology and metabolism: *TEM.* 2010, 21, 214-222.

[11] Bird, A. DNA methylation patterns and epigenetic memory. *Gene Dev.* 2002, 16, 6-21.

[12] Gardinergarden, M.; Frommer, M. Cpg Islands in Vertebrate Genomes. *J Mol Biol.* 1987, 196, 261-282.

[13] Reik, W.; Dean, W. DNA methylation and mammalian epigenetics. *Electrophoresis.* 2001, 22, 2838-2843.

[14] Kacem, S.; Feil, R. Chromatin mechanisms in genomic imprinting. *Mamm Genome.* 2009, 20, 544-556.

[15] Jaenisch, R.; Bird, A. Epigenetic regulation of gene expression: how the genome integrates intrinsic and environmental signals. *Nat Genet.* 2003, 33, 245-254.

[16] Reddy, M.A.; Natarajan, R. Epigenetic mechanisms in diabetic vascular complications. *Cardiovascular research.* 2011, 90, 421-429.

[17] Forsdahl, A. Are poor living conditions in childhood and adolescence an important risk factor for arteriosclerotic heart disease? *British journal of preventive & social medicine.* 1977, 31, 91-95.

[18] Barker, D.J.; Winter, P.D.; Osmond, C.; Margetts, B.; Simmonds, S.J. Weight in infancy and death from ischaemic heart disease. *Lancet.* 1989, 2, 577-580.

[19] Painter, R.C.; Roseboom, T.J.; Bleker, O.P. Prenatal exposure to the Dutch famine and disease in later life: an overview. *Reproductive toxicology.* 2005, 20, 345-352.

[20] Curhan, G.C.; Willett, W.C.; Rimm, E.B.; Spiegelman, D.; Ascherio, A.L.; Stampfer, M.J. Birth weight and adult hypertension, diabetes mellitus, and obesity in US men. *Circulation.* 1996, 94, 3246-3250.

[21] Hales, C.N.; Barker, D.J.P.; Clark, P.M.S.; Cox, L.J.; Fall, C.; Osmond, C.; Winter, P.D. Fetal and Infant Growth and Impaired Glucose-Tolerance at Age 64. *Brit Med J. 1991*, 303, 1019-1022.

[22] Tobi, E.W.; Lumey, L.H.; Talens, R.P.; Kremer, D.; Putter, H.; Stein, A.D.; Slagboom, P.E.; Heijmans, B.T. DNA methylation differences after exposure to prenatal famine are common and timing- and sex-specific. *Human molecular genetics.* 2009, 18, 4046-4053.

[23] Steegers-Theunissen, R.P.; Obermann-Borst, S.A.; Kremer, D.; Lindemans, J.; Siebel, C.; Steegers, E.A.; Slagboom, P.E.; Heijmans, B.T. Periconceptional maternal folic acid use of 400 microg per day is related to increased methylation of the IGF2 gene in the very young child. *PloS one.* 2009, 4, e7845.

[24] Godfrey, K.M.; Sheppard, A.; Gluckman, P.D.; Lillycrop, K.A.; Burdge, G.C.; McLean, C.; Rodford, J.; Slater-Jefferies, J.L.; Garratt, E.; Crozier, S.R.; Emerald, B.S.; Gale, C.R.; Inskip, H.M.; Cooper, C.; Hanson, M.A. Epigenetic gene promoter methylation at birth is associated with child's later adiposity. *Diabetes.* 2011, 60, 1528-1534.

[25] Mayor-Lynn, K.; Toloubeydokhti, T.; Cruz, A.C.; Chegini, N. Expression profile of microRNAs and mRNAs in human placentas from pregnancies complicated by preeclampsia and preterm labor. *Reproductive sciences (Thousand Oaks, Calif.).* 2011, 18, 46-56.

[26] Maccani, M.A.; Padbury, J.F.; Lester, B.M.; Knopik, V.S.; Marsit, C.J. Placental miRNA expression profiles are associated with measures of infant neurobehavioral outcomes. *Pediatric research.* 2013, 74, 272-278.

[27] Huang, L.; Shen, Z.; Xu, Q.; Huang, X.; Chen, Q.; Li, D. Increased levels of microRNA-424 are associated with the pathogenesis of fetal growth restriction. *Placenta.* 2013, 34, 624-627.

[28] Tang, Q.; Wu, W.; Xu, X.; Huang, L.; Gao, Q.; Chen, H.; Sun, H.; Xia, Y.; Sha, J.; Wang, X.; Chen, D.; Xu, Q. miR-141 contributes to fetal growth restriction by regulating PLAG1 expression. *PloS one.* 2013, 8, e58737.

[29] Ferland-McCollough, D.; Fernandez-Twinn, D.S.; Cannell, I.G.; David, H.; Warner, M.; Vaag, A.A.; Bork-Jensen, J.; Brons, C.; Gant, T.W.; Willis, A.E.; Siddle, K.; Bushell, M.; Ozanne, S.E. Programming of adipose tissue miR-483-3p and GDF-3 expression by maternal diet in type 2 diabetes. *Cell death and differentiation.* 2012, 19, 1003-1012.

[30] Jiang, H.; Wu, W.; Zhang, M.; Li, J.; Peng, Y.; Miao, T.T.; Zhu, H.; Xu, G. *Aberrant upregulation of miR-21 in placental tissues of macrosomia.* 2014. 34, 658-663.

[31] Lucchini, R.; Barba, G.; Giampietro, S.; Trivelli, M.; Dito, L.; De Curtis, M. [Macrosomic infants: clinical problems at birth and afterward]. *Minerva pediatrica.* 2010, 62, 65-66.

[32] Bocock, P.N.; Aagaard-Tillery, K.M. Animal models of epigenetic inheritance. *Seminars in reproductive medicine.* 2009, 27, 369-379.

[33] Rees, W.D.; Hay, S.M.; Brown, D.S.; Antipatis, C.; Palmer, R.M. Maternal protein deficiency causes hypermethylation of DNA in the livers of rat fetuses. *The Journal of nutrition.* 2000, 130, 1821-1826.

[34] Burdge, G.C.; Slater-Jefferies, J.; Torrens, C.; Phillips, E.S.; Hanson, M.A.; Lillycrop, K.A. Dietary protein restriction of pregnant rats in the F0 generation induces altered methylation of hepatic gene promoters in the adult male offspring in the F1 and F2 generations. *The British journal of nutrition.* 2007, 97, 435-439.

[35] Lillycrop, K.A.; Phillips, E.S.; Jackson, A.A.; Hanson, M.A.; Burdge, G.C. Dietary protein restriction of pregnant rats induces and folic acid supplementation prevents epigenetic modification of hepatic gene expression in the offspring. *The Journal of nutrition.* 2005, 135, 1382-1386.

[36] Sandovici, I.; Smith, N.H.; Nitert, M.D.; Ackers-Johnson, M.; Uribe-Lewis, S.; Ito, Y.; Jones, R.H.; Marquez, V.E.; Cairns, W.; Tadayyon, M.; O'Neill, L.P.; Murrell, A.; Ling, C.; Constancia, M.; Ozanne, S.E. Maternal diet and aging alter the epigenetic control of a promoter-enhancer interaction at the Hnf4a gene in rat pancreatic islets. *Proceedings of the National Academy of Sciences of the United States of America.* 2011, 108, 5449-5454.

[37] Jousse, C.; Parry, L.; Lambert-Langlais, S.; Maurin, A.C.; Averous, J.; Bruhat, A.; Carraro, V.; Tost, J.; Letteron, P.; Chen, P.; Jockers, R.; Launay, J.M.; Mallet, J.; Fafournoux, P. Perinatal undernutrition affects the methylation and expression of the leptin gene in adults: implication for the understanding of metabolic syndrome. *FASEB journal : official publication of the Federation of American Societies for Experimental Biology.* 2011, 25, 3271-3278.

[38] Vo, T.X.; Revesz, A.; Sohi, G.; Ma, N.; Hardy, D.B. Maternal protein restriction leads to enhanced hepatic gluconeogenic gene expression in adult male rat offspring due to impaired expression of the liver X receptor. *The Journal of endocrinology.* 2013, 218, 85-97.

[39] Cong, R.; Jia, Y.; Li, R.; Ni, Y.; Yang, X.; Sun, Q.; Parvizi, N.; Zhao, R. Maternal low-protein diet causes epigenetic deregulation of HMGCR and CYP7alpha1 in the liver of weaning piglets. *The Journal of nutritional biochemistry.* 2012, 23, 1647-1654.

[40] Vucetic, Z.; Kimmel, J.; Totoki, K.; Hollenbeck, E.; Reyes, T.M. Maternal high-fat diet alters methylation and gene expression of dopamine and opioid-related genes. *Endocrinology.* 2010, 151, 4756-4764.

[41] Vucetic, Z.; Kimmel, J.; Reyes, T.M. Chronic high-fat diet drives postnatal epigenetic regulation of mu-opioid receptor in the brain. *Neuropsychopharmacology: official publication of the American College of Neuropsychopharmacology.* 2011, 36, 1199-1206.

[42] Khalyfa, A.; Carreras, A.; Hakim, F.; Cunningham, J.M.; Wang, Y.; Gozal, D. Effects of late gestational high-fat diet on body weight, metabolic regulation and adipokine expression in offspring. *International journal of obesity* (2005). 2013, 37, 1481-1489.

[43] Sharpe, R.M. Environmental/lifestyle effects on spermatogenesis. Philosophical transactions of the Royal Society of London. Series B, *Biological sciences.* 2010, 365, 1697-1712.

[44] Robertson, S.A. Seminal plasma and male factor signalling in the female reproductive tract. *Cell and tissue research.* 2005, 322, 43-52.

[45] Ng, S.F.; Lin, R.C.; Laybutt, D.R.; Barres, R.; Owens, J.A.; Morris, M.J. Chronic high-fat diet in fathers programs beta-cell dysfunction in female rat offspring. *Nature.* 2010, 467, 963-966.

[46] Carone, B.R.; Fauquier, L.; Habib, N.; Shea, J.M.; Hart, C.E.; Li, R.; Bock, C.; Li, C.; Gu, H.; Zamore, P.D.; Meissner, A.; Weng, Z.; Hofmann, H.A.; Friedman, N.; Rando, O.J. Paternally induced transgenerational environmental reprogramming of metabolic gene expression in mammals. *Cell.* 2010, 143, 1084-1096.

[47] Wei, Y.; Yang, C.R.; Wei, Y.P.; Zhao, Z.A.; Hou, Y.; Schatten, H.; Sun, Q.Y. Paternally induced transgenerational inheritance of susceptibility to diabetes in mammals. *Proceedings of the National Academy of Sciences of the United States of America.* 2014, 111, 1873-1878.

[48] Benatti, R.O.; Melo, A.M.; Borges, F.O.; Ignacio-Souza, L.M.; Simino, L.A.; Milanski, M.; Velloso, L.A.; Torsoni, M.A.; Torsoni, A.S. Maternal high-fat diet consumption modulates hepatic lipid metabolism and microRNA-122 (miR-122) and microRNA-370 (miR-370) expression in offspring. *The British journal of nutrition.* 2014, 111, 2112-2122.

[49] Dumortier, O.; Hinault, C.; Gautier, N.; Patouraux, S.; Casamento, V.; Van Obberghen, E. Maternal protein restriction leads to pancreatic failure in offspring: role of misexpressed microRNA-375. *Diabetes.* 2014, 63, 3416-3427.

[50] Lie, S.; Morrison, J.L.; Williams-Wyss, O.; Suter, C.M.; Humphreys, D.T.; Ozanne, S.E.; Zhang, S.; Maclaughlin, S.M.; Kleemann, D.O.; Walker, S.K.; Roberts, C.T.; McMillen, I.C. Periconceptional undernutrition programs changes in insulin-signaling molecules and microRNAs in skeletal muscle in singleton and twin fetal sheep. *Biology of reproduction.* 2014, 90, 5.

[51] Lie, S.; Morrison, J.L.; Williams-Wyss, O.; Suter, C.M.; Humphreys, D.T.; Ozanne, S.E.; Zhang, S.; MacLaughlin, S.M.; Kleemann, D.O.; Walker, S.K.; Roberts, C.T.; McMillen, I.C. Impact of embryo number and maternal undernutrition around the time of conception on insulin signaling and gluconeogenic factors and microRNAs in the liver of fetal sheep. American journal of physiology. *Endocrinology and metabolism.* 2014, 306, E1013-1024.

[52] Vickers, M.H.; Breier, B.H.; Cutfield, W.S.; Hofman, P.L.; Gluckman, P.D. Fetal origins of hyperphagia, obesity, and hypertension and postnatal amplification by hypercaloric nutrition. American journal of physiology. *Endocrinology and metabolism* 2000, 279, E83-87.

[53] Vickers, M.H.; Gluckman, P.D.; Coveny, A.H.; Hofman, P.L.; Cutfield, W.S.; Gertler, A.; Breier, B.H.; Harris, M. Neonatal leptin treatment reverses developmental programming. *Endocrinology.* 2005, 146, 4211-4216.

[54] Gluckman, P.D.; Lillycrop, K.A.; Vickers, M.H.; Pleasants, A.B.; Phillips, E.S.; Beedle, A.S.; Burdge, G.C.; Hanson, M.A. Metabolic plasticity during mammalian development is directionally dependent on early nutritional status. *Proceedings of the*

National Academy of Sciences of the United States of America. 2007, 104, 12796-12800.

[55] Weaver, I.C.; Cervoni, N.; Champagne, F.A.; D'Alessio, A.C.; Sharma, S.; Seckl, J.R.; Dymov, S.; Szyf, M.; Meaney, M.J. Epigenetic programming by maternal behavior. *Nature neuroscience.* 2004, 7, 847-854.

[56] Weaver, I.C.; Champagne, F.A.; Brown, S.E.; Dymov, S.; Sharma, S.; Meaney, M.J.; Szyf, M. Reversal of maternal programming of stress responses in adult offspring through methyl supplementation: altering epigenetic marking later in life. *The Journal of neuroscience : the official journal of the Society for Neuroscience.* 2005, 25, 11045-11054.

[57] Weaver, I.C. Epigenetic programming by maternal behavior and pharmacological intervention. *Nature versus nurture: let's call the whole thing off. Epigenetics-Us.* 2007, 2, 22-28.

In: Epigenetics: Mechanisms and Clinical Perspectives
Editor: Yildiz Dincer
ISBN: 978-1-63484-490-1

Chapter 2

EFFECTS OF OXIDATIVE STRESS ON EPIGENETIC MECHANISMS

Yıldız Dincer[1,*], PhD, and Onur Baykara[2], PhD

[1]Istanbul University, Cerrahpasa Medical Faculty,
Department of Medical Biochemistry, Istanbul, Turkey
[2]Istanbul University, Cerrahpasa Medical Faculty,
Department of Medical Biology, Istanbul, Turkey

ABSTRACT

Oxidative stress is a state in which production of reactive oxygen species exceeds the capacity of antioxidant systems. Reactive oxygen species have one or more unpaired electrons, making them highly reactive with other cellular molecules such as protein, lipid, and nucleic acid. The peroxidation of polyunsaturated fatty acids in biological membranes results in impaired membrane integrity. Oxidatively modified proteins lose *their* capacity to carry out the *physiological functions* and they may form intracellular aggregates. Attacks of reactive oxygen species to DNA results in strand breakages and base oxidation. Major DNA oxidation product is 8-hydroxydeoxyguanosine which has a pro-mutagenic potential. Due to these damaging effects, oxidative stress plays an important role in various pathologies such as cancer, diabetes, chronic inflammatory diseases and neurodegenerative disorders. Epigenetic changes are regular and natural events which regulate gene expression without changing base sequences on DNA. Dysregulation of regular epigenetic mechanisms is a contributory factor for many of human pathologies. Recently, reactive oxygen species have been shown to cause epigenetic dysregulations that play a pivotal role in human disorders. The basic epigenetic mechanisms and their dysregulation by reactive oxygen species have been reviewed in this chapter.

Keywords: epigenetic modifications, DNA methylation, histone acetylation, miRNAs, oxidative stress, 8-hydroxydeoxyguanosine

[*] Correspondence author: E-mail:yldz.dincer@gmail.com.

ABBREVIATIONS

5-hmC	5-hydroxymethylcytosine
5-mC	5-methylcytosine
8-OHdG	8-hydroxydeoxy guanosine
CoA	Acetyl co-enzyme A
ATG12	Autophagy-12
ATG8	Autophagy-8
CAT	Catalase
CNS	Central Nervous System
CpG sites	Cytosine- phosphate-guanine sites
DNA	Deoxyribonucleic acid
DGCR8	DiGeorge Sydrome Critical Region 8
DNMT	DNA methyltransferase
Exp5	Exportin 5
GSH-Px	Glutathione Peroxidases
GST	Glutathione S-Transferases
GNAT	Gcn5-related N-acetylases
ELP3	Gnc5, PCAF and elongator complex protein 3
HAT	Histone Acetyl Transferases
HDACi	Histone Deacetylase Inhibitors
HDAC	Histone Deacetylases
HMTS	Histone Methyl Transferases
FAT10	Human leukocyte antigen F-associated
IUPAC	International Union of Pure and Applied Chemistry
LINE-1	Long-interspersed nuclear element-1
MDBP	Methyl-CpG-binding domain proteins
miRNA	MicroRNA
MDS	Myelodysplastic syndrome
NAD^+	Nicotinamide adenine dinucleotide
ncRNA	Noncoding RNA
NF-κb	Nuclear factor-κb
OFR	Oxygen free radicals
PAZ	Piwi, Argonaut and Zwille
piRNA	Piwi-interacting RNA
RNA	Ribonucleic acid
rRNA	Ribosomal RNA
RISC	RNA-induced silencing complex
SAM	S-adenosyl-L-methionine
SET	Su(var)3-9, Enhancer of Zeste, Trithorax
siRNA	Small interfering RNA
snoRNA	Small nucleolar RNA
Sumo	Small Ubiquitin-like Modifier
SOD	Superoxide dismutase
tRNA	Transfer RNA

UCRP	Ubiquitin cross-reactive protein
UFM-1	Ubiquitin fold modifier-1
UBL5	Ubiquitin like protein-5
UBLs	Ubiquitin like proteins
Ub	Ubiqutin

1. Epigenetic Mechanisms

The term epigenetics was first coined by C.H Waddington in 1942, a word derived from epigenesis and genetics. Epigenesis is the word for describing live organisms differentiating from a single cell into an organ or organism. Epigenetic mechanisms modulate gene expression patterns without affecting the deoxyribonucleic acid (DNA) sequence. To stop confusions, a consensus definition was put on use of the epigenetic trait: "stably heritable phenotype resulting from changes in a chromosome without alterations in the DNA sequence" [1]. These mechanisms are acquired throughout life and depend on environmental clues such as diet, lifestyle and toxin exposure. They are heritable mechanisms so epigenetic changes can be transferred to the new cell when cells are divided with mitosis. If this process takes place in germ cells and meiosis, genomic imprinting may occur. In some occasions, such as differentiation of embryonic stem cells to specific tissues, the expression is stabilized so that the cells cannot go back to their previous state. Although all cells in one individual have the same DNA sequence, epigenetic regulation occurs at the specific gene loci in the specific cells to yield specific cellular phenotypes. A change in the phenotype does not usually effect the genotype. Gene expression can be activated or silenced via epigenetic regulations. Epigenetic changes can alter transcriptional activity of genes, and may mediate the differences in risk for certain diseases. Methylation, phosphorylation, acetylation, ubiquitination, sumoylation are included in epigenetic processes.

Chromatins are complex macromolecules in cells. DNA, RNA, histones and non-histone proteins form chromatin structure. DNA is tightly packed into the nucleus in the chromatin structure, and this complex can be exposed to modification with acetyl groups (acetylation), non-coding RNAs like microRNAs (miRNAs) and small interfering RNAs (siRNAs). These modifications may cause changes in chromatin structure which in turn effect the expression patterns of the genes. So to summarize, epigenetic modifications are typically divided into three categories: (1) DNA methylation (2) histone post-translational modifications and (3) microRNAs.

1.1. DNA Methylation

DNA methylation can be briefly explained as covalent addition of a methyl group (CH_3) to the 5-carbon of cytosine forming 5-methylcytosine (5-mC) mostly located at cytosine-phosphate-guanine sites (CpG sites) that are present in the 5'-untranslated regions of gene promoters. Although DNA methylation usually ends up with suppression of gene transcription due to binding of methyl groups into the major groove of DNA, this process is essential for development and directly associated with processes in the cell such as X-chromosome inactivation, genomic imprinting and in carcinogenesis. 5-mC can be hydroxylated forming 5-hydroxymethylcytosine (5-hmC) which consists of approximately 1.5% of human DNA [2]. DNA methylation is carried out by DNA methyltransferase (DNMT) activity, and methyl group donor is S-adenosyl-L-methionine (SAM). DNMTs are family of enzymes which catalyze the transfer of methyl groups to DNA. DNMT in mammals contains two main regions, C terminal region and N-terminal region. The C-terminal is the catalytic part containing approximately 500 amino acids, and this is the part which center of enzyme is located. The N terminal includes 621 amino acids, and is not vital for DNMT1 activity [3].

Although DNMTs have been classified in four groups since many years including DNMT1, DNMT2, DNMT3a, and DNMT3b, in mammals, there are actually three types of methyltransferases: DNMT1, DNMT3a, and DNMT3b [4]. DNMT1 is the major methyltransferase for maintenance of methylation, for this reason it is usually called as maintenance methyltransferase [5]. It catalyzes the transfer of a methyl group from SAM on to the 5' position of the cytosine ring, generating 5-methylcytosine. DNMT2 which used to be known as a DNMT, now termed as TRDMT1, is not a methyltransferase, even though its sequence is strongly associated with other DNMTs. Its DNA methylation activity is very low [6]. It only has a C domain and probably has a role in recognition of DNA damage and mutation repair [7]. DNMT2 (TRDMT1-tRNA aspartic acid methyltransferase 1) has both ribonucleic acid (RNA) methyltransferase activity and 5-cytosine DNA methyltransferase activity [8]. Previous studies have shown that DNMT2 functions as tRNA methyl-transferase by methylating cytosine-38 in aspartic acid of tRNA [8, 9].

DNMT3 can methylate both hemimethylated and unmethylated CpG at the same rate. DNMT3 family has three members which are DNMT3a, DNMT3b and DNMT3L. The structure of DNMT3 family is quite similar to DNMT1 and they have a catalytic domain with a regulatory region. DNMT3a and DNMT3b have different functions from those of DNMT1, they serve as *de novo* methyltransferases. For establishment and mitotic inheritance of tissue-specific DNA methylation patterns in normal development, de novo methylation is essential. It occurs frequently during early development and gametogenesis in mice to remethylate the repetitive sequences those are demethylated during early stage of embryogenesis [10].

DNA methylation results in gene silencing by the recruitment of methyl CpG-binding transcriptional repressors and by interfering with the DNA binding of transcriptional activators. Hypomethylated DNA is associated with active chromatin. Transcription level of DNA is affected by DNA methylation at the promoter region via two ways: (1) Methylated DNA prevents the binding of transcription factors to the gene promoter (2) methylated DNAs are occupied by methyl-CpG-binding domain proteins (MBDPs). MBDPs bring together the other epigenetic components consequently forming compact and inactive heterochromatin, therefore cause decreased or silenced gene transcription [11].

Aberrant DNA methylation on disease-related gene promoters was determined in various diseases, especially in cancer. DNMT inhibitors are promising therapeutic targets. Both Decitabine (5-aza-2'-deoxycytidine) and Azacytidine are hypomethylating agents. They show their function by inhibiting DNMTs. Decitabine can be incorporated into DNA where Azacytidine incorporates both into DNA and RNA [12], as azacytidine is the chemical analogue of DNA and RNA. Both of the drugs have been approved by the FDA for treatment of myelodysplastic syndrome.

1.2. Histone Post-translational Modifications

In eukaryotic organisms, DNA is wrapped around a protein octamer. This octamer is composed of two units of H2A, H2B, H3 and H4 histone proteins which are linked to each other through a linker protein called H1 histone. This complex unit is called as nucleosome and is the basic unit of the chromosomes Histones can undergo post-translational covalent modifications which are phosphorylation on serine or threonine residues, methylation on lysine or arginine, acetylation and deacetylation on lysines, ubiquitylation of lysines and sumoylation on lysines [13, 14]. The best defined histone modification is acetylation. Histone modifications are proposed to affect chromosome structure and function, especially during transcription and chromatin remodeling processes. Core modifications of histones can also act on DNA templated processes, like replication and transcriptional processes. They can also effect nucleosomal structure [15, 16]. Simon et al. have shown that histone posttranslational modifications within distinct structured regions of the nucleosome directly regulate the inherent dynamic properties of the nucleosome such as unwrapping and disassembly [17]. North et al. have shown that phosphorylation of H3(T118) alters nucleosome dynamics and remodeling [18]. Histone modifications are catalyzed by histone modification enzymes such as histone acetyltransferases, histone deacetylases, histone metyltransferases and histone demethylases.

Histone acetyltransferases (HAT) acetylate conserved lysine residues on histone tails. HATs transfer an acetyl group from acetyl co-enzyme A (CoA) forming ε-N-acetyllysine. Depending on their localization, HATs (EC 2.3.1.48) are classified into two groups, the ones localized in the nucleus and carrying a conserved motif are Type A HATs, and the ones localized in the cytoplasm are Type B HATs [19]. Type A HATs also have three primary subclasses which are 1) GNAT (Gcn5-related N-acetylases) (Gnc5, PCAF and elongator complex protein 3(ELP3)) 2) MYST (HMOF/MYST1, HBO1/MYST2, MOZ/MYST3, MORF/MYST4, Type60 and Type p300/CBP (p300 and CBP) families.

The acetylated lysine is neutralized from positive charges (normally, the histone tails are positively charged which help these tails interact with negatively charged phosphate groups found on DNA backbone), and this acetylation decreases the electrostatic interaction between DNA backbone and histone tails. This mechanism is used to control gene expression by activating and inactivating genes on DNA. Simply, HATs are transcriptional co-activators and can promote transcription factors for activation of transcription. Acetylation of nucleosomes by HATs helps transcription factors to reach the DNA more easily so that transcription can increase.

Histone deacetylases (EC 3.5.1.98, HDAC) remove acetyl groups from an ε-N-acetyl lysine amino acid on histone tails, therefore the histones can wrap tighter around the DNA which helps to regulate the expression of a gene. The density of wrapping can alter expression levels. HDACs are transcriptional co-repressors. HDACs usually inactivate gene expression [20]. Depending on their sequence homology and subcellular localization, HDACs can be classified into four classes as I, II (a & b), III and IV. HDAC. Class I, (HDACs 1, 2, 3, and 8) are localized in the nucleus. Class II (HDACs 4, 5, 6, 7, 9, and 10), Class III (SirT 1, 2, 3, 4, 5, 6 and 7) and Class IV HDACs (HDAC 11) can be localized either on nucleus or cytoplasm. Class III HDACs can also be localized in the mitochondria [21, 22]. HDAC Class I, II and IV are zinc-dependent which are all expressed in the brain [23], whereas HDAC III is nicotinamide adenine dinucleotide (NAD^+) dependent. Although Class I HDACs are known as nuclear enzymes, they were shown to play a role in unfolded protein response by localizing to the endoplasmic reticulum [24]. In addition to their gene silencing function, sirtuins are the key regulators of multiple biological and metabolic events including energy metabolism, glucose and lipid metabolisms, circadian rhythms, inflammation, stress resistance, apoptosis, and autophagy. Many scientists have performed studies on HDACs and their functions in various diseases to find out the underlying epigenetic mechanism.

HDAC 2 is expressed in the central nervous system (CNS) and has a negative effect on memory and synaptic plasticity. Guan et al. have shown that binding of HDAC2 to the promoters of synaptic-plasticity-related genes and thereby negatively regulates their transcriptions [25]. Graff et al. have knocked down HDAC2 by short-hairpin-RNA and they have shown that structural synaptic plasticity and memory impairments were restored in mice [26]. These findings support the information that HDAC2 is associated with disrupted brain functions in neurodegenerative diseases.

Histone deacetylase inhibitors (HDACi) cause to an increase in histone acetylation and reduce the affinity of the histone to the negatively charged DNA backbone. By this method, chromatin gets less dense so that the transcription factors have the chance of reaching the DNA and increasing gene expression. Certain non-histone proteins (DNA repair proteins, DNA binding proteins, transcription factors, chaperones) are also substrate for HDACs. Therefore HDACs take place in various pathways including inhibition of DNA repair, induction of apoptosis, generation of reactive oxygen species (ROS), inhibition of cell cycle arrest and inhibition of antiangiogesis [27]. In this context HDACi are considered as an emerging therapeutic target for cancer [28, 29] and HIV [30].

Histone methyl transferases (HMTs) (2.1.1.43 EC) are the enzymes (e.g., histone-lysine N-methyltransferases and histone-arginine N-methyltransferases) which modify the histones. They function in adding methyl groups to lysine (K) and arginine (R) residues. Lysine methyltransferases catalyze mono-, di-, and trimethylation of the lysine ε-amino group. Histone methylation can be managed by SAMs as the methyl group donor [31]. HMTs are composed of two classes, lysine specific or arginine specific. Lysine-specific HMTs have two subgroups which carry a special type of protein domain called SET (Su(var)3-9, Enhancer of Zeste, Trithorax) domain containing and non-SET domain containing. Histone methylation controls histone-protein interaction. It mediates the binding of HDACs to histones, thus silences the gene indirectly.

Ubiquitination (Ubiquitylation) of histones is required for histone methylation. Ubiquitin (Ub) is a regulatory protein and is found almost in any tissue of eukaryotes. Ubiquitination may cause the proteins to degrade with proteasomes, change their locations in the cell and increase or decrease protein-protein interactions [32]. Ubiquitination may also play major roles in apoptosis, cell cycle, DNA transcription and repair, response to oxidative stress and many other mechanisms. The protein to be ubiquinated must be activated, conjugated and ligated. These steps can be managed by ubiquitin-activating enzymes (E1s), ubiquitin-conjugating enzymes (E2s), and ubiquitin ligases (E3s), respectively. Ubiqutin first activates E1 and then it is transferred to E2. E3 Ubiquitin ligases are activated by a E2-Ub complex [33]. At the end of this process, ubiquitin is bound to lysine residues on the protein's N-terminus or the protein substrate. The binding can be either with a peptide bond or an isopeptide bond, respectively [34]. In general, defects in ubiquitination mechanism can alter the functions of the genes and proteins. Ubiquitination is associated with many types of disease including cancer, neurodegenerative and inflammatory diseases. [35, 36].

Ubiquitination of histone proteins is required for histone methylation, therefore silences the gene indirectly.

Although ubiquitin is the most studied and commonly known protein, another family of ubiquitin like proteins (UBLs) also exists. These proteins act like ubiquitin and modify proteins but they have distinct functions. One of the major members of this family is Sumos. Small Ubiquitin-like Modifier (Sumo) proteins are a family of small proteins comprising approximately 100 amino acids. Sumoylation is directed by an enzymatic cascade analogous to that involved in ubiquitination but SUMO modification usually functions on the opposite direction of ubiquitination and can help stabilizing protein substrates. As a post-translational modification, sumoylation plays different roles in cellular mechanisms, transportations, apoptosis and response to oxidative stress. Sumoylation regulates protein-protein interaction and localization, and enhances the interaction between histones and HDACs. Therefore it is involved in gene silencing, indirectly. Other members of UBLs include, but not limited to, ubiquitin cross-reactive protein (UCRP), ubiquitin fold modifier-(UFM-1), ubiquitin like protein-5 (UBL5), human leukocyte antigen F-associated (FAT10), autophagy-8 (ATG8), autophagy-12 (ATG12) and many others.

Carbonylation, mostly occurring on lysine and arginine residues, is a consequence of oxidative stress and it is mostly irreversible. Carbonylation causes loss of protein function, the modified proteins are aggregated and they accumulate as unfolded or damaged proteins. [37]. Sharma et al. have shown histone carbonylation to occur on histones H1, H2A, H2B and H3 but not H4 in rat liver. Contrary the wide belief that carbonylation increases by age, they have found the carbonylation was lower in older rats [38].

Besides these most frequently investigated post translational modifications, there are many other modifications such as *acylation* (*alkanoylation*- addition of functional groups on proteins such as addition of fatty acids to amino acids), *alkylation* (transfer of an alkyl group to another molecule, most common form is methylation with a distinction, in methylation only one carbon is transferred while in alkylation long chain of carbon groups are transferred), *glycosylation* (enzymatic addition of sugar molecule to a protein substrate to help proteins folding correctly, stabilize proteins, prevent protein degradation), *glycation* (non-enzymatic glycosylation, covalent binding of a protein with sugar molecule), *biotinylation* (covalent binding of biotin to a protein target), *neddylation* (conjugation of ubiquitin-like protein NEDD8 to protein substrates), *succinylation* (addition of succinyl group to a lysine residue), *sulfation* (addition of a sulfate group to a tyrosine residue).

1.3. Micro RNAs and Epigenetic

Small noncoding RNAs (ncRNAs) are functional RNA molecules which do not encode a protein. Even though they are not translated into a protein, they play major roles in cellular and molecular mechanisms. The family of ncRNAs include micro RNAs (miRNAs), transfer RNAs (tRNAs), ribosomal RNAs (rRNAs), small interfering RNAs (siRNAs), Piwi-interacting RNAs (piRNAs), and small nucleolar RNA (snoRNAs).

miRNAs are a family of short non-coding RNAs of approximately 19-25 long nucleotides which modulates the expression of mRNA. They play crucial roles in cell proliferation, apoptosis, cell cycle control, differentiation and cellular metabolism. miRNAs are found in serum, plasma, saliva, and other body fluids of humans and also in other organisms such as plants, animals and some viruses. Their regulation of gene expression is at post-transcriptional level. miRNAs bind to target mRNAs at their 3'-untranslated regions and suppress their translation and/or promote their degradation, thus sequence-specifically control the translation of target mRNAs [39]. In addition, miRNAs bind complementary sequences on DNA and block access for transcription factors, which in turn results in turned off genes. miRNAs may show their functions on many different targets and a gene can be regulated by different miRNAs [40].

A miRNA is produced from a RNA-coding gene or from introns. It is thought that about 40% of all miRNA genes are located in the introns of protein coding or non-coding genes [41]. miRNAs start their life as primary transcripts known as a pri-miRNAs which are transcribed by RNA polymerase II in the nucleus and they carry a 5' cap and a poly-A-tail. The pri-miRNAs are then transcribed to pre-miRNA, a 70 nucleotide long structure, in the microprocessor complex consisting of RNAse III enzyme Drosha and dsDNA binding protein DGCR8 (DiGeorge Sydrome Critical Region 8- Pasha in invertebrates). The pre-miRNAs are imported out of the nucleus into the cytoplasm by RAN-GTP complex and exporting 5 (Exp5) protein. This complex plays an important role in translocation of the RNA and proteins through the nuclear pores. When pre-miRNAs are out of the nucleus, Dicer, a RNAse III enzyme containing an ATPase/RNA helicase, a DUF283 (Domain of unknown function) domain, a PAZ (Piwi, Argonaut and Zwille) domain, two catalytic RNase III domains (RIIIa and RIIIb), and a C-terminal double-stranded RNA-binding domain (dsRBD), cleaves the pre-miRNA stem loop into two complementary RNA molecules but only one of these molecules integrate with the RNA-induced silencing complex (RISC), resulting with mature miRNA.

Even though the biogenesis of miRNA has been extensively studied, the regulation of miRNA expression still needs to be clarified. A miRNA can be localized in various regions of the genome. If they are localized between the genes, they are called intragenic, if the localization is in the gene they are called intergenic. In an in silico analysis by Sato et al. it has been shown that of 939 miRNAs, 293 (31.2%) were intergenic, whereas 317 (44.4%), 119 (12.7%) and 110 (11.7%) were overlapped by RNA transcripts in the same, opposite and both directions, respectively [42].

There are numerous studies investigating the relationship between miRNA and epigenetic regulations in various diseases such as cancer, diabetic, neurologic and inflammatory related diseases, aging and this number is still growing. Even though the underlying mechanism is not clearly understood, miRNA dysregulation is determined in all of these pathologies. It is thought that miRNAs contribute to formation of diseases via epigenetic regulations, and miRNA modulation is a new therapeutic approach that is extensively investigated in various diseases.

2. Oxidative Stress

Living organisms mostly rely on oxygen. Oxygen is like a double-edged knife. The cells of living organisms cannot live without oxygen but oxygen and its radical species can also harm cells by reacting with cellular components. Impaired balance between the oxidants and the anti-oxidants may cause a damage to the cells and this is called the oxidative stress. If the body cannot fight against the harmful reactive oxygen species, with another word oxygen free radicals (OFR), such as O_2^- (superoxide radical), $OH^\cdot$ (hydroxyl radical), triplet oxygen, triplet carbine (:CH2) and H_2O_2 (hydrogen peroxide), they may cause irreversible damage on the cellular components such as DNA, proteins, lipids. This damage may even cause stop cellular signaling which is essential for the viability, motility and production of the cell.

According to IUPAC (International Union of Pure and Applied Chemistry) a free radical can be an atom, a molecule or an ion that has unpaired valence electrons [43]. These free radicals are highly reactive with other molecules. As they have an unbalanced structure and lacking of electrons, they are very prone to interact with other molecules in order to steal their electrons from them. They can even react with each other to form dimers, trimers or polymers. The most infamous radicals are superoxide anion (O_2^-) and hydroxyl radical ($OH^\cdot$). O_2^- is formed after one electron reduction of dioxygen (O_2). Superoxides are toxic substances and produced by the immune system to be used against foreign substances. In case of a dismutation of superoxide, molecular oxygen (O_2) and hydrogen peroxide (H_2O_2) are formed and this reaction is catalyzed by superoxide dismutase enzyme (SOD-E.C 1.15.1.1).

$$2\,O_2^- + 2\,H^+ \xrightarrow{SOD} H_2O_2 + O_2$$

The main source of OFR in the cells is the leakage from mitochondrial electron transport chain especially from complex I (NADH dehydrogenase) and III (coenzyme Q and cytochrome C oxidoreductase) [44]. When components of the oxidative phosphorylation located in the inner membrane of the mitochondria arc reduced, production of mitochondrial superoxide radical increases. Superoxide radical and hydroxyl radical start the lipid peroxidation in cytoplasm, nucleus, mitochondria and endoplasmic reticulum which increases membrane permeability. Due to effect of OFR proteins, nuclear and mitochondrial DNA are oxidized. Oxidation of proteins may result in functional loss. OFR attacks to DNA causes DNA strand breaks and base modifications. Among the oxidized bases, 8-hydroxydeoxyguanosine (8-OHdG) is the most mutagenic lesion. By mispairing with A residues, 8-OHdG leads to an increased frequency of spontaneous G:C → T:A transversion. This mutation is recorded in many mutated proto-oncogenes and tumor suppressor, and 8-OHdG is strongly implicated in the initiation of carcinogenesis.

As mentioned above, free radicals can steal electrons from other atoms, destabilizing and perturbing them. In order to overcome this attack, the body uses antioxidants. Antioxidants are used to neutralize OFR before they harm the cells. They can be classified in two main groups, endogenous and exogenous antioxidants. Endogenous antioxidants can be sub-grouped into two as enzymatic antioxidants and non-enzymatic antioxidants. The first one includes the enzymes such as Superoxide dismutase (SOD), Glutathione S-Transferases (GST), Glutathione Peroxidases (GSH-Px), Catalase (CAT), mitochondrial cytochrome oxidase system, whereas the latter group consists of selenium, melatonin, hemoglobin, glutathione, bilirubin, methionine, retinal, thiols, and many others. Exogenous antioxidants can be classified as vitamins (ascorbic acid, alpha tocoferol, beta carotene, folic acid), drugs (xanthine oxidase inhibitors, NAPDH oxidase inhibitors etc.) and nutrient antioxidants (sodium benzoate, catechins, anthocyanins, tannins, lycopene, etc.)

Due to the damaging effects on cellular macromolecules, OFR contribute to pathogenesis of many diseases including cancer, ageing, neurodegenerative diseases, obesity, cardiovascular diseases, neuroinflammation, etc. [45-49]. It has been recently recognized that OFR also modulate epigenetic mechanisms.

3. Interactions between Oxidative Stress and Epigenetic Mechanisms

OFR may modulate epigenetic mechanisms. They can effect epigenetic mechanisms in various ways: 1) OFR changes the pattern of DNA methylation 2) Oxidative stress reduces methyl-accepting ability of DNA 3) OFR cause changes in histone modifications [50].

3.1. Oxidative Stress - DNA Methylation

Distruption of DNA methylation patterns by OFR is the best defined in carcinogenesis. Briefly, the hypermethylation of the DNA can cause gene silencing resulting with decreased gene expression, uncontrolled cell proliferation and growth which may end up with cancer. On the other hand, hypomethylation may cause loss of imprinting and chromosomal instability [51, 52]. Hydroxyl radical-induced DNA lesions such as 8-hydroxy-2-deoxyguanosine (8- OHdG) have been shown to contribute to decreased DNA methylation. 8- OHdG adducts interfere with the ability of DNA to function as a substrate for DNMTs. This results in global hypomethylation of the genome which in turn leads to oncogene activation and chromosomal instability. In addition, OFR are responsible for gene silencing by leading aberrant hypermethylation in CpG island-rich promoter region of tumor suppressor genes [53]. Many tumor suppressor genes have been found to be silenced via OFR-mediated aberrant methylation of promoter regions [54-57]. Recent studies have shown a further modification on the methylated CpGs which may make the gene less prone to transcription than those in the methylated state. This further modification is hydroxylation of 5-methylcytosine. Oxidative stress results in hydroxymethylation of DNA, especially in neurons [58]. Regarding other diseases, Patchsung et al. have demonstrated a relationship between increased oxidative stress and hypomethylation of the transposable long-interspersed nuclear element-1 (LINE-1) in patients with bladder cancer. They have found that LINE-1 hypomethylation levels and the number of hypomethylated loci in both the blood- and urine-derived cells are increased which is also causing an increase in oxidative stress in the bladder cancer patients [59]. In patients with myelodysplastic syndrome (MDS), two tumor suppressor genes, P15 and P16, have been found to be hypermethylated due to oxidative stress caused by peroxides, superoxide anion in leukocytes (CD45+ cells) from bone marrows of the patients [60]. This situation may be interpreted as hypermethylation of tumor suppressor genes cause them stop functioning thus increase the chance of cancer progression. Gu et al. have studied histone acetylation and DNA methylation, involved in the transcription of Alzheimer's disease-related genes in human neuroblastoma cells which is treated with hydrogen peroxide. They have found that expression of amyloid-β precursor protein and β-site amyloid-β precursor protein-cleaving enzyme 1 is upregulated due to demethylation in the gene promoters associated with the reduction of methyltransferases. They have also shown down-regulation of HDACs in the same cells. They have concluded that oxidative stress causes an imbalance between DNA methylation and demethylation, and histone acetylation and deacetylation [61]. In fact, both DNMTs and HDACs are vulnerable to oxidative stress due to their protein structure. Increased oxidative stress may inactivate them that results in impaired epigenetic pattern.

3.2. Oxidative Stress - Histone Post-Translational Modifications

Addition of a methyl or phosphate group to histones does not directly affect chemistry of histone. However, acetylation of histone tails directly alter gene expression that higher acetylation of the histones ends up with higher level of transcription. Probably due to this reason, studies examining the effect of oxidative stress on histone post-translational modifications have been focused on acetylation/deacetylation rather than other modifications. OFR- dependent signaling pathways prompt transcriptional and epigenetic dysregulation. Oxidative stress induces various kinase signaling pathways leading to histone acetylation/deacetylation. Activation of upstream kinases, such as nuclear factor-κB (NF-κB)-inducing kinase, mitogen and stress-activated kinase-1 by OFR results in downstream chromatin remodeling, which in turn alters the function of gene promoters [62]. In this context, OFR-mediated epigenetic alterations are especially shown to modulate transcription of specific genes that control apoptosis, autophagy, senescence, proliferation, transformation, and differentiation [63]. Although all histone modifiying enzymes may be oxidized by OFR, sirtuins are more vulnerable due to their dependency to NAD^+. When OFR interact with DNA, strand breakage occurs. This is a signal for polyADP ribose polymerase activation, the enzyme that uses NAD^+. Increased oxidative stress may lead intracellular NAD^+ depletion and decreased sirtuin activity. Since sirtuins have a regulator function on various biological pathways, its low activity results in impaired cellular function. p53 is one of the major targets of SIRT1. Decreasing SIRT1 activity increases p53 acetylation, which in turn promotes apoptosis and senescence [64]. Oxidative stress has been shown to accelerate cellular senescence, due to increased p53 acetylation via decreasing the function of SIRT1 through NAD^+ depletion [65].

3.3. Oxidative Stress – miRNAs

miRNAs are not translated into a protein but play a crucial role in regulation of gene expression via translational repression or mRNA degradation. Although OFR-induced gene regulation has been extensively studied at epigenetic level, the effects of OFR on gene expression regulation by miRNAs is currently uncertain. Certain studies have shown that miRNAs are involved in regulation of cell survival in response to oxidative stress [66, 67]. The activation of apoptosis by Fas death receptor is stimulated by OFR. In an in vitro study Lin et al. have demonstrated that Fas is a functional target gene for miR-23a, forced overexpression of miR-23a reduces cell death induced by H_2O_2 incubation in macular retinal pigment epithelial cells, and this protective effect of miR-23a is disappeared by miR-23a inhibition [68].

In fact, it should be kept in mind that miRNAs are encoded by miRNA genes and their expressions are mostly regulated by epigenetic mechanisms. Epigenetic mechanisms may affect the expression of miRNAs in both physiological and pathologic conditions in a tissue-specific manner [69]. Hyper or hypo methylation of miRNA coding gene promoters leads to altered miRNA expression patterns, and this event may result in different diseases including cancer, glioblastoma, Parkinson's disease and Alzheimer's disease [70, 71]. Furthermore, recent studies have revealed that miRNAs are transcriptionally regulated by a variety of DNA methylation-specific methyl-CpG-binding domain proteins (MBDs) [72]. Besides DNA methylation, histone post-translational modifications and associated changes in chromatin structure can affect miRNAs at the transcriptional level leading their both up- and down regulation [73].

CONCLUSION

Eukaryotic organisms including human beings are very complex structures and a variety of mechanisms interact with each other. Genetic and epigenetic changes work hand by hand to keep this fabulous factory working. Aberrant epigenetic modifications and regulations in post-transcriptional and post-translational processes make either direct or indirect contributions to the diseases such as cancer, neurodegenerative and inflammatory diseases. Oxidative stress is a state of persistent generation of OFR overwhelming cellular antioxidant defense, and it has long been known to contribute to pathogenesis of various diseases. Although harmful effects of oxidative stress on genes have been discovered a long time ago, OFR-epigenome interaction and its consequences have gained interest in the last decade. OFR-induced epigenetic mechanisms involve changes in DNA methylation profiles, histone acetylations and miRNA expression. Epigenetic modifications are potentially reversible and there is a great potential for the development of effective strategies to cure diseases. Scientist all over the world are studying on different mechanisms and interactions of genes and proteins to find cures for diseases. Understanding the role of oxidative stress on epigenetic mechanisms can help to see light at the end of the tunnel.

REFERENCES

[1] Berger SL, Kouzarides T, Shiekhattar R, Shilatifard A. An operational definition of epigenetics. *Genes Dev.* 2009; 23(7):781–3.

[2] Lister R, Pelizzola M, Dowen RH, Hawkins RD, Hon G, Tonti-Filippini J, Nery JR, Lee L, Ye Z, Ngo QM, Edsall L, Antosiewicz-Bourget J, Stewart R, Ruotti V, Millar AH, Thomson JA, Ren B, Ecker JR. Human DNA methylomes at base resolution show widespread epigenomic differences. *Nature.* 2009; 462(7271):315-22.

[3] Ghoshal K, Bai S. DNA methyltransferases as targets for cancer therapy. *Drugs Today.* 2007; 43(6):395–422.

[4] Gopalakrishnan S, Van Emburgh BO, Robertson KD. *Mutat Res.* 2008; 647(1-2):30-8.

[5] Miremadi A, Oestergaard MZ, Pharoah PD, Caldas C. Cancer genetics of epigenetic genes. *Hum Mol Genet.* 2007; 16:R28–49.

[6] Jeltsch A, Nellen W, Lyko F. Two substrates are better than one: dual specificities for Dnmt2 methyltransferases. *Trends Biochem Sci.* 2006; 31(6):306-8.
[7] Hermann A, Schmitt S, Jeltsch A. The human Dnmt2 has residual DNA-(cytosine-C5) methyltransferase activity. *J Biol Chem.* 2003; 278(34):31717–21.
[8] Schaefer M, Lyko F. Solving the Dnmt2 enigma. *Chromosoma.* 2010; 119(1):35-40.
[9] Goll MG, Kirpekar F, Maggert KA, Yoder JA, Hsieh CL, Zhang X, Golic KG, Jacobsen SE, Bestor TH. Methylation of tRNAAsp by the DNA methyltransferase homolog Dnmt2. *Science.* 2006; 311(5759):395-8.
[10] Okano M, Bell DW, Haber DA, Li E. DNA methyltransferases Dnmt3a and Dnmt3b are essential for de novo methylation and mammalian development. *Cell.* 1999; 99(3):247-57.
[11] Suzuki MM, Bird A. DNA methylation landscapes: provocative insights from epigenomics. *Nat Rev Genet.* 2008; 9(6):465-76.
[12] Cihák A. Biological effects of 5-azacytidine in eukaryotes. *Oncology.* 1974; 30(5):405–422.
[13] Vaquero A, Loyola A, Reinberg D. The constantly changing face of chromatin. *Sci Aging Knowledge Environ.* 2003; 2003(14):RE4.
[14] Bowman GD, Poirier MG. Post-translational modifications of histones that influence nucleosome dynamics. *Chem Rev.* 2015; 115(6):2274-95.
[15] Tessarz P, Kouzarides T. Histone core modifications regulating nucleosome structure and dynamics. *Nat Rev Mol Cell Biol.* 2014; 15(11):703-8.
[16] Jack AP, Hake SB. Getting down to the core of histone modifications. *Chromosoma.* 2014; 123(4):355-71.
[17] Simon M, North JA, Shimko JC, Forties RA, Ferdinand MB, Manohar M, Zhang M, Fishel R, Ottesen JJ, Poirier MG. Histone fold modifications control nucleosome unwrapping and disassembly. *Proc Natl Acad Sci U S A.* 2011; 108(31):12711-6.
[18] North JA, Javaid S, Ferdinand MB, Chatterjee N, Picking JW, Shoffner M, Nakkula RJ, Bartholomew B, Ottesen JJ, Fishel R, Poirier MG. Phosphorylation of histone H3(T118) alters nucleosome dynamics and remodeling. *Nucleic Acids Res.* 2011; 39(15):6465-74.
[19] Lee KK, Workman JL. Histone acetyltransferase complexes: one size doesn't fit all. *Nat. Rev. Mol. Cell Biol.* 2007; 8 (4):284–95.
[20] Bannister AJ, Kouzarides T. Regulation of chromatin by histone modifications. *Cell Res.* 2011; 21(3):381-95.
[21] Gibson CL, Murphy SP. Benefits of histone deacetylase inhibitors for acute brain injury; a systematic review of animal studies. *J. Neurochem.* 2010; 115:806–813.
[22] Volmar CH, Wahlestedt C. Histone deactylases (HDACs) and brain function. *Neuroepigenetics.* 2015; 1:20-27.
[23] Broide RS, Redwine JM, Aftahi N, Young W, Bloom FE, Winrow CJ. Distribution of histone deacetylases 1-11 in the rat brain. *J. Mol. Neurosci.* 2007; 31:47–58.
[24] Kahali S, Sarcar B, Prabhu A, Seto E, Chinnaiyan P. Class I histone deacetylases localize to the endoplasmic reticulum and modulate the unfolded protein response. *FASEB J.* 2012; 26(6):2437-45.
[25] Guan JS, Haggarty SJ, Giacometti E, Dannenberg JH, Joseph N, Gao J. HDAC2 negatively regulates memory formation and synaptic plasticity. *Nature.* 2009; 459; 55–60.

[26] Graff J, Rei D, Guan JS, Wang WY, Seo J, Hennig KM. An epigenetic blockade of cognitive functions in the neurodegenerating brain. *Nature.* 2012; 483:222–226.

[27] Lakshmaiah KC, Linu AJS. Aparna DL, Smitha CS. Epigenetic therapy of cancer with histone deacetylase inhibitors. *J Cancer Res Thera.* 2014; 10(3):469.

[28] Ellis L, Hammers H, Pili R. Targeting tumor angiogenesis with Histone Deacetylase Inhibitors. *Cancer Letters.* 2009; 280(2):145.

[29] Li L, Sun Y, Liu J, Wu X, Chen L, Ma L, Wu P. Histone deacetylase inhibitor sodium butyrate suppresses DNA double strand break repair induced by etoposide more effectively in MCF-7 cells than in HEK293 cells. *BMC Biochem.* 2015; 16:2.

[30] Campbell GR, Bruckman RS, Chu YL, Spector SA Autophagy induction by histone deacetylase inhibitors inhibits HIV type 1. *J Biol Chem.* 2015; 290(8):5028-40.

[31] Smith BC, Denu JM. Chemical mechanisms of histone lysine and arginine modifications. *Biochim Biophys Acta.* 2009; 1789:45–57.

[32] Schnell JD, Hicke L. Non-traditional functions of ubiquitin and ubiquitin-binding proteins. *J Biol Chem.* 2003; 278(38):35857–60.

[33] Popovic D, Vucic D, Dikic I. Ubiquitination in disease pathogenesis and treatment. *Nature Med* 2014; 20:1242–1253.

[34] Komander D, Rape M. The ubiquitin code. *Annu Rev Biochem.* 2012; 81:203–29.

[35] Dawson TM. Parkin and defective ubiquitination in Parkinson's disease. *J Neural Transm Suppl.* 2006; (70):209-13.

[36] Guo G, Gui Y, Gao S, Tang A, Hu X, Huang Y, Jia W, Li Z, He M, Sun L. Frequent mutations of genes encoding ubiquitin-mediated proteolysis pathway components in clear cell renal cell carcinoma. *Nat Genet.* 2011; 44(1):17-9.

[37] Dalle-Donne I, Aldini G, Carini M, Colombo R, Rossi R, Milzani A. Protein carbonylation, cellular dysfunction, and disease progression. *J Cell Mol Med.* 2006; 10(2):389–406.

[38] Sharma R, Nakamura A, Takahashi R, Nakamoto H, Goto S. Carbonyl modification in rat liver histones: decrease with age and increase by dietary restriction, *Free Radic Biol Med.* 2006; 40:1179–1184.

[39] Kim, VN. Small RNAs: classification, biogenesis, and function, *Mol Cells.* 2005; 19:1–15.

[40] Friedman RC, Farh KK, Burge CB, Bartel DP; Farh BB. Most mammalian mRNAs are conserved targets of microRNAs. *Genome Res.* 2009; 19 (1): 92–105.

[41] Rodriguez A, Griffiths-Jones S, Ashurst JL, Bradley A, Griffiths J, Ashurst B. Identification of mammalian microRNA host genes and transcription units. *Genome Res.* 2004; 14(10A): 1902–10.

[42] Sato F, Tsuchiya S, Meltzer SJ, Shimizu K. MicroRNAs and epigenetics. *FEBS J.* 2011; 278(10):1598-609.

[43] IUPAC Gold Book. Revised nomenclature for radicals, ions, radical ions and related species. *Pure&App/Chern.* 1993; 65(6):1357-1455.

[44] Muller FL, Liu Y, van Remmen H. Complex III releases superoxide to both sides of the inner mitochondrial membrane. *J Biol Chem.* 2004; 279:49064–49073.

[45] Mikhed Y, Daiber A, Steven S. Mitochondrial Oxidative Stress, Mitochondrial DNA Damage and Their Role in Age-Related Vascular Dysfunction. *Int J Mol Sci.* 2015; 16(7):15918-15953.

[46] Zawia NH, Lahiri DK, Cardozo-Pelaez F. Epigenetics, oxidative stress, and Alzheimer disease. *Free Radic Biol Med.* 2009; 46(9):1241-9.

[47] Kwok JB. Role of epigenetics in Alzheimer's and Parkinson's disease. *Epigenomics.* 2010; 2(5):671-82.

[48] Prattichizzo F, Giuliani A, Ceka A, Rippo MR, Bonfigli AR, Testa R, Procopio AD, Olivieri F. Epigenetic mechanisms of endothelial dysfunction in type 2 diabetes. *Clin Epigenetics.* 2015; 7(1):56.

[49] Yara S, Lavoie JC, Levy E. Oxidative stress and DNA methylation regulation in the metabolic syndrome. *Epigenomics.* 2015; 7(2):283-300.

[50] Ziech D, Franco R, Pappa A. Reactive oxygen species (ROS)--induced genetic and epigenetic alterations in human carcinogenesis. *Mut Res-Gen Tox En,* 2011; 711:167-73.

[51] Esteller M. Cancer epigenomics: DNA methylomes and histone modification maps. *Nat Rev Genet.* 2007; 8(4):286-98.

[52] Feinberg AP, Tycko B. The history of cancer epigenetics. *Nat Rev Cancer.* 2004; 4(2):143-53.

[53] Franco R, Schoneveld O, Georgakilas A, Panayiotidis MI. Oxidative stress, DNA methylation and carcinogenesis. *Cancer Lett.* 2008; 266:6–11.

[54] Ushijima T, Takada EO. Aberrant methylations in cancer cells: where do they come from? *Cancer Sci,* 2005; 96:206–11.

[55] Ushijima T. Detection and interpretation of altered methylation patterns in cancer cells. *Nat Rev Cancer.* 2005; 5:223–31.

[56] Toyokuni A. Molecular mechanisms of oxidative stress-induced carcinogenesis: from epidemiology to oxygenomics. *IUMBM Life.* 2008; 60:441–7.

[57] Georgakilas AG, Aziz K, Ziech D. BRCA1 involvement in toxicological responses and human cancer etiology. *Toxicol Lett,* 2009; 188:77–83.

[58] Kriaucionis S, Heintz N. The nuclear DNA base 5-hydroxymethylcytosine is present in Purkinje neurons and the brain. *Science.* 2009; 324(5929):929-30.

[59] Patchsung M, Boonla C, Amnattrakul P, Dissayabutra T, Mutirangura A, Tosukhowong P. Long interspersed nuclear element-1 hypomethylation and oxidative stress: correlation and bladder cancer diagnostic potential. *PLoS One.* 2012; 7(5):e37009.

[60] Gonçalves AC, Cortesão E, Oliveiros B, Alves V, Espadana AI, Rito L, Magalhães E, Pereira S, Pereira A, Costa JM, Mota-Vieira L, Sarmento-Ribeiro AB. Oxidative stress levels are correlated with P15 and P16 gene promoter methylation in myelodysplastic syndrome patients. *Clin Exp Med.* 2015. [Epub ahead of print].

[61] Gu X, Sun J, Li S, Wu X, Li L. Oxidative stress induces DNA demethylation and histone acetylation in SH-SY5Y cells: potential epigenetic mechanisms in gene transcription in Aβ production. *Neurobiol Aging.* 2013; 34(4):1069-79.

[62] Sundar IK, Yao H, Rahman I. Oxidative stress and chromatin remodeling in chronic obstructive pulmonary disease and smoking-related diseases. *Antioxidant Redox Signal.* 2013; 18(15):1956-71.

[63] Hoshino S, Yoshida M, Inoue K, Yano Y, Yanagita M, Mawatari H, Yamane H, Kijima T, Kumagai T, Osaki T, Tachiba I, Kawase I. Cigarette smoke extract induces endothelial cell injury via JNK pathway. *Biochem Biophys Res Commun.* 2005; 329:58–63.

[64] Vaziri H, Dessain SK, Ng Eaton E, Imai SI, Frye RA, Pandita TK, Guarente L, Weinberg RA. hSIR2(SIRT1) functions as an NAD-dependent p53 deacetylase. *Cell.* 2001; 107:149–159.

[65] Furukawa A, Tada-Oikawa S, Kawanishi S, Oikawa S. H2O2 accelerates cellular senescence by accumulation of acetylated p53 via decrease in the function of SIRT1 by NAD+ depletion. *Cell Physiol Biochem.* 2007; 20:45–54.

[66] Cheng Y, Liu X, Zhang S, Lin Y, Yang J, Zhang C. MicroRNA-21 protects against the H(2)O(2)-induced injury on cardiac myocytes via its target gene PDCD4. *J Mol Cell Cardiol.* 2009; 47:5–14.

[67] Tang Y, Zheng J, Sun Y, Wu Z, Liu Z, Huang G. MicroRNA-1 regulates cardiomyocyte apoptosis by targeting Bcl-2. *Int Heart J.* 2009; 50:377–387.

[68] Lin H1, Qian J, Castillo AC, Long B, Keyes KT, Chen G, Ye Y. Effect of miR-23 on oxidant-induced injury in human retinal pigment epithelial cells. *Invest Ophthalmol Vis Sci.* 2011; 52(9):6308-14.

[69] Sato F, Tsuchiya S, Meltzer SJ, Shimizu K. MicroRNAs and epigenetics. *FEBS J.* 2011; 278:1598-1609).

[70] Baer C, Claus R, Frenzel LP, Zucknick M, Park YJ, Gu L, Weichenhan D, Fischer M, Pallasch CP. Extensive promoter DNA hypermethylation and hypomethylation is associated with aberrant microRNA expression in chronic lymphocytic leukemia. *Cancer Res.* 2012; 72:3775-3785.

[71] Li J, Harris RA, Cheung, SW, Coarfa C, Jeong M, Goodell MA, White LD, Patel A, Kang SH. Genomic hypomethylation in the human germline associates with selective structural mutability in the human genome. *PLoS Genet.* 2012; 8; e1002692.

[72] Saito Y, Saito H., MicroRNAs in cancers and neurodegenerative disorders. *Front Genet.* 2012; 3:194).

[73] Brait M, Sidransky D. Cancer epigenetics: above and beyond. *Toxicol Mech Methods.* 2011; 21:275e288.

In: Epigenetics: Mechanisms and Clinical Perspectives
Editor: Yildiz Dincer
ISBN: 978-1-63484-490-1

Chapter 3

EPIGENETIC MODIFICATIONS AND THEIR POTENTIAL ROLE IN TUMORIGENESIS

Muthu K. Shanmugam[1], PhD, Frank Arfuso[2], PhD, and Gautam Sethi[1,2,*], PhD

[1]Department of Pharmacology, Yong Loo Lin School of Medicine, National University of Singapore, Singapore
[2]School of Biomedical Sciences, CHIRI Biosciences Research Precinct, Curtin University, Western Australia, Australia

ABSTRACT

Cancer is one of the deadliest malignancies that have plagued mankind for decades. Epigenetic mechanism(s) play a central role in the homeostasis of normal cell proliferation and differentiation. Global epigenetic modifications are frequently associated with cancer initiation, progression, as well as metastasis. These changes include DNA methylation, histone lysine methylation/demethylation, acetylation/ deacetylation, including methylation and acetylation of non-histone proteins, and can alter the expression of various oncogenic signaling cascades, which in-turn can lead to uncontrolled proliferation. In this chapter, we primarily focus on the major epigenetic changes that occur in oncogenes, tumor suppressor genes, transcription factors and cancer stem cells, which in turn mediate tumor growth. These modifications are controlled by regulatory enzymes such as DNA methyltransferase, histone acetyltransferases, histone deacetylase, lysine acetyltransferase, and arginine and lysine methyl transferases. In addition, we also describe a few selected pharmacological agents that can modulate the action of these enzymes and display significant potential for cancer therapy.

Keywords: cancer, epigenetics, transcription factors, oncogenes

* Correspondence author: gautam_sethi@nuhs.edu.sg.

ABBREVIATIONS

ALL	acute lymphocytic leukemia
AML	acute myeloid leukemia
APC	adenomatosis polyposis coli
ASH2L	absent, small or homeotic discs–like 2
AR	androgen receptor
BRCA1	breast cancer 1, early onset
COX	cyclooxygenase
CML	chronic myelogenous leukemia
CARM1	coactivator associated arginine methyltransferase 1
Cdk9	cyclin dependent kinase 9
CDKN2	cyclin dependent kinase 2
CpG	cytosine-phosphate-guanine
CREB	c-AMP response element binding protein
CSC	cancer stem cell
CDH1	E-cadherin
CREB	cAMP response element binding
CBP	cAMP response element binding (CREB) protein
CTL	cytotoxic T-Lymphocytes
DAPK1	death associated protein kinase 1
DLBCL	diffuse lager B-cell lymphoma
DNMT	DNA methyltransferase
DR	death receptor
EZH2	enhancer of zeste 2 polycomb repressive complex 2 subunit
ERRB2	human epidermal growth factor receptor 2
EMT	epithelial-to-mesenchymal transitions
ER	estrogen receptor
FDA	food and drug administration
FBXL11	F-box and leucine-rich repeat protein 11
GNAT5	Gcn5-related N-acetyltransferase
GSTP1	glutathione S-transferase-pi
GM-CSF	granulocyte macrophage colony stimulating factor
GSK3β	glycogen synthase kinase 3β
HCC	hepatocellular carcinoma
HIF1	hypoxia-inducible factor
HRE	HIF1α binding to the HIF1α-response element
H3K27me3	trimethylated lysine 27 on histone H3
H3K4me 1/2/3	mono-di-tri-methylated histone H3 at lysine 4
H3K64me3	trimethylated histone 3 at lysine 64
H3K9ac	acetylated histone H3 at lysine 9
H3K9me 1/2/3	mono-di-tri-methylated histone H3 at lysine 9
H4K20me3	trimethylated histone 4 at lysine 20
HMG	high mobility group
HSC	hematopoietic stem cells

IGF	insulin like growth factor
IKK	IκB kinase
IL	interleukin
JAK	janus kinase
JmjC/N	jumonji-domain-containing protein histone lysine demethylases
KAT	histone lysine acetyltransferases
KATi	histone lysine acetyltransferases inhibitor
KDAC	histone lysine deacetylase
KDACi	histone lysine deacetylase inhibitor
KMT	histone lysine methyltransferases
KDMs	histone lysine demethylases
MBD	methyl-binding domain
MBD2	methyl-CpG-binding domain protein 2
MBD3	methyl-CpG binding domain protein 3
MHC	major histocompatibility complex
MLL	histone methyl transferase
MOZ	monocytic leukaemia zinc finger protein
MKP1	mitogen activated protein kinase (MAPK) phosphatase-1
NSD1	nuclear receptor binding Su(var)3–9, enhancer of zeste and Trithorax (SET) domain protein 1
iNOS	inducible nitric oxide synthase
NF-κB	nuclear factor kappaB
NFAT	nuclear factor of activated T cell
PG	prostaglandins
PPAR-γ	proliferator-activated receptor gamma
PTEN	phosphatase and tensin homolog
PARP-1	poly-ADP-ribose-polymerase 1
PCAF	p300/CBP associated factor
PcG	polycomb group of proteins
PR	progesterone receptor
PRC2	polycomb repressive complex 2
PRMT	potein arginine methyl transferases
PHF	plant homeodomain (PHD)-finger protein
PKMT	protein lysine methyl transferase
PKDM	protein lysine demethylases
RASSF1	ras association domain-containing gene
RelA	v-rel avian reticuloendotheliosis viral oncogene homolog A
SAM	s-adenosyl methionine
SMYD	SET and MYND-domain containing
SIRT	sirtuin
Skp2	S-phase kinase associated protein 2
STAT3	signal transducer and activator of transcription
TLR	toll like receptor
TGFβ	transforming growth factor beta
TIMP-3	tissue inhibitor of metalloproteinase 3

TNFα	tumor necrosis factor alpha
TSA	trichostatin A
VHL	Von Hippel–Lindau

1. Introduction

Historically, epigenetics is defined as heritable changes in gene expression patterns that occur without any changes to the DNA but involve changes that are adequate to regulate activation or repression of gene expression. The term epigenetics was first introduced by developmental biologist Conrad H. Waddington in 1942. The latest revised definition of epigenetics includes the transient changes that occur during gene expression [1]. Epigenetics is an important physiological process that occurs all through the development and differentiation of any organisms that have the same set of DNA sequences, and drives them to diversify into different cell types [2]. In addition to its pivotal role in the development of an organism, deregulated epigenetics mechanisms are often encountered in the development and severity of disease. It is now increasingly apparent that chronic inflammation-driven diseases occur due to both genetic and epigenetic alterations [3, 4]. It was Rudolf Virchow in 1870 who linked inflammation with cancer, diabetes, obesity, arthritis, allergy, and atherosclerosis [5]. From then on the connection between inflammation and cancer development has been frequently observed in chronic inflammatory bowel disease and colorectal cancer [6], Barrett's esophagitis and esophageal cancer [7], and in virus-induced cancers such as human papilloma virus-induced cervical cancer, hepatitis B or C virus-induced liver cancer [8]. It has been estimated that 15% of all cancers are linked to inflammation [6]. In addition, environmental factors such as exposure to radiation are also key contributors in cancer progression [7]. The first and foremost defense response to any type of injury to the human body is the recruitment of immune cells such as leukocytes, lymphocytes, neutrophils, monocytes, and macrophages. These cells reduce the inflammatory reaction in an acute attack. However, in chronic inflammation, pro-inflammatory molecules are produced and secreted by these cells, which produces a cascading effect and thus, has been implicated in the development of various major human diseases including malignancies [9]. Thus, excessive production of pro-inflammatory molecules such as cytokines, chemokines, prostaglandins (PGs), nitric oxide (NO), and leukotrienes deregulate the cellular signaling and transcriptional mechanism(s), which ultimately leads to the development of neoplasms [10]. Thus, changes to the epigenome mediated by various enzymes are now been recognized as being important mediators in the activation or silencing of genes, the impact of which are as significant as heritable permanent genetic mutations in the DNA [11]. The impact of epigenetic alterations in disease progression has led pharmaceutical companies to target these histone and non-histone modifying enzymes for anti-cancer drug development [12-14].

In the eukaryotic cell nucleus, DNA is wrapped around the core histone octamer H2A, H2B, H3 and H4, thus forming the fundamental unit of chromatin, the nucleosome [15]. Post-translation modifications on core histone tails, along with cytosine methylation of the DNA, determines the accessibility of the chromatin and thereby enhancing the ability of transcription factors to bind to DNA and initiate gene transcription [15]. These transient changes are unstable and disappear very rapidly, and can be easily modulated by any external stimulus [15, 16]. In contrast to transient change, permanent change to the DNA often leads to underdeveloped or defective organs and to the development of diseases. These heritable permanent DNA changes often occur at specific dinucleotide sites along the genome such as the cytosine 5' of guanine (CpG) sites. It has been observed that 40% of genes contain clusters of CpG dinucleotide upstream of the transcription start site and are often found to be 70-80% methylated [17, 18]. Escalation in DNA methylation is often seen in chronic inflammation-driven diseases, viral infections, and cancers. Thus, aberrant DNA methylation now serves as a biomarker for the early detection of cancer and can also be related to the therapeutic effectiveness of drugs affecting DNA methylation and histone acetylation [4]. Epigenetic modifications of chromatin and DNA have in recent times been acknowledged as being either the expressive or suppressive factors in directly controlling gene transcription [18]. The post-translation modification on histone proteins in chromatin and methylation of DNA are regulated by two distinct pathways [18]. Environmental factors influence the epigenome at all times during the development of human life, especially nutritional factors, which can have profound effects in the expression of specific genes by epigenetic modifications [19]. The epigenome is characterized by global hypomethylation, but distinct CpG islands located in the regulatory regions of genes can be hypermethylated, leading to repression of gene expression [20]. Based on the levels of methylation status of CpG islands, monoallelic silencing and other epigenetic regulatory mechanisms have been reported in key inflammatory response genes [11, 18]. Histone hypermethylation silences genes involved in cell cycle control (*CDKN2*) [21], apoptosis (*DAPK1*) [22], DNA repair (*BRCA1*) [23], metastasis (*TIMP3*) [24], drug resistance, tumor heterogeneity, stem cell-like state [25], differentiation, epithelial-mesenchymal transition, metabolism [26], and VEGFR2 [27]. This global phenomenon of CpG hypermethylation is a common biological pathway to switching off genes. While DNA mutations have been linked to chronic inflammation-driven cancers, it is was also observed that epigenetic alterations can contribute to familial cancer risk [28]. Chronic inflammation mediates the initiation of a variety of diseases including cancer progression, which can be explained on the basis of epigenetic dysregulation [29-32]. Recent studies have indicated that DNA hypermethylation and histone modifications play important roles in gene silencing [33].

Selective silencing of DNA methyltransferase leads to acute fatty acid-induced, non-CpG methylation of proliferator-activated receptor gamma (PPAR-γ) co-activator 1 alpha promoter [34]. In this chapter, we primarily focus on the major epigenetic changes that occur in oncogenes, tumor suppressor genes, transcription factors, and cancer stem cells, which in turn mediate tumor growth. These modifications are controlled by regulatory enzymes such as DNA methyltransferase, histone methylases, histone demethylases, histone lysine acetyltransferases, histone deacetylases, and arginine and lysine methyl transferases. In addition, we also describe a few selected pharmacological agents that can modulate the action of these enzymes and display significant potential for cancer therapy.

2. The Dysregulation of Epigenetic Factors in Various Cancers

2.1. The Deregulation of DNA Methylation in Cancer

DNA methylation is a widespread epigenetic modification in eukaryotic cells. DNA methylation occurs at the 5' end of the CpG dinucleotide of the cytosine ring, with S-adenosyl-L-methionine as its methyl donor [35, 36]. This reaction is catalyzed by a family of closely related DNA methytransferases (DNMT), DNMT1, DNMT3A, and DNMT3B [37]. DNMT1 is the most abundant DNMT in mammals and maintains the methylation status on CpG dinucleotides [38]. In normal eukaryotic cells, the CpG islands are usually found in the unmethylated state to maintain euchromatin structure, and thereby allowing gene expression. However, during cancer development, many of the genes are hypermethylated at their CpG islands and silence gene transcription by changing the open euchromatic structure to a compact heterochromatic structure that is repressive for transcription [39]. DNMT1 predominantly adds a methyl group to DNA when one strand is already methylated (hemi-methylated) [40]. Once methylated they recruit the methyl-CpG-binding domain (MBD) proteins such as Kaiso, MeCp2, and members of the MBD family, which recognize methylated CpG islands and initiate chromatin silencing [41].

In cancer, promoter DNA hypermethylation has been observed for over two decades with the subsequent silencing of tumor suppressor genes [42, 43]. Global hypomethylation 'shores' also occur in cancer cells and are associated with genomic instability [44]. Both hypomethylation and hypermethylation have been seen across different types of cancers [45]. In cancer cells, hypermethylation are often observed in the promoter regions of tumor suppressor genes such as retinoblastoma gene (RB1), phosphatase and tensin homolog (PTEN), and DLC-1, glutathione S-transferase-pi (GSTP1), and E-cadherin (CDH1), thereby resulting in loss-of-function in colon cancers [46, 47]. Hypermethylation targeting of specific genes such as p16, p15, MINT1, MINT2, MINT31, MLH1, MSH2, MGMT, DAPK, APC, LKB1, IGF2, and COX2 has also been detected in colon cancer [47-49]. Global methylations have been observed in tumors of breast, prostate, lung, as well as in hematological malignancies [50, 51]. Overexpression of DNMT1 is widely reported in leukemias [51], and mutation and hypermethylation of this gene is observed in a variety of hematological malignancies [52]. Hypermethylation of genes such as GSTP1, RASSF1A, APC, MGMT, CCNA1, CDH1, RUNX3, ESR2, CDKN2A, RARβ2, IGFBP7, and SFRP2 are all implicated in prostate cancer development and progression [53, 54], including E-cadherin, CD44, galectins [55], ER, androgen receptor, retinoic acid receptor, and genes involved in apoptosis [56], tissue inhibitors of metalloproteinases (TIMPs) [57], and cyclinD kinases [58]. In breast cancer, more than 100 genes have been hypermethylated in all type of cell processes including cell cycle regulation genes CCND2, p16ink4A/CDKN2A [59], apoptosis genes such as APC, TWIST, and HOXA5 [60, 61], metastasis, invasion genes such as RARb2, CDH1, ER, BRCA1, CCND2, p16 and TWIST, angiogenic genes VEGFR2, eNOS and in cell signaling pathways [62]. The estrogen receptor (ER)α and progesterone receptor (PR), which are critical in hormone regulation, are also frequently methylated in breast cancer [63]. DNA hypomethylation is also frequently reported in breast cancer, especially on FEN1 BCSG1, PLAU, IGF2, and CDH3 [64, 65]. In addition, recent research has shown that miRNA that function as tumor suppressors can be silenced via DNA methylation [66]. Lower levels of genome-wide DNA hypomethylation have been observed in metastatic prostate cancer in contrast to non-metastatic prostate cancer [67, 68]. Gene-specific hypomethylation also occurs in urokinase plasminogen activator (uPa), cell proliferation genes, and in various cellular processes such as invasion and metastasis [69], cancer/testis antigen gene [70], hydroxylation of estrogens [71], and X-chromosome inactivation [72]. In pancreatic carcinoma, DNA hypomethylation occurs in the promoter regions of the RASSF1A gene and K-Ras mutations [73]. Similarly, global hypomethylation and hypermethylation are also observed in lung cancer [74]. DNA hypomethylation in H19, IGF2, and MEST genes results in abnormal lung cancer growth. In NSCLC cancer, testis antigen and melanoma-associated antigen were upregulated due to hypomethylation [75]. Hypermethylation of tumor suppressor genes such as *p53, CDKN2, p16INK4* [21], *MGMT* [76], *DAPK, caspase 8, ARF, FAS, TRAILR1* [76, 77], *RASSF1A, NORE1A, G0S2, cadherins, TIMP3* [78, 79] may also result in their silencing.

2.2. The Deregulation of Histone Lysine Methylation in Cancer

ε-*N*-methyl-lysine was first found in a bacterial flagellar protein in 1959 [80] and subsequently, in 1964, this post-translational modification (PTM) was also identified in histones [81]. Other PTMs include acetylation, ubiquitination, phosphorylation, sumoylation, and ribosylation at histone tails [82-84]. Several reports have indicated that PTM do influence the density of chromatin and thus accessibility to DNA and gene transcription [84]. Core histones H2A, H2B, H3, and H4 combine to form the basic structure of the histone octamer, the nucleosome [84, 85]. Mono, di-, or tri-methylation of histones takes place in the amino acid of the N-terminal tail of histones that are prone to covalent modifications. It has been shown that methylation predominantly occurs in histone H3 or H4 lysine residues such as H3K4, H3K9, H3K27, H3K36, and H4K20, and one lysine residue located in H3 (H3K79) [86, 87]. Nuclear receptor binding Su (var)3–9, Enhancer of zeste and Trithorax (SET) domain protein 1 (NSD1) is the largest protein of the NSD family of histone lysine methyltransferases. It is located on chromosome 5q35 and consists of 2696 aa [88]. In the multi-domain NSDs, it is the SET domain that is responsible for methyltransferase activity. The NSD protein lysine methyltransferases (KMTs) family is composed of three members, NSD1 (KMT3B), NSD2 (WHSC1/MMSET), and NSD3 (WHSC1L1), which regulate gene expression through methylation of lysine 36 on histone H3 (H3K36) [89]. An accumulation of evidence has unveiled the role of NSDs in a variety of cancers due to genetic alterations or aberrant overexpression. Numerous studies have reported the role of NSD1 with a distinct *HOX* gene expression pattern in acute myeloid leukemia (AML) [90-93]. In addition to NSD1, NSD2 has also been implicated in cancer pathogenesis. NSD2 mRNA was found to be overexpressed in 13 types of cancers [94], and overexpression of NSD2 protein in 3774 tumor samples of colon, small cell lung, skin, multiple myeloma, and bladder cancer [88, 95]. In breast, prostate, oligodendroglioma, and head and neck cancers, NSD2 overexpression was associated with higher tumor grade [96]. A high level of NSD2 is an indicator of poor overall and disease free survival in hepatocellular carcinoma and endometrial cancer [97, 98]. In addition, NSD3 has been identified as the fourth most frequently amplified methyltransferase in breast cancer [99, 100]. However, histone tri-methylation on the lysine 4 residue of histone

3 (H3K4me3) leads to gene activation while tri-methylation on lysine 9 (H3K9me3) or on lysine 27 (H3K9me3) leads to gene repression [19]. It is of note that histone methylation often leads to a condensed chromatin structure and thereby suppress gene expression [101]. EZH2 (KMT6) catalyzes the trimethylation of histone H3K27me3 [102] and initiates gene silencing via local chromatin reorganization. Several studies suggest that EZH2 plays a critical role in initiation, progression, and metastasis of prostate, breast, bladder, ovarian, renal carcinoma, lung, liver, brain, gastric, esophageal, pancreatic, and melanoma cancer, and also in cancer stem cells [102, 103]. In solid tumors, a hypoxic microenvironment induces hypoxia-inducible factor (HIF1)α binding to the HIF1α-response element (HRE) and transactivates EZH2 transcription [104]. There are also other methyltrasferases that are overexpressed in variety of cancers, such as KMT2A in human lymphoid and myeloid leukemias [105, 106], KMT1A in colon cancer [107], KMT1C in lung carcinoma [108, 109], Ash2L in hematological malignancies, hepatocellular carcinoma (HCC) and gastrointestinal cancers [110, 111], and SET and MYND-domain containing (SMYD)-3 in breast, colon, and HCC [112, 113]. Other lysine methyltransferases that are either mutated or downregulated in cancers are menin, KMT3B, and KMT8 [39, 114, 115].

2.3. The Deregulation of Histone Lysine Demethylases in Cancer

Lysine demethylases are a diverse group of demethylases. The lysine-specific histone demethylase (LSD) family and JmjC-domain containing histone demethylases (JHDMs) protein family have been identified as the two major families of demethylases that regulate histone demethylation, which are responsible for removing the methylation mark on lysines [82, 116-124]. LSD1 (KDM1A) and LSD2 (KDM1B) are flavin-dependent monoamine oxidases that can only demethylate mono- and di-methyl lysine and not tri-methyl lysine residues [117, 122, 125]. LSD1 demethylates H3K4me1/2 and H3K9me1/2, and non-histone proteins such as p53, E2F1, and DNMT1 [124]. LSD1 is overexpressed in prostate, breast, bladder, lung, colon, and neuroblastoma cancers [126], and is under expressed in HCC [111, 127]. LSD1 is also associated with mixed lineage leukemia (MLL) [128], and sustains the growth of MLL-AF9 acute myeloid leukemia cells by preventing differentiation and apoptosis [129], and serves as a biomarker for aggressive malignancies and tumor relapse [130]. It is of note that LSD1 upregulates cyclin A2 *(CCNA2)* and human epidermal growth factor receptor 2 *(ERRB2)* in aggressive breast cancer [130]. LSD2 is highly overexpressed in breast cancer and in invasive tumors, in contrast, a decreased expression of LSD2 is seen in leukemia and in seminoma [131]. The JmjC domain containing demethylases can demethylate mono-, di-, and tri-methyl lysine residues [120, 121, 124, 132]. Around 30 JmjC domain containing proteins have been identified in the human genome [133] and are divided into six sub-families based on histone lysine sites and demethylation states. They are KDM2, KDM3, KDM4, KDM5, KDM6, and plant homeo domain (PHD)-finger protein (PHF) [134]. JmjC domain-containing proteins belong to the Fe (II) and 2-oxoglutarate (2-OG)-dependent dioxygenases. They demethylate di- or tri-methylation marks on H3K4, H3K9, H3K27, H3K36, and non-histone proteins [124]. The KDM2 subfamily is comprised of KDM2A and KDM2B, which can demethylate H3K36A/B and the non-histone protein nuclear factor kappa-light-chain-enhancer of activated B cells (NF-κB) [135]. KDM2A/B is also involved in cellular processes such as ribosomal RNA transcription, cell proliferation, apoptosis, differentiation, and carcinogenesis [136-138]. Overexpression of KDM2A/B is observed in bladder cancer, T-acute lymphocytic leukemia (T-ALL), B-ALL, AML, breast, pancreatic adenocarcinoma, and seminoma. In contrast, reduced expression of KDM2A/B is observed in prostate cancer and glioblastoma [124, 139, 140]. In the human genome, the KDM4 family is sub-divided into two major groups based on their domain structure. KDM4A, KDM4B, and KDM4C form the first group while KDM4D, KDM4E, and KDM4F form the second group. The first group contains two PHD domains and two tudor domains in addition to the JmjC and JmjN domains, whereas the second group has only JmjC and JmjN domains [141]. KMD4A/B/C demethylates both histone and non-histone protein methylation such as WIZ, CDYL1, and G9a [142]. Therefore, KDM4 demethylases act as either repressors or activators of transcription and are involved in regulating various cellular processes including cell cycle and DNA damage repair, as well as tumor initiation and progression [124, 143, 144]. Numerous evidence suggests that overexpression of KDM4 demethylase is involved in breast, prostate, lymphoma, gastric, lung, bladder, medulloblastomas, and colorectal cancers, whereas mutational deletion or knockdown of KDM4A/B/C inhibits cancer cell proliferation and growth [133, 144-146]. The KDM5 family contains KDM5A, KDM5B, KDM5C, and KDM5D. This enzyme removes the methylation marks on H3K4me2/3 [132]. KDM5s are also implicated in cancer development. KDMs are either overexpressed (KDM5A) or downregulated (KDM3B) in breast cancers [147] Differences in KDMs have been shown to be associated with aberrant histone methylation and demethylation of H3K4 and shorter relapse free survival [147]. The abundance of methyl moieties on the lysine residue determines cancer prognosis. For example, increased H3K4me2 is found in cancerous cells but not in normal cells, and is a biomarker for tumor recurrence in prostate cancer [148-150]. Overexpression of KDM5A has been observed in lung and gastric cancers [151, 152], while overexpression of KDM5B was detected in bladder, AML, lung, breast, chronic myelogenous leukemia (CML), cervical, renal, metastatic prostate cancer, and testicular cancer [153]. Knockdown of KDM5B by siRNA results in inhibition of cancer growth [153, 154]. The KMD6 subfamily contains KDM6A, KDM6B, and UTY (KDM6C), which all exhibit high levels of catalytic activity on H3K27me2/3 [155, 156]. Several oncogenic mutations on KDM6A are observed in the cells of multiple myeloma, renal cell carcinoma, and esophageal squamous cell carcinoma [157, 158]. In addition, KMD6B can activate the tumor suppressor *p16INK4A* and *p14ARF*; hence, any mutation or deregulation in these genes might lead to carcinogenesis [159].

3. Histone Lysine Acetylation in Cancer

In addition to the other post-translational modifications, the physiological roles of N-ε-lysine protein acetylation have been identified in tumorigenesis [160]. Some histone acetylation marks have been associated with compaction of chromatin [161], DNA repair [162], protein stability, and protein-protein interaction [82], and lysine acetylation has emerged as the ubiquitous post-translational modification that is found across the entire proteome [163, 164]. Public repositories such as phosphositeplus database show approximately 15,000 lysine acetylation sites in human cells, with prominent roles in signaling networks [165]. Acetylation was initially identified 45 years ago primarily in histones, but lysine acetylation also occurs in non-histone proteins such as transcription factors, chromatin regulators, cytoplasmic proteins involved in the regulation of cytoskeleton dynamics, and in proteins involved with energy metabolism and endocytosis [163, 164, 166, 167].

3.1. The Deregulation of Histone Lysine Acetyltransferases in Tumor Proliferation and Metastasis

Access to DNA that is wrapped around histones occurs via post-translational modifications such as lysine acetylation. Approximately 10-20 lysine residues get acetylated per histone by lysine acetylatransferases (KATs), which add an acetyl group to the lysine residue on histone. Acetylation of histone increases the negative charge on DNA, thus facilitating the DNA accessibility to proteins involved in replication, transcription, or DNA repair [168-170]. The most understood set of KATs are GNAT5 (Gcn5-related N-acetyltransferase), PCAF (p300/CBP associated factor), Hat1, Elp3, and Hpa2, all of which are involved in transcription initiation. cAMP response element binding (CREB) protein (CBP) and p300 are the two most abundant KATs, both proteins are regulators of RNA polymerase II-mediated transcription, and are ~300 kDa proteins possessing structural homologues with high sequence similarity [171]. Additionally, both KATs acetylate at multiple sites on each of the four core histones H2A, H2B, H3, and H4, and regulate human growth and development [172]. Inactivation or mutation on either of both KATs leads to abnormal organ development and diseases in humans, while insufficiency in p300 leads to aberrant levels of acetylation in multiple cancers [173-177]. The germ line mutation of CBP results in Rubinstein-Taybi syndrome with increased susceptibility to childhood malignancies [171]. Therefore, it is vital to understand the divergent functions such as specificity and selectivity for lysine residues on core histones in treating the disease that they cause. Constitutive activation of transcription factors are observed in all types of cancers. In particular, acetylation of K9, K14, K18, and K56 of histone H3, and K5, K8, K13, and K16 of histone H4 are strongly involved in transcriptional activation [15, 178]. Histone lysine acetyltransferase is known to play important roles in normal and malignant hematopoiesis by acetylating histone and non-histone proteins. In normal hematopoietic stem/progenitor cells, myeloid progenitor cells, and lymphoid cells, protein acetylation by KATs such as p300/CBP, MOZ, HBO1, and GCN5 regulates hematopoietic stem cell (HSC) self-renewal, proliferation, and their differentiation into committed hematopoietic progenitors [179]. Several studies have identified aberrant KAT activity on oncogenes and tumor suppressor proteins that have been shown to be involved in the development of hematological malignancies. In AML, T-cell leukemia, lymphoma, B-cell ALL, and CML leukemogenic proteins physically interact with the KATs p300/CBP, PCAF, MOZ, Tip60, and GCN5, and target cell cycle genes, oncogenes, and tumor suppressor genes such as *Id1, p21, Egr1, c-Myc, p53, p53, RARβ, PU.1, Syk, Btk*, and *AML1* that promote malignant transformation of cells [179]. Over a dozen human lysine acetyltransferases have been identified. Two noted KATs are MOZ (monocytic leukemia zinc finger protein; a.k.a. MYST3 and KAT6A) and its paralog MORF (a.k.a. MYST4 and KAT6B) form tetrameric complexes with BRPF1 (bromodomain- and PHD finger-containing protein 1) and two small non-catalytic subunits. These two acetyltransferases and BRPF1 are found to be recurrently mutated in leukemia, non-hematologic malignancies, and in multiple developmental disorders. The BRPF1 gene is mutated in childhood leukemia and adult medulloblastoma [180]. In addition to hematologic malignancies, the MOZ and MORF gene is mutated in esophageal adenocarcinoma [181] and metastatic medulloblastoma [182]. The MORF gene is altered in leiomyomata [183, 184], breast cancer [185], and castration-resistant prostate cancer [186]. Moreover, both the MOZ and MORF genes have been identified as very common KATs amplified in different types of cancers [187], suggesting that they have an oncogenic role. Deletion of CBP/p300 specifically reduces acetylation on H3K18ac and H3K27ac in mouse embryonic fibroblasts [188]. Mice heterozygous for CBP spontaneously develop myelodysplastic and/or myeloproliferative tumors [189]. Genetic mutations/deletions of CBP/p300 are correlated with the development of transitional cell bladder cancer [190], follicular lymphoma, and DLBCL [191], relapsed ALL [192] and colon cancer [193]. Overexpression of CBP/p300 is a poor prognostic indicator in small-cell lung cancer [194], and is associated with higher grade tumor with distant metastasis of nasopharyngeal tumor [195]. In melanoma patients, decreased expression of p300 is associated with poor prognosis [196]. CBP/p300 can also induce tumorigenesis independently by acetylating S-phase kinase associated protein 2 (Skp2) [197]. Reduced expression of Tip60 is an indicator for poor prognosis in primary and metastatic melanoma patients [198], and it is also downregulated in colon, lung, head and neck carcinoma, metastatic prostate cancer, and lymphoma [199-204]. In hormone refractory or androgen independent prostate cancer, Tip60 accumulates in the nucleus [202]. In castration resistant prostate cancer LNCaP derived CxR cells; Tip60 is upregulated in the nucleus, which results in the activation of transcription even in the absence of androgen [205]. ERα and PRs have been implicated in breast cancer initiation and progression, and increased activation of ERα is a risk factor for the development of breast cancer by increasing cell proliferation and growth of abnormal breast tissue. ERα is a transcription factor that is activated by hormonal and growth factor signals. Several reports have highlighted the importance of ERα post-translational modification. It has been shown that ERα is acetylated by co-activator p300 at lysine in the hinge/ligand domain at K229, K299, K302, and K303 [206]. In atypical breast hyperplasia samples, lysine to arginine substitution was found at residue K303R in ERα, with increased estradiol dependent activation [207]. Another study demonstrated that, in the presence of p160 and p300 co-activators, ERα was acetylated at K266 and K268 [208]. In breast cancer cells where the tumor suppressor gene BRCA1 is mutated, p300/CBP acetylates ERα and drives cell proliferation, while reintroduction of BRCA1 downregulates p300 and not CBP [209]. Prostate cancer is one of the most frequently diagnosed cancers in men and a leading cause of death worldwide [210, 211]. The androgen receptor (AR) plays a crucial role in prostate cancer growth and progression, and mutations in the AR often lead to tumors becoming resistant to androgen ablation therapy [212]. Post-translational modifications of the AR result in enhanced activity. The AR is acetylated in response to dehydrotestosterone. The AR is acetylated at a conserved lysine residue at K630, K632, and K633, which increases its transcriptional activity [213-217]. It was found that a gain of function mutation increased cell proliferation and colony size in soft agar, and accelerated prostate tumor growth *in vivo* [214]. Histone acetylation was detected in prostate cancer patient tumors and was a positive predictor for tumor recurrence [218]. Deacetylation of ERα and AR by SIRT1 inactivates both the receptors' activity, and thus serves as an alternative approach in the prevention and therapy of breast and prostate cancer [212].

Oncogenic KATs p300 and TIP60 acetylate the AR in a ligand-independent manner and induce transcription of target genes involved in cell proliferation, metastasis, and apoptosis [173, 219]. In resected prostate cancer patients treated with endocrine therapy, p300 and CBP expression is increased. Inhibition of p300 but not CBP using targeted siRNA showed a significant decrease in proliferation rate and a significant increase in the induction of apoptosis [173]. This might explain why p300 is upregulated in highly metastatic prostate cancer and is an indicator of poor prognosis.

3.2. The Deregulation of Histone Lysine Deacetylases (KDACs) in Cancer

Addition or removal of the acetyl group is catalyzed by acetylases and deacetylases respectively. Aberrant expression due to mutation or deletions of various KDACs is commonly encountered in chronic inflammation associated diseases, in particular cancer. Thus, KDACs are an attractive target for cancer therapy. Numerous studies have indicated global histone acetylation is deregulated in human cancers, and absence of the acetylation mark in H4K16 and H4K20 is critical in establishing the tumor phenotype [220]. The KDACs family of enzymes is divided into four classes and 18 members, based on their sequence homology to yeast KDAC [221, 222]. KDACs deacetylate the acetyl moiety on lysines and thereby can interact with numerous transcription factors to regulate a large number of genes [221]. KDACs are transcriptional repressors because they condense the chromosomes and make them less accessible to transcription initiation [223]. Class 1 KDAC1, 2, 3, and 8 are large protein complexes that silence genes by removing the acetyl moiety on lysine residues on histone tails. High expression of KDAC1, 2, and 3 is seen in urothelial bladder cancer [224]; it is also associated with shorter survival of colorectal cancer patients, and is highly expressed in prostate cancer. Interestingly, KDAC2 is the single prognostic factor that is associated with shorter prostate-specific antigen relapse time following radical prostatectomy [225]. In pancreatic cancers, this class of KDAC is associated with dedifferentiation and rapid proliferation of cells [226-228]. KDAC1 was found to be overexpressed in several organ specific cancers such as gastric, HCC, lung, prostate, breast, and pancreatic cancer samples derived from patients and is a predictor of poor prognosis [228, 229]. Other studies have also reported the overexpression of KDAC1, 2, and 3 in renal cell carcinoma, gastric cancer, and Hodgkin's lymphoma [228, 230, 231]. KDAC8 overexpression is associated with high grade tumor and poor overall survival in childhood neuroblastoma [232]. In a different study analyzing breast tumors, KDAC1 and KDAC3 expression correlates with ER and PR expression, pointing to KDAC1 as an independent prognostic marker [233]. In a tissue microarray representing 44 categories of malignant and borderline mesenchymal tumors, KDAC2 was highly expressed compared to KDAC1, indicating that KDAC2 more likely contributes to the pathogenesis of the disease [234]. KDAC2 is also aberrantly expressed in lung cancer tissues [235]. Class II KDACs are sub-divided into two groups: class IIa consists of KDAC4, KDAC5, KDAC7, and KDAC9; and class IIb consists of KDAC6 and KDAC10. Numerous studies both *in vitro* and *in vivo* have established an association between class IIa KDAC expression and tumor development. KDAC4 is overexpressed in Walsdenstrom's macroglobulinemia [236], breast, renal, and colorectal cancer [237].

Table 1. List of synthetic inhibitors of histone lysine deacetylase either approved or undergoing clinical trials

Compound	KDAC targets	Current status	References
Compound 6j and SEN196	SIRT2	Phase II	[439]
Panobinostat (LBH-589)	KDAC1, 2, 3 and 6	FDA approved	[440]
Vorinostat (SAHA)	KDAC1, 2, 3 and 6	FDA approved	[439]
Romidepsin (FK228)	KDAC1, 2, 3 and 8	FDA approved	[439]
Valproic acid	Class I and Class IIa	FDA approved	[438]
Belinostat (PXD101)	KDAC1, 2, and 3	Phase II	[438]
Entinostat (MS275)	KDAC1, 2, 3 and 8	Phase II	[438]
Mocetinostat (MGCD01030)	KDAC1, 2 and 3	Phase II	[438]
Givinostat (ITF2357)	KDACs	Phase II	[438]
Practinostat (SB939)	KDACs	Phase II	[438]
Chidamide (CS055/HBI-8000)	KDACs	Phase II	[438]
Quisinostat (JNJ26481585)	KDACs	Phase II	[438]
CHR-5154	Macrophage targeted KDACs	Phase I	[441]
CUDC-907	KDACs	Phase I	[441]
Abexinostat (PCI-24781)	KDAC1, 2, 3, 6 and 10	Phase II	[438]
Trichostatin A	KDAC1, 2, 3 and 6	Preclinical	[442]

In addition, KDAC4 dysfunction is also implicated in cancer development. For example, KDAC4 homozygous deletion was observed in melanoma cells [238], and KDAC4 mutations have also been observed in breast cancer [239]. In renal cell carcinoma, KDAC4 is associated with repression of *Pax7* and regulates HIF1α transcriptional activity. KDAC5 and KDAC9 are overexpressed and are poor prognostic indicators in medulloblastoma patients, demonstrating an intricate relationship in their expression and poor survival [240]. KDAC5 and KDAC3 overexpression is correlated with copy number gains in hepatocellular carcinoma [241]. In colorectal cancer, KDAC5 and KDAC7 are overexpressed and interact with the transcription factor GATA-1 [226, 227]. In pancreatic cancer, KDAC7 is overexpressed [242, 243]. In ALL, high levels of KDAC7 and KDAC9 are associated with poor prognosis [244]. In contrast, KDAC7 is not overexpressed in breast, renal, and bladder cancer, and myeloproliferative neoplasms [245]. KDAC9 has been reported to be overexpressed in cervical cancer [246], whereas in glioblastoma, KDAC9 is downregulated [247]. Class IIb KDAC6 is overexpressed in breast and oral squamous cell carcinoma [228]. Deacetylation is also catalyzed by another class of enzymes called sirtuins (Class III KDACs). Sirtuins are NAD+ dependent enzymes [248-250]. Sirtuins consist of seven members (Sir1-7) homologous to the yeast KDAC silent information regulator 2 (Sir2) that was cloned and characterized in 1984 as a gene required for maintaining silent chromatin in yeast [251]. Sirtuins are regulators of various cellular and pathological processes such as cell proliferation and differentiation, cell survival, stress response and DNA damage, genome stability, metabolism, energy homeostasis, organ development, aging, and cancer. Sirtuins are overexpressed in variety of cancers. SIRT1 is upregulated in prostate, acute myeloid leukemia, and colorectal cancer. SIRT3 and SIRT7 are overexpressed in breast cancer whereas SIRT2 is downregulated in gastric cancer and glioma [252, 253]. KDAC11 belongs to the Class 4 KDACs, and it is overexpressed in colon, prostate, breast, and ovarian cancer [254]. In Hodgkin lymphoma, KDAC11 regulates OX40, a mediator of anti-tumor immunity [228]. However, inhibition of KDACs activity by sodium butyrate or trichostatin A induces apoptosis by upregulating pro-apoptotic protein Bad in human glioma T98G, U251MG, and U877MG cells [255]). There are several FDA approved KDACs inhibitors that are used to treat multiple myeloma, hematological malignancies and solid tumors (Table 1).

4. Histone Acetylation "Code" as a Diagnostic Marker for Cancer Progression

Heritable epigenetic marks on histones, such as acetylation, methylation, ubiquitination, and phosphorylation, are maintained with high fidelity during cell division. This phenomenon of histone modifications that act in a dependable manner is called the "histone code" [116, 256]. These post-translational modifications present on the histone tail "codes" are decoded by various cellular machinery regulating cellular processes [256]. Histone post-translational modifications have been detected in almost all cancers types. Loss of H4K16 acetylation and H4K20Me3 is one of the hallmarks of cancers [220]. H3K4Me, H3K9Me2, H3K9Me3, and H3Ac were all reduced in prostate cancer compared to normal prostate tissue [150]. Furthermore, a decrease in H3K4Me2, H3K9Me2, and H3K18Ac levels are all biomarkers of lung cancer, kidney cancer, and pancreatic adenocarcinoma [257, 258]. In lung cancer, aberrant hyperacetylation of H4K5 and H4K8 was observed compared to normal lung epithelium, where hypoacetylation was observed [259]. In addition, H4K20Me3 was shown to be a biomarker for early detection of squamous cell carcinoma [259]. Barlesi et al., 2007 showed that low levels of H2AK5Ac indicated poor prognosis in lung cancer patients [260]. Similarly, adenocarcinoma patients with reduced H3K9Ac exhibited better survival [260]. DNA mismatch repair (MMR) is observed in several cancers and is characterized by frequent alterations in simple repetitive DNA sequences, which is a phenomenon called microsatellite instability and plays a role in apoptosis. In a recent review by Guo Min Li et al., 2013, it was reported that the H3K6Me3 histone biomarker plays an important role in the regulation of MMR *in vivo* in hereditary nonpolyposis colorectal cancer or Lynch syndrome [261]. Taken together, these studies show that histone post-translational modifications can serve as biomarkers for the early diagnosis of cancer and as prognostic predictors.

5. Epigenetic Modifications in Oncogenes and Tumor Suppressor Genes

Oncogenes or proto-oncogenes transcribe oncogenic proteins that alter normal cell proliferation to an abnormal and uncontrolled proliferation of transformed cells [262]. Oncogenes were discovered three decades ago and have, over the years, provided valuable information in understanding the genetic landscape of cancers, differentiation, and apoptosis [262]. One of the most commonly encountered oncogenes in cancer is the *Myc* oncogene. *Myc* regulates a complex inflammatory milieu [263] and, upon activation in B cells, it rapidly induces synthesis and release of IL1β [264]. The pleiotropic effects of oncogenes include creating a pro-tumor microenvironment, and a persistent, constitutive activation of pro-inflammatory transcritption factors [262, 265, 266]. To date approximately 20 transforming oncogenes including *Ras, Raf, Myc, c-Src, EGFR,* and *HER-2*, and a number of tumor suppressor genes such as *p53, VHL,* and *PTEN* are directly implicated in angiogenesis [267-269]. For example, vascular endothelial growth factor production is upregulated in cancer cells expressing a mutant *Ras* oncogene [268]. Global hypomethylation of genomic DNA is one of the main contributors in the development of aggressive cancers by upregulating oncogenes such as *c-Myc* or *h-Ras* [20]. In addition, promoter hypomethylation can also cause activation of aberrant expression of oncogenes and induce loss of imprinting in some loci. The tumor suppressor gene, *MASPIN* (also known as *SERPINB5*) becomes hypermethlyated in breast and prostate cancer cells [270], whereas *MASPIN* is hypomethylated in other tumor types. *MASPIN* hypomethylation and its subsequent expression increases with the degree of dedifferentiation of some types of cancer cells [271, 272]. The *Myc* oncogene is deregulated in most cancers, where it exhibits abnormally high levels of expression. In a recent study by Wasylishen AR et al., 2014, they show that *Myc* signaling is regulated by six lysine residues near the highly conserved *Myc* homology box IV. These lysine residues are important for the negative regulation of *Myc*-induced transformation and tumorigenesis, which may expose novel therapeutic strategies to target *Myc* in cancer [273]. This region has been previously associated with *Myc* acetylation [274, 275].

The Ras small GTPase family of proteins regulates cellular responses to external stimuli by post-translational modifications. Disruptions to these signals through mutations of *Ras* are involved in 90% of pancreatic cancer and 50% of colon cancer. *Ras* has been shown to be acetylated at lysine 104 and is a negative regulatory modification of *Ras* oncogenicity [276]. In hematological malignancies such as AML and ALL, low frequency mutation in the KAT genes leads to poor prognosis in patients [277]. Chromosomal translocations during mitosis and meiosis generate fusion proteins and chimeric proteins with oncogenic potential. In some rare cases, KAT gene translocations form chimeric proteins that retain KAT catalytic activity and bromodomains. For example, this is seen in mixed lineage leukemia (MLL)-CBP [t(11;16) (q23;p13)] or in MLL-p300 [t(11;22)(q23;q13)] fusions [278]. In addition, monocytic leukemia zinc finger protein (MOZ) fusions such as the MOS-CBP [t(8;16)(p11;p13)] or MOZ-p300 [t(8;22)(p11;q13)] are also reported for AML [174, 279, 280]. These fusions are frequently encountered in patients who have been treated with topoisomerase II inhibitors for the treatment of other cancers, and who can develop secondary therapy-related leukemia [280]. Other oncogenic fusions, in particular the most frequent in AMLs, AML1-ETO [t(8;21)(q22;q220], require p300 mediated acetylation to induce leukemogenesis [281]. This clearly indicates the importance of transcriptional co-activators p300/CBP and their oncogenic role in cancer progression. In another oncogenic fusion between nuclear receptor co-activator TIF2 and MOZ is associated with AML [inv (8)(p11q13)] and recruits CBP to function as an oncoprotein [180, 282, 283]. Indeed, altered expression of the KATs is found in various cancers [237]. Overexpression of p300 is observed in HCC [284], breast cancer [285], prostate cancer [219], and non-small cell lung cancer [286]. In human skin cancer, upregulation of the enzyme ornithine decarboxylase leads to histone H4 acetylation and skin tumor progression [287]. All these studies suggest KATs as being an integral part of carcinogenesis. p300/CBP acetylates lysine residues K372, K373, and K382 at the carboxy terminal of p53. In addition, DNA damage induces acetylation of

K320 and K373, which regulates interactions at DNA binding sites of pro-apoptotic genes [288].

Tumor suppressor *p53* is the most extensively studied gene that is mutated in almost all types of cancers [289]. Under normal physiological conditions, *p53* induces cell cycle arrest, induces apoptosis, and upsets other cellular processes such as senescence and differentiation [290]. The frequent post-translation modification observed in *p53* is acetylation by p300 on 5 lysine residues at the C-terminal regulatory region [291]. Acetylation of p53 is triggered by ultraviolet light and ionizing radiation [292]). In addition, the *Ras* oncogene and other stress factors also induce acetylation of p53 [293]. PCAF acetylates K320, while TIP60 (HIV Tat-interacting protein; 60 kDa) and MOF (males absent on the first)-two members of the MYST family-acetylated K120 are located within the DNA binding domain of *p53* [294, 295]. K120 acetylation induces apoptotic target genes but with minimal or no effect on cell cycle genes. Acetylation of *p53* neutralizes the positive charge on lysine and impairs its ability to form hydrogen bonds. Acetylation also regulates both loss of function and gain of function such as enhancing DNA binding [296] and preventing nonspecific DNA binding by the *p53* C-terminal domain [297]. In addition, acetylation forms docking sites for transcription co-activators, where K382 increases *p53* affinity to CBP bromodomain [298, 299] whereas deacetylation of K373 and K382 enables interaction with the tandem bromodomains of TAF1, a TFIID subunit [299]. Acetylation of *p53* blocks ubiquitination and K320 acetylation but PCAF prevents ubiquitination by E4F1 [300]. Remarkably, both acetylation and ubiquitination activate *p53*. Ubiquitination of K370, K372, K373, K381, and K382 indicates *p53* nuclear export and proteasomal degradation [296]. Thereby, acetylation of these residues stabilizes *p53* and promotes its nuclear localization. Interestingly, in human cancer, p300 and PCAF acetylation sites are not mutated, except K120 which is mutated in human cancer and is linked with cell cycle control and apoptosis [290, 294, 295, 301]. PC4 is a transcriptional coactivator that is acetylated by p300. PC4 interacts with p53, and both proteins become acetylated, which enhance their DNA binding ability and leads to the expression of p53 responsive genes, especially during DNA damage [302, 303].

p53 is also methylated at lysine residues [304]. Lysine mono-methylation of *p53* at K372 by protein lysine methyl transferase (PKMT) and NSD1 regulates its stability and transcriptional activity [305]. This methylation pattern is restricted to the nucleus and is not seen in any of the cytosolic fractions [305]. Subsequently, Huang et al., 2006, showed that PKMT SMYD2 methylates *p53* at K370 and impairs the expression of *CDKN1A* [306]. Inhibiting SMYD2 by small interfering RNA activates *p53*-mediated apoptosis of cancer cells [306]. Interestingly, mono- or di-methylation at K370 of *p53* has contrasting functions. Mono-methylation results in repression of *p53* activity, whereas di-methylation increases the association of *p53* with the co-activator tandem tudor domain containing the p53-binding protein1 [126]. LSD1 preferentially demethylates di-methylated *p53* at K370 [126]. The complex nature of methylation or demethylation by SYMD2 or LSD1 respectively can promote oncogenesis, as evidenced by the overexpression of both the enzymes [126]. In another study by Shi et al., 2007, they show that PKMT SETD8 monomethylates *p53* at K382 and suppresses *p53* transcriptional activity in cancer cells [307], indicating that methylation is also an important regulator of *p53* function.

E2F is a family of transcription factors that modulates important cellular events, including cell cycle progression, apoptosis, and DNA damage response, and also functions as a tumor suppressor gene [153, 308]. RB1 is a cell cycle regulator. Both *E2F* and *RB1* genes are exposed to PTMs in several types of cancer [309]. *RB1* is methylated at K810 by SMYD2, which has been identified as being pivotal for cell cycle regulation [310-312]. Methylated *RB1* subsequently enhances phosphorylation of serine 807 and 811 [312]. Unmethylated *RB1* is a suppressor of *E2F* transcriptional activities but, upon RB1 methylation at K810, it transactivates E2F transcriptional activity, and promotes RB1 phosphorylation and cell cycle progression. Furthermore, MYPT1 K442 is a target for methylation. Demethylation of MYPT1 by SETD7 and LSD1 enhances polyubiquitination and subsequent degradation [313], and thereby releases E2F, which transcribes genes that are involved in S-phase to promote cell cycle progression [313]. Taken together, methylation and demethylation regulates cell cycle progression by modulation RB1 activity. Deregulated *E2F* has been detected in a variety of human cancers [308]. In p53 deficient cells, methylation of K185 or E2F1 by SETD7 induces polyubiquitination and degradation of E2F1, and thus prevents E2F1 accumulation upon DNA damage [314]. In addition, LSD1 also demethylates *E2F1* at K185 and maintains an adequate amount of unmethylated *E2F1* in cancer cells [314].

6. Epigenetic Modifications in Cellular Signaling Pathways

6.1. NF-κB Signaling Pathway

The master transcription factor NF-κB was first discovered by David Baltimore in 1986. It is present in the cytoplasm in a hetrodimeric form bound to its inhibitor, and upon phosphorylation of its inhibitor, NF-κB translocates to the nucleus of the cell and binds to the promoter region in the immunoglobulin kappa chain in B cells [315]. NF-κB regulates multiple biological functions, including inflammation, immunity, cell proliferation, and apoptosis [316]. NF-κB is activated by many diverse agents such as interleukins, cytokines, T and B-cell mitogens, gram negative bacterial lipopolysaccharide, viruses and viral proteins, double-stranded DNA, and chemical and physical stress [317-319]. NF-κB regulates more than 400 genes involved in inflammation, immunoregulation, tumor cell proliferation, invasion, metastasis and angiogenesis, chemoresistance and radioresistance [320-322]. Activation of an inflammatory response is mainly mediated through NF-κB, which is susceptible to post-translational modifications such as acetylation, lysine methylation, and arginine methylation, in addition to its primary phosphorylation modification [323]. Acetylation modification has been shown to be a major player in inflammatory gene expression [324] in T-cells and monocytes [325-327]. Acetylation occurs in the lysine residue of RelA, which regulates NF-κB transcription activation, DNA binding affinity, IκB-α assembly, and subcellular localization [328, 329]. Inhibitory kappa kinase (IKKα), in association with polymerase II and CREB binding protein (CBP), binds to the NF-κB dependent promoter, where it acetylates H3K9 [330] and also phosphorylates H3Ser10 [331]. This IKKα-dependent and cytokine-mediated phosphorylation is essential for CBP-mediated acetylation of H3K14 [332]. One of the common features in cytokine-mediated inflammation and in autocrine and paracrine activation is that it leads to increased NF-κB driven inflammatory gene expression [333]. However, glucocorticoids and KDAC2 can reverse NF-κB activation and thus the inflammatory reaction [334]. Poly (ADP-ribose) polymerase -1 (PARP-1), in association with NF-κB, contributes to the progression of chronic inflammation driven diseases [335]. PARP-1 is a promoter-specific co-activator of NF-κB *in vivo* and is independent of its enzymatic activity [336]. PARP-1 directly interacts with p300, p50, and p65, and synergistically activates NF-κB [337]. Under inflammatory conditions, p300 can acetylate PARP-1 at specific lysine residues in a variety of cell types. NF-κB mediated H3 and H4 acetylation induced by IL1β, TNFα, or endotoxins can induce the production of granulocyte colony stimulation factor (GM-CSF) [338]. Thus, it is obvious that reversible acetylation of NF-κB, PARP1, and histones plays a prominent role in regulating gene expression during inflammation. Another important post-translational modification is methylation. An altered lysine methylation pattern is often observed during inflammatory diseases. The H3K4 methyltransferase SET7/9 can also regulate NF-κB(p65) to its promoters, thereby activating inflammatory gene expression [339]. In addition, arginine methyltransferases PRMT1 and PRMT4 are transcriptional co-activators of NF-κB, although they have not been directly implicated in inflammation [340-342]. Recent studies have implicated NF-κB hyperactivity upon deletion of H3K9 methyl marks from its promoter region, especially in hyperglycemic memory [343, 344], suggesting a strong link between epigenetic modification and NF-κB in the development of chronic inflammation-driven disease [345, 346]. Constitutive activation of transcription factors is observed in all types of cancers. In particular, acetylation of K9, K14, K18, and K56 of histone H3, and acetylation of K5, K8, K13, and K16 of histone H4 are strongly involved in transcriptional activation [15, 178]. The master transcription factor NF-κB is another non-histone protein and a key regulator of the cellular immune response that is also regulated by reversible acetylation at K218, K221, and K310. Acetylation of NF-κB regulates different functions such as DNA binding affinity, IκBα assembly and subcellular localization, and in the recruitment of NF-κB co-activators such as BRD4 and PTEFB [347], which stimulates the transcription of NF-κB target genes. In turn, acetylated NF-κB is deacetylated by KDAC3 [348]. Acetylated NF-κB and its subsequent recruitment of bromodomain-containing factor Brd4, which is a member of the bromodomain extraterminal (BET) family, leads to sustained activation of NF-κB in adult T-cell leukemia [349]. Acetylation of p65 at K122 and K123 reduces its affinity for DNA and increases IκB interaction, while suppressing downstream gene expression [350]. In addition, acetylation of p50 at K431, K440, and K441 enhances transcriptional activation [351]. In breast cancer, the ER along with NF-κB plays an important role in cell proliferation and survival. Pradhan et al., 2012 show that there is a major increase in NF-κB dependent histone acetylation around the estrogen response element in the BIRC3 promoter region in aggressive luminal B breast tumors [352], which is in contrast to earlier reports that suggested repressive interactions between ER and NF-κB [352].

The importance of protein lysine methylation in the pathogenesis of cancer is broadly divided into 5 categories. First, lysine methylation affects other post-translational modifications directly or indirectly. Second, lysine methylation regulates protein-protein interaction. Third, lysine methylation inhibits polyubiquitination of lysine residues. Fourth, lysine methylation regulates subcellular localization of proteins. Fifth, lysine methylation regulates promoter binding affinity of transcription factors, thereby controlling the expression levels of target genes [353, 354]. Lysine di-methylation of RelA by PKMT and NSD1 increases its promoter binding ability and transcriptional activity [88, 135, 355]. Upon TNFα stimulation, SETD7 methylates K37 RelA in the nucleus and increases the promoter binding affinity of RelA. Lu et al., 2013 reported that NSD1 methylates RelA at two sites, K218me1 and K221me2, to increase NF-κB transcription of target genes, while any mutations on these two sites diminish its ability to initiate transcription [355]. In addition, SETD6 methylates RelA at K310 in primary immune cells and completely abolishes the NF-κB mediated inflammatory response [356]. RelA methylation is demethylated by PKDM F-box and leucine-rich repeat protein 11 (FBXL11). Overexpression of FBXL11 inhibits NF-κB activation and impedes the growth of HT29 colon cancer cells [135]. Yang et al., 2012 found that, in castration-resistance prostate cancer, NSD2 interacts with NF-κB and acetyltransferase p300 and di or tri-methylates H3K36 at the promoter regions of NF-κB target genes [357]. Overall, transcriptional activation or inhibition of NF-κB is intricately linked to the site of methylation on RelA.

6.2. STAT3 Signaling Cascade

Signal transducer and activator of transcription (STAT) is another important transcription factor that is constitutively activated in cancer cells. The primary route of signaling in the STAT signaling pathway is via cytokine ligand-induced receptor dimerization, which activates receptor-associated tyrosine kinases and in-turn phosphorylates STATs, induces dimerization, nuclear translocation, DNA binding, and cytokine responsive genes [358]. Acetylation of STAT3 is a cytokine-induced, post-translational modification. STAT1, STAT2, STAT3, STAT5, and STAT6 have been shown to be acetylated by CBP/p300 [359]. Of all the STAT family of transcription factors, the most prominent is STAT3. STAT3 gets acetylated by histone acetyl transferase CBP/p300 at K685. Acetylation enhances its DNA binding affinity, increases transcriptional activation, promotes protein–protein interaction, and modulates dimerization [360]. Deacetylation by KDAC3, and to a lesser extent by KDAC1 and KDAC2, can lead to loss of DNA binding and suppression of transcription [360]. Thus, the acetylation and deacetylation reaction is a novel signaling mechanism that regulates the IL6/STAT3 signaling pathway in hepatocellular carcinoma. In addition to K685 acetylation, STAT3 also gets acetylated by p300 on aa49 and aa87 at its NH2 terminal domain, but the exact role of this in transcriptional initiation in still unclear [361]. When lysine residues were substituted by arginine at K679, K685, and K707, it reduced STAT3 phosphorylation compared to a K685R mutant alone, signifying the importance of the regulatory role of multiple acetylations [362].

In another report by Ohbayashi et al., 2007, they showed that IL-6 or leukemia inhibitory factor induced acetylation of STAT3 at K685 in 293T and Hep3B cells. This effect was abolished by the PI3K inhibitor, LY294002 [363]. STAT1 also gets acetylated and binds to the NF-κB subunit p65, and negatively regulates NF-κB target genes in cancer cells [364]. In the majority of published studies, cytokines have been shown to be the major inducers of acetylation; however, KDAC inhibitors have also been shown to hyperacetylate STAT3. In diffuse, large B-cell lymphoma (DLBLC), the KDAC inhibitor LBH589 leads to STAT3 hyperacetylation and dephosphorylation, and thereby inhibiting STAT3 transcription in DLBCL cells [365]. Acetylation of the transcription factor E2F1, which regulates cell cycle progression, at the N-terminal promotes DNA binding activity [366, 367] while acetylation attenuates Foxo1 and p65 transcription factors' ability to bind to DNA [350, 368]. The innate immune response activated by toll like receptors (TLRs) upon exposure to lipopolysaccharide often leads to chronic inflammation. Stimulation of TLRs induces expression of mitogen activated protein kinase (MAPK) phosphatase-1 (MKP1), which is acetylated at a key lysine residue K57, and promotes the dephosphorylation of p38MAPK and c-jun N-terminal kinase (JNK), resulting in the attenuated production of pro-inflammatory cytokines [369]. STAT3 is a latent transcription factor that is constitutively activated in numerous cancer cells [370]. STAT3 is di-methylated by SETD7 at K140 following phosphorylation, dimerization, nuclear translocation, and promoter binding; however, this reversible methylation is demethylated by LSD1 [371]. This transient dimethylation has a minimal repressive effect and contributes to specific transcriptional regulation. Furthermore, STAT3 is trimethylated at K180 by EZH2, which increases the phosphorylation of STAT3 in glioblastoma cells [372]. Interestingly, inhibiting EZH2 directly inhibits Polycomb gene expression and indirectly blocks STAT3 activation, which serves as a strategic target for glioblastoma therapy.

6.3. Wnt/β-Catenin Signaling Pathway

The Wnt signaling pathway is evolutionarily conserved and plays important roles in the development of organ systems. The important mediators in this pathway are adenomatous polyposis coli (APC), glycogen synthase kinase 3β (GSK3β), axin, and the transcriptional cofactor β-catenin (encoded by *CTNNB1*) [373]. Under normal circumstances, β-catenin levels remain low due to constitutive phosphorylation of β-catenin by GSK3β and subsequent polyubiquitination and degradation of β-catenin [374]. Wnt ligand binding to the frizzled (Fz) receptor activates the Disheveled (Dvl) adaptor protein, which in turn inhibits GSK3β activity and thereby inhibits β-catenin degradation. β-catenin translocates to the nucleus where it binds to the T-cell factor (Tcf)/lymphoid-enhancing factor (Lef) family of transcription factors and induces the expression of genes such as *CMYC* and *CCND1* [375]. A high level of nuclear β-catenin is often observed in a variety of cancers [374]. Aberrant Wnt signaling is a very common phenomenon in colon and hepatocellular carcinoma [376]. Epigenetic silencing of genes involved in Wnt signaling (by DNA methylation), and histone modification in tumor suppressor genes is a major contributor in aberrant Wnt/β-catenin signaling in cancers [377]. Promoter hypermethylation of Wnt inhibitors such as secreted frizzled-related protein (SFRP), Wnt inhibitory factor 1 (WIF1), HDPR1, and Dickkopf 3 (DKK3) is seen in a variety of tumors [378-380]. Schiefer et al., 2014 recently demonstrated that hypermethylation of the SFRP family was a common feature of human glioblastoma multiforme [381].

Overexpression of Wnt1, 2, 3A, and 5A leads to oncogenic activation of canonical Wnt signaling in cancers of the breast, colon, lung, and prostate [382, 383]. Wnt5A, a tumor suppressor, is silenced by methylation in hematological malignancies. Inactivation of Wnt7A and Wnt9A by promoter hypermethylation has been reported in ALL, pancreatic, and colon cancer [384, 385]. Promoter hypermethylation of SFRP1and WIF1 is a poor prognostic indicator in breast cancer and acute promyelocytic leukemia respectively [386-388]. In gastric cancer, DKK3 hypermethylation is strongly correlated with a shorter survival time for patients [389]. Epigenetic silencing of *CDH1* and *APC* by promoter methylation leads to aberrant activation of Wnt/β-catenin signaling in invasive ductal carcinoma of the breast [390]. NSD2 methyltransferase may activate the Wnt signaling pathway through interaction with β-catenin [391]. NSD2 was shown to interact with β-catenin and its co-activators IQ motif containing GTPase activating protein 1 (IQGAP1) and T-cell lymphoma invasion and metastasis 1 (TIAM1) proteins and increase motility, adhesion, and migration properties [88]. Inhibiting NSD2 concomitantly decreases the expression of cyclin D1 (*CCND1*), a target of the β-catenin/Tcf4 complex, through H3K39 trimethylation [88]. Interestingly, Ge et al., 2009 reported that PCAF acetylation stabilizes β-catenin in colon cancer cells [392].

7. Epigenetic Modification in Cancer Stem Cells

Cancer stem cells (CSCs) represent a small sub-population in the whole tumor population that represents the tumor initiating clone. They are very similar to normal stem cells in that they have unlimited self-renewal capacities with the potential to differentiate into many cancer cell types. CSCs are resistant to chemotherapy and may be responsible for the distant site metastasis of cancers [393, 394]. EZH2 catalyzes methylation of histone H3K27me3 to cause the silencing of tumor suppressor genes [395, 396]. Extensive reports have shown that EZH2 plays a critical role in cancer initiation, progression, and metastasis, and also in cancer stem cell biology [102, 397]. EZH2 plays a pivotal role in the maintenance and self-renewal of embryonic and adult stem cells [398-400]. EZH2 has been implicated in the maintenance of CSC, promotion of breast cancer progression [401], and glioblastoma CSC self-renewal and progression [402, 403]. In hematological malignancies, DNA hypermethylation by DNMTs of oncogenic pathway regulatory genes may contribute to the differentiation of hematopoietic stem cells to leukemia stem cells (LSCs) and leukemogenesis [51, 52, 404-406].

Activating mutations of the DNMT3A gene is a poor prognosis indicator in AML, with leukemogenic mutations of Flt3-ITD, NPM1, PTPN11, and IDH2 genes being associated with intermediate risk [407]. In addition, other mechanisms mediated by KDMs and KATs play key roles in the development and maintenance of LSCs [52]. In AML-CSCs and non-CSCs there is a difference in the histone methylation pattern, especially in H3K4me3 and H3K27me3, whereas it is the same in DNA methylation patterns [408]. In glioblastoma multiforme (GBM)-CSCs, the loss of polycomb mark H3K27me3 from the promoters is the primary cause for the activation of multiple transcription factors [409]. In particular, activation of the achaete-scute complex homolog-like 1 (Ascl1) transcription factor causes activation of the Wnt signaling pathway, which is required for the maintenance of stemness and its tumorigenicity [410, 411].

8. Epigenetic Modifications Associated with Cancer Chemo- and Radio-Resistance

Epigenetic alterations are now increasingly recognized as major factors in tumor biology and are integral for their role in treatment resistance. Epigenetic modulators have made a profound impact in preventing, delaying, or reversing chemo-resistance, as seen with the use of natural product molecules or synthetic molecules that can reverse DNA methylation or histone deacetylation [412, 413]. The emergence of a CSC population is now recognized as one of the causative factors contributing to multi-therapy resistance, which is further compounded by new mutations arising in genes associated with oncogenic signaling pathways [414-416]. For example, *K-Ras* mutation in multiple tumor types, *APC* in colon cancer, and *VHL* in renal cancer empower the cancer cells with self-renewal and tumorigenic properties [25]. CBP/p300 is implicated in resistance in certain cancer types. P300 expression is reduced in doxorubicin-resistant bladder cancer cells and knockdown of p300 leads to bladder cancer cells being resistant to doxorubicin [417].

In addition, CBP/p300 is required in the interleukin-4-mediated castration resistance of prostate cancer cells. Downregulation of CBP/p300 abolished interleukin-4-mediated AR activation [418]. Overall, the changes in genetic alterations, expression level, and subcellular localization of CBP/p300 are associated with progression, prognosis, and resistance of many types of cancers. Acquired drug resistance is the most common phenomenon in cancer therapy [419].

Overexpression of the hMLH1 DNA mismatch repair enzyme in ovarian cancer confers resistance to platinum drugs; 25% of patients acquired hMLH1 methylation during chemotherapy, and hMLH1 methylation is poor prognosis indicator [420].

Treatment with the DNMT inhibitor azacitidine reactivates hMLH1, reverses resistance [421, 422], and sensitizes ovarian cancer cells to platinum drugs *in vitro* [423]. The methylation status of the cells serves as a biomarker to monitor progression in patients under treatment with platinum drugs [413, 424, 425]. KAT Tip60 is overexpressed in cisplatin-resistant human lung cancer cells, and knockdown of Tip60 expression sensitizes the cells to cisplatin therapy [426]. ABCB1 (MDR1) is a multidrug resistance p-glycoprotein that is often induced in traditional chemotherapy. However, upon chemotherapy, hypermethylated promoter regions decrease in these cells while, in contrast, epigenetic methylation is implicated in the development of multidrug resistance [427, 428]. These findings suggest that epigenetic methylation independently could contribute to multidrug resistance and not as a part of accumulating genetic mutations in cancer development. In cisplatin resistant ovarian cancer cells, KDAC4 is overexpressed and promotes cancer cell survival by deacetylating the transcription factor STAT1 [429]. The promoter methylation-specific silencing of gene O-6-methylguanine-DNA methyltransferase (*MGMT*) in glioma serves as a biomarker to the alkylating agent carmustine or temozolomide [430-433]. Romidepsin, a KDAC inhibitor, increases sensitivity to EGFR tyrosine kinase inhibitors in lung cancer [434]. In a combination therapy with the KDAC inhibitors entinostat and erlotinib, patients experienced disease progression due to erlotinib resistance that was due to epigenetic changes [435]. In another combination therapy, all-trans-retinoic acid with the LSD1/KDM1A inhibitor tranylcypromine has been shown to upregulate the myeloid differentiation pathway [129, 436, 437].

9. PHARMACOLOGICAL INHIBITORS OF EPIGENETIC MODIFICATIONS

Especially in malignant epithelial and hematological tumors, deregulation of cellular signaling pathways is often due to overexpression of epigenetic modifying enzymes that either methylate or acetylate the DNA or the lysine residues on the histone. KDAC inhibitors are classified into groups based on their chemical structure, including hydroxamic acids (trichostatin A, vorinostat), carboxylic acids (valproate, butyrate), aminobenzamides (entinostat, mocetinostat), cyclic peptides (apicidin, romidepsin), epoxyketones (trapoxins), and hybrid molecules [438]. Table 1 lists the KDAC inhibitors that have been approved by the FDA and the other drugs that are in clinical trials for the treatment of cancers. Table 2 depicts the DNA methyltransferase inhibitors that are in clinical trials. Table 3 indicates the histone methyltransferase and histone demethylase inhibitors currently undergoing clinical trials. The histone acetyl transferase inhibitors currently in use are presented in Table 4.

Table 2. A list of selected pharmacological DNA methyltransferase inhibitors

Compound	Mechanism(s) of action	Current status	Reference
5-Azacytidine (Vidaza)	Nucleoside DNMT inhibitor	FDA approved	[441]
Decitabine (Dacogen)	Nucleoside DNMT inhibitor	FDA approved	
5-fluor-2'-deoxycitidine (FdCyd) with tetrahydrouridine (THU)	FdCyd metabolite inhibits DNMT	In Phase I and Phase II clinical trials	
ASTX-727	Nucleoside DNMT inhibitor	In phase I clinical trials	
CC-486	Nucleoside DNMT inhibitor	In phase I-III clinical trials	
Hydralazine	Non-nucleoside DNMT inhibitor	In Phase I-III clinical trials	
MG-98	Non-nucleoside DNMT inhibitor	In phase I clinical trials	
RRx-001	Non-nucleoside DNMT inhibitor	In Phase I and phase II clinical trials	
SGI-110	Nucleoside DNMT inhibitor	In phase I and Phase II clinical trials	
Zebularine	Nucleoside DNMT inhibitor	Preclinical	
SGI-1027	Non-nucleoside DNMT inhibitor	Preclinical	
RG108	Non-nucleoside DNMT inhibitor	Preclinical	
Psammaplin A	Non-nucleoside DNMT inhibitor	Preclinical	
Procain and procainamide	Non-nucleoside DNMT inhibitor	Preclinical	
CP-4200	Nucleoside DNMT inhibitor	Preclinical	
Epigallocatechin-3-gallate	Nucleoside DNMT inhibitor	Preclinical	

Table 3. A list of selected Histone lysine methyl transferase (KMTi) and histone lysine demethylase inhibitors (KDMi)

Compound	Target	Reference
Chaetocin	KMTi	[442]
DZNep	KMTi	
BIX-01294	KMTi	
AMI-1	KMTi	
AMI-5	KMTi	
Allantodapsone	KMTi	
RM-65	KMTi	
Tranylcypromine	KMTi	
Pargyline	KMTi	
Phenelzine	KMTi	

Table 4. A list of selected Histone lysine acetyltransferase inhibitors (KATi)

Compound	Target(s)	References
Garcinol	PCAF/p300	[442, 443]
Isogarcinol/LTK14	PCAF/p300	[439, 443]
Curcumin	CBP/p300	[439, 443]
Anacardic acid	PCAF/p300	[439, 443]
Lys–CoA	p300	[439]
Lys–CoA–CH3	PCAF	[439]
Ph–Lys-CoA	P300	[439]

PERSPECTIVES AND CONCLUSION

Epigenetic changes such as DNA methylation and histone modification by enzyme DNA methylases, histone lysine methyltransferases, demethylases, histone lysine acetyltransferases and deacetylases control critical gene transcription mechanisms. Their deregulation is often seen in diseases and malignancies. Detailed investigations of the changes that regulate disease progression are essential, and epigenetic alterations constitute valuable therapeutic targets. Large scale screening of the epigenome has helped in identifying several new epigenetic targets that have enormous potential in epigenetic-specific therapy. Several KDAC and DNA methyltransferase inhibitors have been approved by the FDA for the treatment of cancers. While natural product agents have shown significant progress in preclinical studies, well defined human clinical trials are essential to validate their potential as a therapeutic modality. Overall, existing evidence(s) clearly suggest that drugs targeting the epigenome have made significant inroads in cancer therapy, especially in the management of hematological malignancies and in other aggressive solid tumors.

ACKNOWLEDGMENTS

This work was supported by NUHS Bench-to- Bedside grant to GS.

Conflict of Interest Statement

The authors declare no competing financial interests.

REFERENCES

[1] Hirst M, Marra MA. Epigenetics and human disease. *Int J Biochem Cell Biol.* 2009;41(1):136-46.
[2] Feinberg AP. Phenotypic plasticity and the epigenetics of human disease. *Nature.* 2007;447(7143):433-40.
[3] Egger G, Liang G, Aparicio A, Jones PA. Epigenetics in human disease and prospects for epigenetic therapy. *Nature.* 2004;429(6990):457-63.
[4] Esteller M. Epigenetics in cancer. *N Engl J Med.* 2008;358(11):1148-59.
[5] Heidland A, Klassen A, Rutkowski P, Bahner U. The contribution of Rudolf Virchow to the concept of inflammation: what is still of importance? *J Nephrol.* 2006;19 Suppl 10:S102-9.
[6] Balkwill F, Mantovani A. Inflammation and cancer: back to Virchow? *Lancet.* 2001;357(9255):539-45.
[7] Schottenfeld D, Beebe-Dimmer J. Chronic inflammation: a common and important factor in the pathogenesis of neoplasia. *CA Cancer J Clin.* 2006;56(2):69-83.
[8] Saigo K, Yoshida K, Ikeda R, Sakamoto Y, Murakami Y, Urashima T, et al. Integration of hepatitis B virus DNA into the myeloid/lymphoid or mixed-lineage leukemia (MLL4) gene and rearrangements of MLL4 in human hepatocellular carcinoma. *Hum Mutat.* 2008;29(5):703-8.
[9] Aggarwal BB, Shishodia S, Sandur SK, Pandey MK, Sethi G. Inflammation and cancer: how hot is the link? *Biochem Pharmacol.* 2006;72(11):1605-21.
[10] Surh YJ, Kundu JK, Na HK, Lee JS. Redox-sensitive transcription factors as prime targets for chemoprevention with anti-inflammatory and antioxidative phytochemicals. *J Nutr.* 2005;135(12 Suppl):2993S-3001S.
[11] Portela A, Esteller M. Epigenetic modifications and human disease. *Nat Biotechnol.* 2010;28(10):1057-68.
[12] Eliseeva ED, Valkov V, Jung M, Jung MO. Characterization of novel inhibitors of histone acetyltransferases. *Mol Cancer Ther.* 2007;6(9):2391-8.
[13] Selvi BR, Mohankrishna DV, Ostwal YB, Kundu TK. Small molecule modulators of histone acetylation and methylation: a disease perspective. *Biochim Biophys Acta.* 2010;1799(10-12):810-28.
[14] Wagner JM, Hackanson B, Lubbert M, Jung M. Histone deacetylase (HDAC) inhibitors in recent clinical trials for cancer therapy. *Clin Epigenetics.* 2010;1(3-4):117-36.

[15] Berger SL. The complex language of chromatin regulation during transcription. *Nature.* 2007;447(7143):407-12.
[16] Bird A. Perceptions of epigenetics. *Nature.* 2007;447(7143):396-8.
[17] Bird A. DNA methylation patterns and epigenetic memory. *Genes Dev.* 2002;16(1):6-21.
[18] Wilson AG. Epigenetic regulation of gene expression in the inflammatory response and relevance to common diseases. *J Periodontol.* 2008;79(8 Suppl):1514-9.
[19] Bishop KS, Ferguson LR. The interaction between epigenetics, nutrition and the development of cancer. *Nutrients.* 2015;7(2):922-47.
[20] Das PM, Singal R. DNA methylation and cancer. *J Clin Oncol.* 2004;22(22):4632-42.
[21] Otterson GA, Khleif SN, Chen W, Coxon AB, Kaye FJ. CDKN2 gene silencing in lung cancer by DNA hypermethylation and kinetics of p16INK4 protein induction by 5-aza 2'deoxycytidine. *Oncogene.* 1995;11(6):1211-6.
[22] Katzenellenbogen RA, Baylin SB, Herman JG. Hypermethylation of the DAP-kinase CpG island is a common alteration in B-cell malignancies. *Blood.* 1999;93(12):4347-53.
[23] Dobrovic A, Simpfendorfer D. Methylation of the BRCA1 gene in sporadic breast cancer. *Cancer Res.* 1997;57(16):3347-50.
[24] Kang SH, Choi HH, Kim SG, Jong HS, Kim NK, Kim SJ, et al. Transcriptional inactivation of the tissue inhibitor of metalloproteinase-3 gene by dna hypermethylation of the 5'-CpG island in human gastric cancer cell lines. *Int J Cancer.* 2000;86(5):632-5.
[25] Easwaran H, Tsai HC, Baylin SB. Cancer epigenetics: tumor heterogeneity, plasticity of stem-like states, and drug resistance. *Mol Cell.* 2014;54(5):716-27.
[26] Li L, Li W. Epithelial-mesenchymal transition in human cancer: comprehensive reprogramming of metabolism, epigenetics, and differentiation. *Pharmacol Ther.* 2015;150:33-46.
[27] Chen LJ, Wei SY, Chiu JJ. Mechanical regulation of epigenetics in vascular biology and pathobiology. *J Cell Mol Med.* 2013;17(4):437-48.
[28] Fleming JL, Huang TH, Toland AE. The role of parental and grandparental epigenetic alterations in familial cancer risk. *Cancer Res.* 2008;68(22):9116-21.
[29] Brigati C, Noonan DM, Albini A, Benelli R. Tumors and inflammatory infiltrates: friends or foes? *Clin Exp Metastasis.* 2002;19(3):247-58.
[30] Coussens LM, Werb Z. Inflammation and cancer. *Nature.* 2002;420(6917):860-7.
[31] Mantovani A, Allavena P, Sica A, Balkwill F. Cancer-related inflammation. *Nature.* 2008;454(7203):436-44.
[32] Rosin MP, Anwar WA, Ward AJ. Inflammation, chromosomal instability, and cancer: the schistosomiasis model. *Cancer Res.* 1994;54(7 Suppl):1929s-33s.
[33] Vaissiere T, Sawan C, Herceg Z. Epigenetic interplay between histone modifications and DNA methylation in gene silencing. *Mutat Res.* 2008;659(1-2):40-8.
[34] Barres R, Osler ME, Yan J, Rune A, Fritz T, Caidahl K, et al. Non-CpG methylation of the PGC-1alpha promoter through DNMT3B controls mitochondrial density. *Cell Metab.* 2009;10(3):189-98.
[35] Herman JG, Baylin SB. Gene silencing in cancer in association with promoter hypermethylation. *N Engl J Med.* 2003;349(21):2042-54.
[36] Weber M, Schubeler D. Genomic patterns of DNA methylation: targets and function of an epigenetic mark. *Curr Opin Cell Biol.* 2007;19(3):273-80.

[37] Okano M, Xie S, Li E. Cloning and characterization of a family of novel mammalian DNA (cytosine-5) methyltransferases. *Nat Genet.* 1998;19(3):219-20.
[38] Bernstein BE, Meissner A, Lander ES. The mammalian epigenome. *Cell.* 2007;128(4):669-81.
[39] Chen QW, Zhu XY, Li YY, Meng ZQ. Epigenetic regulation and cancer (review). *Oncol Rep.* 2014;31(2):523-32.
[40] Jeltsch A. Molecular enzymology of mammalian DNA methyltransferases. *Curr Top Microbiol Immunol.* 2006;301:203-25.
[41] McCabe MT, Brandes JC, Vertino PM. Cancer DNA methylation: molecular mechanisms and clinical implications. *Clin Cancer Res.* 2009;15(12):3927-37.
[42] Galm O, Herman JG, Baylin SB. The fundamental role of epigenetics in hematopoietic malignancies. *Blood Rev.* 2006;20(1):1-13.
[43] Weisenberger DJ, Velicescu M, Preciado-Lopez MA, Gonzales FA, Tsai YC, Liang G, et al. Identification and characterization of alternatively spliced variants of DNA methyltransferase 3a in mammalian cells. *Gene.* 2002;298(1):91-9.
[44] Schoofs T, Muller-Tidow C. DNA methylation as a pathogenic event and as a therapeutic target in AML. *Cancer Treat Rev.* 2011;37 Suppl 1:S13-8.
[45] Hansen KD, Timp W, Bravo HC, Sabunciyan S, Langmead B, McDonald OG, et al. Increased methylation variation in epigenetic domains across cancer types. *Nat Genet.* 2011;43(8):768-75.
[46] Schuebel KE, Chen W, Cope L, Glockner SC, Suzuki H, Yi JM, et al. Comparing the DNA hypermethylome with gene mutations in human colorectal cancer. *PLoS Genet.* 2007;3(9):1709-23.
[47] Yuan BZ, Durkin ME, Popescu NC. Promoter hypermethylation of DLC-1, a candidate tumor suppressor gene, in several common human cancers. *Cancer Genet Cytogenet.* 2003;140(2):113-7.
[48] Fang JY, Lu R, Mikovits JA, Cheng ZH, Zhu HY, Chen YX. Regulation of hMSH2 and hMLH1 expression in the human colon cancer cell line SW1116 by DNA methyltransferase 1. *Cancer Lett.* 2006;233(1):124-30.
[49] Goel A, Arnold CN, Niedzwiecki D, Carethers JM, Dowell JM, Wasserman L, et al. Frequent inactivation of PTEN by promoter hypermethylation in microsatellite instability-high sporadic colorectal cancers. *Cancer Res.* 2004;64(9):3014-21.
[50] Yang L, Rau R, Goodell MA. DNMT3A in haematological malignancies. *Nat Rev Cancer.* 2015;15(3):152-65.
[51] Trowbridge JJ, Sinha AU, Zhu N, Li M, Armstrong SA, Orkin SH. Haploinsufficiency of Dnmt1 impairs leukemia stem cell function through derepression of bivalent chromatin domains. *Genes Dev.* 2012;26(4):344-9.
[52] Vecchio L, Seke Etet PF, Kipanyula MJ, Krampera M, Nwabo Kamdje AH. Importance of epigenetic changes in cancer etiology, pathogenesis, clinical profiling, and treatment: what can be learned from hematologic malignancies? *Biochim Biophys Acta.* 2013;1836(1):90-104.
[53] Ngollo M, Dagdemir A, Karsli-Ceppioglu S, Judes G, Pajon A, Penault-Llorca F, et al. Epigenetic modifications in prostate cancer. *Epigenomics.* 2014;6(4):415-26.
[54] Reibenwein J, Pils D, Horak P, Tomicek B, Goldner G, Worel N, et al. Promoter hypermethylation of GSTP1, AR, and 14-3-3sigma in serum of prostate cancer patients and its clinical relevance. *Prostate.* 2007;67(4):427-32.

[55] Woodson K, Hayes R, Wideroff L, Villaruz L, Tangrea J. Hypermethylation of GSTP1, CD44, and E-cadherin genes in prostate cancer among US Blacks and Whites. *Prostate.* 2003;55(3):199-205.

[56] Phe V, Cussenot O, Rouppret M. Interest of methylated genes as biomarkers in urothelial cell carcinomas of the urinary tract. *BJU Int.* 2009;104(7):896-901.

[57] Pulukuri SM, Patibandla S, Patel J, Estes N, Rao JS. Epigenetic inactivation of the tissue inhibitor of metalloproteinase-2 (TIMP-2) gene in human prostate tumors. *Oncogene.* 2007;26(36):5229-37.

[58] Henrique R, Costa VL, Cerveira N, Carvalho AL, Hoque MO, Ribeiro FR, et al. Hypermethylation of Cyclin D2 is associated with loss of mRNA expression and tumor development in prostate cancer. *J Mol Med* (Berl). 2006;84(11):911-8.

[59] Parrella P, Poeta ML, Gallo AP, Prencipe M, Scintu M, Apicella A, et al. Nonrandom distribution of aberrant promoter methylation of cancer-related genes in sporadic breast tumors. *Clin Cancer Res.* 2004;10(16):5349-54.

[60] Swift-Scanlan T, Vang R, Blackford A, Fackler MJ, Sukumar S. Methylated genes in breast cancer: associations with clinical and histopathological features in a familial breast cancer cohort. *Cancer Biol Ther.* 2011;11(10):853-65.

[61] Fackler MJ, McVeigh M, Evron E, Garrett E, Mehrotra J, Polyak K, et al. DNA methylation of RASSF1A, HIN-1, RAR-beta, Cyclin D2 and Twist in in situ and invasive lobular breast carcinoma. *Int J Cancer.* 2003;107(6):970-5.

[62] Li S, Rong M, Iacopetta B. DNA hypermethylation in breast cancer and its association with clinicopathological features. *Cancer Lett.* 2006;237(2):272-80.

[63] Yan L, Yang X, Davidson NE. Role of DNA methylation and histone acetylation in steroid receptor expression in breast cancer. *J Mammary Gland Biol Neoplasia.* 2001;6(2):183-92.

[64] Dietrich D, Lesche R, Tetzner R, Krispin M, Dietrich J, Haedicke W, et al. Analysis of DNA methylation of multiple genes in microdissected cells from formalin-fixed and paraffin-embedded tissues. *J Histochem Cytochem.* 2009;57(5):477-89.

[65] Yballe CM, Vu TH, Hoffman AR. Imprinting and expression of insulin-like growth factor-II and H19 in normal breast tissue and breast tumor. *J Clin Endocrinol Metab.* 1996;81(4):1607-12.

[66] Lehmann U, Hasemeier B, Romermann D, Muller M, Langer F, Kreipe H. [Epigenetic inactivation of microRNA genes in mammary carcinoma]. *Verh Dtsch Ges Pathol.* 2007;91:214-20.

[67] Kim SJ, Kelly WK, Fu A, Haines K, Hoffman A, Zheng T, et al. Genome-wide methylation analysis identifies involvement of TNF-alpha mediated cancer pathways in prostate cancer. *Cancer Lett.* 2011;302(1):47-53.

[68] Bedford MT, van Helden PD. Hypomethylation of DNA in pathological conditions of the human prostate. *Cancer Res.* 1987;47(20):5274-6.

[69] Pakneshan P, Szyf M, Rabbani SA. Hypomethylation of urokinase (uPA) promoter in breast and prostate cancer: prognostic and therapeutic implications. *Curr Cancer Drug Targets.* 2005;5(7):471-88.

[70] Cho B, Lee H, Jeong S, Bang YJ, Lee HJ, Hwang KS, et al. Promoter hypomethylation of a novel cancer/testis antigen gene CAGE is correlated with its aberrant expression and is seen in premalignant stage of gastric carcinoma. *Biochem Biophys Res Commun.* 2003;307(1):52-63.

[71] Tokizane T, Shiina H, Igawa M, Enokida H, Urakami S, Kawakami T, et al. Cytochrome P450 1B1 is overexpressed and regulated by hypomethylation in prostate cancer. *Clin Cancer Res.* 2005;11(16):5793-801.

[72] Laner T, Schulz WA, Engers R, Muller M, Florl AR. Hypomethylation of the XIST gene promoter in prostate cancer. *Oncol Res.* 2005;15(5):257-64.

[73] Dammann R, Schagdarsurengin U, Liu L, Otto N, Gimm O, Dralle H, et al. Frequent RASSF1A promoter hypermethylation and K-ras mutations in pancreatic carcinoma. *Oncogene.* 2003;22(24):3806-12.

[74] Rauch TA, Zhong X, Wu X, Wang M, Kernstine KH, Wang Z, et al. High-resolution mapping of DNA hypermethylation and hypomethylation in lung cancer. *Proc Natl Acad Sci USA.* 2008;105(1):252-7.

[75] Glazer CA, Smith IM, Ochs MF, Begum S, Westra W, Chang SS, et al. Integrative discovery of epigenetically derepressed cancer testis antigens in NSCLC. *PLoS One.* 2009;4(12):e8189.

[76] Paz MF, Avila S, Fraga MF, Pollan M, Capella G, Peinado MA, et al. Germ-line variants in methyl-group metabolism genes and susceptibility to DNA methylation in normal tissues and human primary tumors. *Cancer Res.* 2002;62(15):4519-24.

[77] Pereira MA, Tao L, Liu Y, Li L, Steele VE, Lubet RA. Modulation by budesonide of DNA methylation and mRNA expression in mouse lung tumors. *Int J Cancer.* 2007;120(5):1150-3.

[78] Katayama H, Hiraki A, Fujiwara K, Matsuo K, Maeda T, Chikamori K, et al. Aberrant promoter methylation profile in pleural fluid DNA and clinicopathological factors in patients with non-small cell lung cancer. *Asian Pac J Cancer Prev.* 2007;8(2):221-4.

[79] Licchesi JD, Westra WH, Hooker CM, Herman JG. Promoter hypermethylation of hallmark cancer genes in atypical adenomatous hyperplasia of the lung. *Clin Cancer Res.* 2008;14(9):2570-8.

[80] Ambler RP, Rees MW. Epsilon-N-Methyl-lysine in bacterial flagellar protein. *Nature.* 1959;184:56-7.

[81] Murray K. The Occurrence of Epsilon-N-Methyl Lysine in Histones. *Biochemistry.* 1964;3:10-5.

[82] Kouzarides T. Chromatin modifications and their function. *Cell.* 2007;128(4):693-705.

[83] Cheung P, Allis CD, Sassone-Corsi P. Signaling to chromatin through histone modifications. *Cell.* 2000;103(2):263-71.

[84] Luger K. Structure and dynamic behavior of nucleosomes. *Curr Opin Genet Dev.* 2003;13(2):127-35.

[85] Luger K, Mader AW, Richmond RK, Sargent DF, Richmond TJ. Crystal structure of the nucleosome core particle at 2.8 A resolution. *Nature.* 1997;389(6648):251-60.

[86] Zhang Y, Reinberg D. Transcription regulation by histone methylation: interplay between different covalent modifications of the core histone tails. *Genes Dev.* 2001;15(18):2343-60.

[87] Margueron R, Trojer P, Reinberg D. The key to development: interpreting the histone code? *Curr Opin Genet Dev.* 2005;15(2):163-76.

[88] Vougiouklakis T, Hamamoto R, Nakamura Y, Saloura V. The NSD family of protein methyltransferases in human cancer. *Epigenomics.* 2015:1-12.

[89] Wagner EJ, Carpenter PB. Understanding the language of Lys36 methylation at histone H3. *Nat Rev Mol Cell Biol.* 2012;13(2):115-26.

[90] Wang GG, Cai L, Pasillas MP, Kamps MP. NUP98-NSD1 links H3K36 methylation to Hox-A gene activation and leukaemogenesis. *Nat Cell Biol.* 2007;9(7):804-12.

[91] Hollink IH, van den Heuvel-Eibrink MM, Arentsen-Peters ST, Pratcorona M, Abbas S, Kuipers JE, et al. NUP98/NSD1 characterizes a novel poor prognostic group in acute myeloid leukemia with a distinct HOX gene expression pattern. *Blood.* 2011;118(13):3645-56.

[92] Thol F, Kolking B, Hollink IH, Damm F, van den Heuvel-Eibrink MM, Michel Zwaan C, et al. Analysis of NUP98/NSD1 translocations in adult AML and MDS patients. *Leukemia.* 2013;27(3):750-4.

[93] Fasan A, Haferlach C, Alpermann T, Kern W, Haferlach T, Schnittger S. A rare but specific subset of adult AML patients can be defined by the cytogenetically cryptic NUP98-NSD1 fusion gene. *Leukemia.* 2013;27(1):245-8.

[94] Toyokawa G, Cho HS, Iwai Y, Yoshimatsu M, Takawa M, Hayami S, et al. The histone demethylase JMJD2B plays an essential role in human carcinogenesis through positive regulation of cyclin-dependent kinase 6. *Cancer Prev Res* (Phila). 2011;4(12):2051-61.

[95] Hudlebusch HR, Skotte J, Santoni-Rugiu E, Zimling ZG, Lees MJ, Simon R, et al. MMSET is highly expressed and associated with aggressiveness in neuroblastoma. *Cancer Res.* 2011;71(12):4226-35.

[96] Saloura V, Cho HS, Kiyotani K, Alachkar H, Zuo Z, Nakakido M, et al. WHSC1 promotes oncogenesis through regulation of NIMA-related kinase-7 in squamous cell carcinoma of the head and neck. *Mol Cancer Res.* 2015;13(2):293-304.

[97] Xiao M, Yang S, Chen J, Ning X, Guo L, Huang K, et al. Overexpression of MMSET in endometrial cancer: a clinicopathologic study. *J Surg Oncol.* 2013;107(4):428-32.

[98] Zhou P, Wu LL, Wu KM, Jiang W, Li JD, Zhou LD, et al. Overexpression of MMSET is correlation with poor prognosis in hepatocellular carcinoma. *Pathol Oncol Res.* 2013;19(2):303-9.

[99] Angrand PO, Apiou F, Stewart AF, Dutrillaux B, Losson R, Chambon P. NSD3, a new SET domain-containing gene, maps to 8p12 and is amplified in human breast cancer cell lines. *Genomics.* 2001;74(1):79-88.

[100] Liu L, Kimball S, Liu H, Holowatyj A, Yang ZQ. Genetic alterations of histone lysine methyltransferases and their significance in breast cancer. *Oncotarget.* 2015;6(4):2466-82.

[101] Pan Y-XZ, Y.; Chen, D. Epigenetic Mechanisms of Colon Cancer Prevention: What Can Nutrition Do? In Molecular Mechanisms and Physiology of Disease. New York, NY, USA: *Springer;* 2014. p. pp. 277–353.

[102] Volkel P, Dupret B, Le Bourhis X, Angrand PO. Diverse involvement of EZH2 in cancer epigenetics. *Am J Transl Res.* 2015;7(2):175-93.

[103] Muller J, Hart CM, Francis NJ, Vargas ML, Sengupta A, Wild B, et al. Histone methyltransferase activity of a Drosophila Polycomb group repressor complex. *Cell.* 2002;111(2):197-208.

[104] Chang CJ, Yang JY, Xia W, Chen CT, Xie X, Chao CH, et al. EZH2 promotes expansion of breast tumor initiating cells through activation of RAF1-beta-catenin signaling. *Cancer Cell.* 2011;19(1):86-100.

[105] Canaani E, Nakamura T, Rozovskaia T, Smith ST, Mori T, Croce CM, et al. ALL-1/MLL1, a homologue of Drosophila TRITHORAX, modifies chromatin and is directly involved in infant acute leukaemia. *Br J Cancer.* 2004;90(4):756-60.

[106] Liu H, Takeda S, Kumar R, Westergard TD, Brown EJ, Pandita TK, et al. Phosphorylation of MLL by ATR is required for execution of mammalian S-phase checkpoint. *Nature.* 2010;467(7313):343-6.
[107] Tell R, Rivera CA, Eskra J, Taglia LN, Blunier A, Wang QT, et al. Gastrin-releasing peptide signaling alters colon cancer invasiveness via heterochromatin protein 1Hsbeta. *Am J Pathol.* 2011;178(2):672-8.
[108] Chen MW, Hua KT, Kao HJ, Chi CC, Wei LH, Johansson G, et al. H3K9 histone methyltransferase G9a promotes lung cancer invasion and metastasis by silencing the cell adhesion molecule Ep-CAM. *Cancer Res.* 2010;70(20):7830-40.
[109] Watanabe H, Soejima K, Yasuda H, Kawada I, Nakachi I, Yoda S, et al. Deregulation of histone lysine methyltransferases contributes to oncogenic transformation of human bronchoepithelial cells. *Cancer Cell Int.* 2008;8:15.
[110] Wang J, Zhou Y, Yin B, Du G, Huang X, Li G, et al. ASH2L: alternative splicing and downregulation during induced megakaryocytic differentiation of multipotential leukemia cell lines. *J Mol Med* (Berl). 2001;79(7):399-405.
[111] Magerl C, Ellinger J, Braunschweig T, Kremmer E, Koch LK, Holler T, et al. H3K4 dimethylation in hepatocellular carcinoma is rare compared with other hepatobiliary and gastrointestinal carcinomas and correlates with expression of the methylase Ash2 and the demethylase LSD1. *Hum Pathol.* 2010;41(2):181-9.
[112] Xi Y, Formentini A, Nakajima G, Kornmann M, Ju J. Validation of biomarkers associated with 5-fluorouracil and thymidylate synthase in colorectal cancer. *Oncol Rep.* 2008;19(1):257-62.
[113] Wang XQ, Miao X, Cai Q, Garcia-Barcelo MM, Fan ST. SMYD3 tandem repeats polymorphism is not associated with the occurrence and metastasis of hepatocellular carcinoma in a Chinese population. *Exp Oncol.* 2007;29(1):71-3.
[114] Morishita M, di Luccio E. Cancers and the NSD family of histone lysine methyltransferases. *Biochim Biophys Acta.* 2011;1816(2):158-63.
[115] Nakshatri H, Appaiah HN, Anjanappa M, Gilley D, Tanaka H, Badve S, et al. NF-kappaB-dependent and -independent epigenetic modulation using the novel anti-cancer agent DMAPT. *Cell Death Dis.* 2015;6:e1608.
[116] Strahl BD, Allis CD. The language of covalent histone modifications. *Nature.* 2000;403(6765):41-5.
[117] Shi Y, Lan F, Matson C, Mulligan P, Whetstine JR, Cole PA, et al. Histone demethylation mediated by the nuclear amine oxidase homolog LSD1. *Cell.* 2004;119(7):941-53.
[118] Shi Y. Histone lysine demethylases: emerging roles in development, physiology and disease. *Nat Rev Genet.* 2007;8(11):829-33.
[119] Tsukada Y, Fang J, Erdjument-Bromage H, Warren ME, Borchers CH, Tempst P, et al. Histone demethylation by a family of JmjC domain-containing proteins. *Nature.* 2006;439(7078):811-6.
[120] Klose RJ, Kallin EM, Zhang Y. JmjC-domain-containing proteins and histone demethylation. *Nat Rev Genet.* 2006;7(9):715-27.
[121] Klose RJ, Yamane K, Bae Y, Zhang D, Erdjument-Bromage H, Tempst P, et al. The transcriptional repressor JHDM3A demethylates trimethyl histone H3 lysine 9 and lysine 36. *Nature.* 2006;442(7100):312-6.

[122] Karytinos A, Forneris F, Profumo A, Ciossani G, Battaglioli E, Binda C, et al. A novel mammalian flavin-dependent histone demethylase. *J Biol Chem.* 2009;284(26):17775-82.

[123] Liu Y, Liu K, Qin S, Xu C, Min J. Epigenetic targets and drug discovery: part 1: histone methylation. *Pharmacol Ther.* 2014;143(3):275-94.

[124] Liu K, Liu Y, Lau JL, Min J. Epigenetic targets and drug discovery Part 2: Histone demethylation and DNA methylation. *Pharmacol Ther.* 2015;151:121-40.

[125] Dong C, Zhang H, Xu C, Arrowsmith CH, Min J. Structure and function of dioxygenases in histone demethylation and DNA/RNA demethylation. *IUCrJ.* 2014;1(Pt 6):540-9.

[126] Hayami S, Kelly JD, Cho HS, Yoshimatsu M, Unoki M, Tsunoda T, et al. Overexpression of LSD1 contributes to human carcinogenesis through chromatin regulation in various cancers. *Int J Cancer.* 2011;128(3):574-86.

[127] Suikki HE, Kujala PM, Tammela TL, van Weerden WM, Vessella RL, Visakorpi T. Genetic alterations and changes in expression of histone demethylases in prostate cancer. *Prostate.* 2010;70(8):889-98.

[128] Guenther MG, Jenner RG, Chevalier B, Nakamura T, Croce CM, Canaani E, et al. Global and Hox-specific roles for the MLL1 methyltransferase. *Proc Natl Acad Sci U S A.* 2005;102(24):8603-8.

[129] Harris WJ, Huang X, Lynch JT, Spencer GJ, Hitchin JR, Li Y, et al. The histone demethylase KDM1A sustains the oncogenic potential of MLL-AF9 leukemia stem cells. *Cancer Cell.* 2012;21(4):473-87.

[130] Lim S, Janzer A, Becker A, Zimmer A, Schule R, Buettner R, et al. Lysine-specific demethylase 1 (LSD1) is highly expressed in ER-negative breast cancers and a biomarker predicting aggressive biology. *Carcinogenesis.* 2010;31(3):512-20.

[131] Katz TA, Vasilatos SN, Harrington E, Oesterreich S, Davidson NE, Huang Y. Inhibition of histone demethylase, LSD2 (KDM1B), attenuates DNA methylation and increases sensitivity to DNMT inhibitor-induced apoptosis in breast cancer cells. *Breast Cancer Res Treat.* 2014;146(1):99-108.

[132] Rotili D, Mai A. Targeting Histone Demethylases: A New Avenue for the Fight against Cancer. *Genes Cancer.* 2011;2(6):663-79.

[133] Cloos PA, Christensen J, Agger K, Helin K. Erasing the methyl mark: histone demethylases at the center of cellular differentiation and disease. *Genes Dev.* 2008;22(9):1115-40.

[134] Pedersen MT, Helin K. Histone demethylases in development and disease. *Trends Cell Biol.* 2010;20(11):662-71.

[135] Lu T, Jackson MW, Wang B, Yang M, Chance MR, Miyagi M, et al. Regulation of NF-kappaB by NSD1/FBXL11-dependent reversible lysine methylation of p65. *Proc Natl Acad Sci U S A.* 2010;107(1):46-51.

[136] Tanaka Y, Okamoto K, Teye K, Umata T, Yamagiwa N, Suto Y, et al. JmjC enzyme KDM2A is a regulator of rRNA transcription in response to starvation. *Embo J.* 2010;29(9):1510-22.

[137] Huang Y, Liu Y, Yu L, Chen J, Hou J, Cui L, et al. Histone demethylase KDM2A promotes tumor cell growth and migration in gastric cancer. *Tumour Biol.* 2015;36(1):271-8.

[138] Wagner KW, Alam H, Dhar SS, Giri U, Li N, Wei Y, et al. KDM2A promotes lung tumorigenesis by epigenetically enhancing ERK1/2 signaling. *J Clin Invest.* 2013;123(12):5231-46.
[139] Curtis C, Shah SP, Chin SF, Turashvili G, Rueda OM, Dunning MJ, et al. The genomic and transcriptomic architecture of 2,000 breast tumours reveals novel subgroups. *Nature.* 2012;486(7403):346-52.
[140] Tzatsos A, Paskaleva P, Ferrari F, Deshpande V, Stoykova S, Contino G, et al. KDM2B promotes pancreatic cancer via Polycomb-dependent and -independent transcriptional programs. *J Clin Invest.* 2013;123(2):727-39.
[141] Katoh M, Katoh M. Identification and characterization of JMJD2 family genes in silico. *Int J Oncol.* 2004;24(6):1623-8.
[142] Ponnaluri VK, Vavilala DT, Putty S, Gutheil WG, Mukherji M. Identification of non-histone substrates for JMJD2A-C histone demethylases. *Biochem Biophys Res Commun.* 2009;390(2):280-4.
[143] Berry WL, Shin S, Lightfoot SA, Janknecht R. Oncogenic features of the JMJD2A histone demethylase in breast cancer. *Int J Oncol.* 2012;41(5):1701-6.
[144] Berry WL, Janknecht R. KDM4/JMJD2 histone demethylases: epigenetic regulators in cancer cells. *Cancer Res.* 2013;73(10):2936-42.
[145] Berdel B, Nieminen K, Soini Y, Tengstrom M, Malinen M, Kosma VM, et al. Histone demethylase GASC1--a potential prognostic and predictive marker in invasive breast cancer. *BMC Cancer.* 2012;12:516.
[146] Rui L, Emre NC, Kruhlak MJ, Chung HJ, Steidl C, Slack G, et al. Cooperative epigenetic modulation by cancer amplicon genes. *Cancer Cell.* 2010;18(6):590-605.
[147] Paolicchi E, Crea F, Farrar WL, Green JE, Danesi R. Histone lysine demethylases in breast cancer. *Crit Rev Oncol Hematol.* 2013;86(2):97-103.
[148] Bianco-Miotto T, Chiam K, Buchanan G, Jindal S, Day TK, Thomas M, et al. Global levels of specific histone modifications and an epigenetic gene signature predict prostate cancer progression and development. *Cancer Epidemiol Biomarkers Prev.* 2010;19(10):2611-22.
[149] Chen Z, Wang L, Wang Q, Li W. Histone modifications and chromatin organization in prostate cancer. *Epigenomics.* 2010;2(4):551-60.
[150] Ellinger J, Kahl P, von der Gathen J, Rogenhofer S, Heukamp LC, Gutgemann I, et al. Global levels of histone modifications predict prostate cancer recurrence. *Prostate.* 2010;70(1):61-9.
[151] Zeng J, Ge Z, Wang L, Li Q, Wang N, Bjorkholm M, et al. The histone demethylase RBP2 Is overexpressed in gastric cancer and its inhibition triggers senescence of cancer cells. *Gastroenterology.* 2010;138(3):981-92.
[152] Qi L, Zhu F, Li SH, Si LB, Hu LK, Tian H. Retinoblastoma binding protein 2 (RBP2) promotes HIF-1alpha-VEGF-induced angiogenesis of non-small cell lung cancer via the Akt pathway. *PLoS One.* 2014;9(8):e106032.
[153] Hayami S, Yoshimatsu M, Veerakumarasivam A, Unoki M, Iwai Y, Tsunoda T, et al. Overexpression of the JmjC histone demethylase KDM5B in human carcinogenesis: involvement in the proliferation of cancer cells through the E2F/RB pathway. *Mol Cancer.* 2010;9:59.
[154] Yamamoto S, Wu Z, Russnes HG, Takagi S, Peluffo G, Vaske C, et al. JARID1B is a luminal lineage-driving oncogene in breast cancer. *Cancer Cell.* 2014;25(6):762-77.

[155] Agger K, Christensen J, Cloos PA, Helin K. The emerging functions of histone demethylases. *Curr Opin Genet Dev.* 2008;18(2):159-68.

[156] Walport LJ, Hopkinson RJ, Vollmar M, Madden SK, Gileadi C, Oppermann U, et al. Human UTY(KDM6C) is a male-specific N-methyl lysyl demethylase. *J Biol Chem.* 2014;289(26):18302-13.

[157] van Haaften G, Dalgliesh GL, Davies H, Chen L, Bignell G, Greenman C, et al. Somatic mutations of the histone H3K27 demethylase gene UTX in human cancer. *Nat Genet.* 2009;41(5):521-3.

[158] Dalgliesh GL, Furge K, Greenman C, Chen L, Bignell G, Butler A, et al. Systematic sequencing of renal carcinoma reveals inactivation of histone modifying genes. *Nature.* 2010;463(7279):360-3.

[159] Agger K, Cloos PA, Rudkjaer L, Williams K, Andersen G, Christensen J, et al. The H3K27me3 demethylase JMJD3 contributes to the activation of the INK4A-ARF locus in response to oncogene- and stress-induced senescence. *Genes Dev.* 2009;23(10): 1171-6.

[160] Stasevich TJ, Hayashi-Takanaka Y, Sato Y, Maehara K, Ohkawa Y, Sakata-Sogawa K, et al. Regulation of RNA polymerase II activation by histone acetylation in single living cells. *Nature.* 2014;516(7530):272-5.

[161] Shogren-Knaak M, Ishii H, Sun JM, Pazin MJ, Davie JR, Peterson CL. Histone H4-K16 acetylation controls chromatin structure and protein interactions. *Science.* 2006;311(5762):844-7.

[162] Celic I, Masumoto H, Griffith WP, Meluh P, Cotter RJ, Boeke JD, et al. The sirtuins hst3 and Hst4p preserve genome integrity by controlling histone h3 lysine 56 deacetylation. *Curr Biol.* 2006;16(13):1280-9.

[163] Choudhary C, Kumar C, Gnad F, Nielsen ML, Rehman M, Walther TC, et al. Lysine acetylation targets protein complexes and co-regulates major cellular functions. *Science.* 2009;325(5942):834-40.

[164] Zhao S, Xu W, Jiang W, Yu W, Lin Y, Zhang T, et al. Regulation of cellular metabolism by protein lysine acetylation. *Science.* 2010;327(5968):1000-4.

[165] Hornbeck PV, Kornhauser JM, Tkachev S, Zhang B, Skrzypek E, Murray B, et al. PhosphoSitePlus: a comprehensive resource for investigating the structure and function of experimentally determined post-translational modifications in man and mouse. *Nucleic Acids Res.* 2012;40(Database issue):D261-70.

[166] Kouzarides T. Acetylation: a regulatory modification to rival phosphorylation? *Embo J.* 2000;19(6):1176-9.

[167] You L, Nie J, Sun WJ, Zheng ZQ, Yang XJ. Lysine acetylation: enzymes, bromodomains and links to different diseases. *Essays Biochem.* 2012;52:1-12.

[168] Unnikrishnan A, Gafken PR, Tsukiyama T. Dynamic changes in histone acetylation regulate origins of DNA replication. *Nat Struct Mol Biol.* 2010;17(4):430-7.

[169] Vo N, Goodman RH. CREB-binding protein and p300 in transcriptional regulation. *J Biol Chem.* 2001;276(17):13505-8.

[170] Capell BC, Berger SL. Genome-wide epigenetics. *J Invest Dermatol.* 2013;133(6):e9.

[171] Henry RA, Kuo YM, Andrews AJ. Differences in specificity and selectivity between CBP and p300 acetylation of histone H3 and H3/H4. *Biochemistry.* 2013;52(34):5746-59.

[172] Kalkhoven E. CBP and p300: HATs for different occasions. *Biochem Pharmacol.* 2004;68(6):1145-55.

[173] Santer FR, Hoschele PP, Oh SJ, Erb HH, Bouchal J, Cavarretta IT, et al. Inhibition of the acetyltransferases p300 and CBP reveals a targetable function for p300 in the survival and invasion pathways of prostate cancer cell lines. *Mol Cancer Ther.* 2011;10(9):1644-55.

[174] Iyer NG, Ozdag H, Caldas C. p300/CBP and cancer. *Oncogene.* 2004;23(24):4225-31.

[175] Gayther SA, Batley SJ, Linger L, Bannister A, Thorpe K, Chin SF, et al. Mutations truncating the EP300 acetylase in human cancers. *Nat Genet.* 2000;24(3):300-3.

[176] Bryan EJ, Jokubaitis VJ, Chamberlain NL, Baxter SW, Dawson E, Choong DY, et al. Mutation analysis of EP300 in colon, breast and ovarian carcinomas. *Int J Cancer.* 2002;102(2):137-41.

[177] Ozdag H, Batley SJ, Forsti A, Iyer NG, Daigo Y, Boutell J, et al. Mutation analysis of CBP and PCAF reveals rare inactivating mutations in cancer cell lines but not in primary tumours. *Br J Cancer.* 2002;87(10):1162-5.

[178] Grunstein M. Histone acetylation in chromatin structure and transcription. *Nature.* 1997;389(6649):349-52.

[179] Sun XJ, Man N, Tan Y, Nimer SD, Wang L. The Role of Histone Acetyltransferases in Normal and Malignant Hematopoiesis. *Front Oncol.* 2015;5:108.

[180] Yang XJ, Ullah M. MOZ and MORF, two large MYSTic HATs in normal and cancer stem cells. *Oncogene.* 2007;26(37):5408-19.

[181] Dulak AM, Stojanov P, Peng S, Lawrence MS, Fox C, Stewart C, et al. Exome and whole-genome sequencing of esophageal adenocarcinoma identifies recurrent driver events and mutational complexity. *Nat Genet.* 2013;45(5):478-86.

[182] Wu X, Northcott PA, Dubuc A, Dupuy AJ, Shih DJ, Witt H, et al. Clonal selection drives genetic divergence of metastatic medulloblastoma. *Nature.* 2012;482(7386):529-33.

[183] Moore SD, Herrick SR, Ince TA, Kleinman MS, Dal Cin P, Morton CC, et al. Uterine leiomyomata with t(10;17) disrupt the histone acetyltransferase MORF. *Cancer Res.* 2004;64(16):5570-7.

[184] Panagopoulos I, Gorunova L, Bjerkehagen B, Heim S. Novel KAT6B-KANSL1 fusion gene identified by RNA sequencing in retroperitoneal leiomyoma with t(10;17)(q22;q21). *PLoS One.* 2015;10(1):e0117010.

[185] Lynch H, Wen H, Kim YC, Snyder C, Kinarsky Y, Chen PX, et al. Can unknown predisposition in familial breast cancer be family-specific? *Breast J.* 2013;19(5):520-8.

[186] Grasso CS, Wu YM, Robinson DR, Cao X, Dhanasekaran SM, Khan AP, et al. The mutational landscape of lethal castration-resistant prostate cancer. *Nature.* 2012;487(7406):239-43.

[187] Zack TI, Schumacher SE, Carter SL, Cherniack AD, Saksena G, Tabak B, et al. Pan-cancer patterns of somatic copy number alteration. *Nat Genet.* 2013;45(10):1134-40.

[188] Jin Q, Yu LR, Wang L, Zhang Z, Kasper LH, Lee JE, et al. Distinct roles of GCN5/PCAF-mediated H3K9ac and CBP/p300-mediated H3K18/27ac in nuclear receptor transactivation. *Embo J.* 2011;30(2):249-62.

[189] Zimmer SN, Lemieux ME, Karia BP, Day C, Zhou T, Zhou Q, et al. Mice heterozygous for CREB binding protein are hypersensitive to gamma-radiation and invariably

develop myelodysplastic/myelo proliferative neoplasm. *Exp Hematol.* 2012;40(4):295-306 e5.

[190] Gui Y, Guo G, Huang Y, Hu X, Tang A, Gao S, et al. Frequent mutations of chromatin remodeling genes in transitional cell carcinoma of the bladder. *Nat Genet.* 2011;43(9):875-8.

[191] Pasqualucci L, Dominguez-Sola D, Chiarenza A, Fabbri G, Grunn A, Trifonov V, et al. Inactivating mutations of acetyltransferase genes in B-cell lymphoma. *Nature.* 2011;471(7337):189-95.

[192] Mullighan CG, Zhang J, Kasper LH, Lerach S, Payne-Turner D, Phillips LA, et al. CREBBP mutations in relapsed acute lymphoblastic leukaemia. *Nature.* 2011;471(7337):235-9.

[193] Ionov Y, Matsui S, Cowell JK. A role for p300/CREB binding protein genes in promoting cancer progression in colon cancer cell lines with microsatellite instability. *Proc Natl Acad Sci U S A.* 2004;101(5):1273-8.

[194] Gao Y, Geng J, Hong X, Qi J, Teng Y, Yang Y, et al. Expression of p300 and CBP is associated with poor prognosis in small cell lung cancer. *Int J Clin Exp Pathol.* 2014;7(2):760-7.

[195] Liao ZW, Zhou TC, Tan XJ, Song XL, Liu Y, Shi XY, et al. High expression of p300 is linked to aggressive features and poor prognosis of nasopharyngeal carcinoma. *J Transl Med.* 2012;10:110.

[196] Rotte A, Bhandaru M, Cheng Y, Sjoestroem C, Martinka M, Li G. Decreased expression of nuclear p300 is associated with disease progression and worse prognosis of melanoma patients. *PLoS One.* 2013;8(9):e75405.

[197] Inuzuka H, Gao D, Finley LW, Yang W, Wan L, Fukushima H, et al. Acetylation-dependent regulation of Skp2 function. *Cell.* 2012;150(1):179-93.

[198] Chen G, Cheng Y, Tang Y, Martinka M, Li G. Role of Tip60 in human melanoma cell migration, metastasis, and patient survival. *J Invest Dermatol.* 2012;132(11):2632-41.

[199] Zhao H, Jin S, Gewirtz AM. The histone acetyltransferase TIP60 interacts with c-Myb and inactivates its transcriptional activity in human leukemia. *J Biol Chem.* 2012;287(2):925-34.

[200] Chevillard-Briet M, Quaranta M, Grezy A, Mattera L, Courilleau C, Philippe M, et al. Interplay between chromatin-modifying enzymes controls colon cancer progression through Wnt signaling. *Hum Mol Genet.* 2014;23(8):2120-31.

[201] Gorrini C, Squatrito M, Luise C, Syed N, Perna D, Wark L, et al. Tip60 is a haplo-insufficient tumour suppressor required for an oncogene-induced DNA damage response. *Nature.* 2007;448(7157):1063-7.

[202] Halkidou K, Gnanapragasam VJ, Mehta PB, Logan IR, Brady ME, Cook S, et al. Expression of Tip60, an androgen receptor coactivator, and its role in prostate cancer development. *Oncogene.* 2003;22(16):2466-77.

[203] ME LL, Vidal F, Gallardo D, Diaz-Fuertes M, Rojo F, Cuatrecasas M, et al. New p53 related genes in human tumors: significant downregulation in colon and lung carcinomas. *Oncol Rep.* 2006;16(3):603-8.

[204] Mattera L, Escaffit F, Pillaire MJ, Selves J, Tyteca S, Hoffmann JS, et al. The p400/Tip60 ratio is critical for colorectal cancer cell proliferation through DNA damage response pathways. *Oncogene.* 2009;28(12):1506-17.

[205] Shiota M, Yokomizo A, Masubuchi D, Tada Y, Inokuchi J, Eto M, et al. Tip60 promotes prostate cancer cell proliferation by translocation of androgen receptor into the nucleus. *Prostate*. 2010;70(5):540-54.

[206] Wang C, Fu M, Angeletti RH, Siconolfi-Baez L, Reutens AT, Albanese C, et al. Direct acetylation of the estrogen receptor alpha hinge region by p300 regulates transactivation and hormone sensitivity. *J Biol Chem*. 2001;276(21):18375-83.

[207] Fuqua SA, Wiltschke C, Zhang QX, Borg A, Castles CG, Friedrichs WE, et al. A hypersensitive estrogen receptor-alpha mutation in premalignant breast lesions. *Cancer Res*. 2000;60(15):4026-9.

[208] Kim MY, Woo EM, Chong YT, Homenko DR, Kraus WL. Acetylation of estrogen receptor alpha by p300 at lysines 266 and 268 enhances the deoxyribonucleic acid binding and transactivation activities of the receptor. *Mol Endocrinol*. 2006;20(7):1479-93.

[209] Fan S, Ma YX, Wang C, Yuan RQ, Meng Q, Wang JA, et al. p300 Modulates the BRCA1 inhibition of estrogen receptor activity. *Cancer Res*. 2002;62(1):141-51.

[210] Amanatullah DF, Reutens AT, Zafonte BT, Fu M, Mani S, Pestell RG. Cell-cycle dysregulation and the molecular mechanisms of prostate cancer. *Front Biosci*. 2000;5:D372-90.

[211] Jemal A, Murray T, Samuels A, Ghafoor A, Ward E, Thun MJ. Cancer statistics, 2003. *CA Cancer J Clin*. 2003;53(1):5-26.

[212] Popov VM, Wang C, Shirley LA, Rosenberg A, Li S, Nevalainen M, et al. The functional significance of nuclear receptor acetylation. *Steroids*. 2007;72(2):221-30.

[213] Fu M, Wang C, Reutens AT, Wang J, Angeletti RH, Siconolfi-Baez L, et al. p300 and p300/cAMP-response element-binding protein-associated factor acetylate the androgen receptor at sites governing hormone-dependent transactivation. *J Biol Chem*. 2000;275(27):20853-60.

[214] Fu M, Wang C, Wang J, Zhang X, Sakamaki T, Yeung YG, et al. Androgen receptor acetylation governs trans activation and MEKK1-induced apoptosis without affecting in vitro sumoylation and trans-repression function. *Mol Cell Biol*. 2002;22(10):3373-88.

[215] Fu M, Rao M, Wang C, Sakamaki T, Wang J, Di Vizio D, et al. Acetylation of androgen receptor enhances coactivator binding and promotes prostate cancer cell growth. *Mol Cell Biol*. 2003;23(23):8563-75.

[216] Fu M, Rao M, Wu K, Wang C, Zhang X, Hessien M, et al. The androgen receptor acetylation site regulates cAMP and AKT but not ERK-induced activity. *J Biol Chem*. 2004;279(28):29436-49.

[217] Gong J, Zhu J, Goodman OB, Jr., Pestell RG, Schlegel PN, Nanus DM, et al. Activation of p300 histone acetyltransferase activity and acetylation of the androgen receptor by bombesin in prostate cancer cells. *Oncogene*. 2006;25(14):2011-21.

[218] Seligson DB, Horvath S, Shi T, Yu H, Tze S, Grunstein M, et al. Global histone modification patterns predict risk of prostate cancer recurrence. *Nature*. 2005;435(7046):1262-6.

[219] Debes JD, Sebo TJ, Lohse CM, Murphy LM, Haugen DA, Tindall DJ. p300 in prostate cancer proliferation and progression. *Cancer Res*. 2003;63(22):7638-40.

[220] Fraga MF, Ballestar E, Villar-Garea A, Boix-Chornet M, Espada J, Schotta G, et al. Loss of acetylation at Lys16 and trimethylation at Lys20 of histone H4 is a common hallmark of human cancer. *Nat Genet*. 2005;37(4):391-400.

[221] Ropero S, Esteller M. The role of histone deacetylases (HDACs) in human cancer. *Mol Oncol.* 2007;1(1):19-25.
[222] Ropero S, Fraga MF, Ballestar E, Hamelin R, Yamamoto H, Boix-Chornet M, et al. A truncating mutation of HDAC2 in human cancers confers resistance to histone deacetylase inhibition. *Nat Genet.* 2006;38(5):566-9.
[223] Swierczynski S, Klieser E, Illig R, Alinger-Scharinger B, Kiesslich T, Neureiter D. Histone deacetylation meets miRNA: epigenetics and post-transcriptional regulation in cancer and chronic diseases. *Expert Opin Biol Ther.* 2015;15(5):651-64.
[224] Poyet C, Jentsch B, Hermanns T, Schweckendiek D, Seifert HH, Schmidtpeter M, et al. Expression of histone deacetylases 1, 2 and 3 in urothelial bladder cancer. *BMC Clin Pathol.* 2014;14(1):10.
[225] Weichert W, Roske A, Gekeler V, Beckers T, Stephan C, Jung K, et al. Histone deacetylases 1, 2 and 3 are highly expressed in prostate cancer and HDAC2 expression is associated with shorter PSA relapse time after radical prostatectomy. *Br J Cancer.* 2008;98(3):604-10.
[226] Witt O, Deubzer HE, Milde T, Oehme I. HDAC family: What are the cancer relevant targets? *Cancer Lett.* 2009;277(1):8-21.
[227] Millard CJ, Watson PJ, Celardo I, Gordiyenko Y, Cowley SM, Robinson CV, et al. Class I HDACs share a common mechanism of regulation by inositol phosphates. *Mol Cell.* 2013;51(1):57-67.
[228] Barneda-Zahonero B, Parra M. Histone deacetylases and cancer. *Mol Oncol.* 2012;6(6):579-89.
[229] Ishihama K, Yamakawa M, Semba S, Takeda H, Kawata S, Kimura S, et al. Expression of HDAC1 and CBP/p300 in human colorectal carcinomas. *J Clin Pathol.* 2007;60(11):1205-10.
[230] Adams H, Fritzsche FR, Dirnhofer S, Kristiansen G, Tzankov A. Class I histone deacetylases 1, 2 and 3 are highly expressed in classical Hodgkin's lymphoma. *Expert Opin Ther Targets.* 2010;14(6):577-84.
[231] Fritzsche FR, Weichert W, Roske A, Gekeler V, Beckers T, Stephan C, et al. Class I histone deacetylases 1, 2 and 3 are highly expressed in renal cell cancer. *BMC Cancer.* 2008;8:381.
[232] Oehme I, Deubzer HE, Wegener D, Pickert D, Linke JP, Hero B, et al. Histone deacetylase 8 in neuroblastoma tumorigenesis. *Clin Cancer Res.* 2009;15(1):91-9.
[233] Krusche CA, Wulfing P, Kersting C, Vloet A, Bocker W, Kiesel L, et al. Histone deacetylase-1 and -3 protein expression in human breast cancer: a tissue microarray analysis. *Breast Cancer Res Treat.* 2005;90(1):15-23.
[234] Pacheco M, Nielsen TO. Histone deacetylase 1 and 2 in mesenchymal tumors. *Mod Pathol.* 2012;25(2):222-30.
[235] Jung KH, Noh JH, Kim JK, Eun JW, Bae HJ, Xie HJ, et al. HDAC2 overexpression confers oncogenic potential to human lung cancer cells by deregulating expression of apoptosis and cell cycle proteins. *J Cell Biochem.* 2012;113(6):2167-77.
[236] Sun JY, Xu L, Tseng H, Ciccarelli B, Fulciniti M, Hunter ZR, et al. Histone deacetylase inhibitors demonstrate significant preclinical activity as single agents, and in combination with bortezomib in Waldenstrom's macroglobulinemia. *Clin Lymphoma Myeloma Leuk.* 2011;11(1):152-6.

[237] Ozdag H, Teschendorff AE, Ahmed AA, Hyland SJ, Blenkiron C, Bobrow L, et al. Differential expression of selected histone modifier genes in human solid cancers. *BMC Genomics*. 2006;7:90.

[238] Stark M, Hayward N. Genome-wide loss of heterozygosity and copy number analysis in melanoma using high-density single-nucleotide polymorphism arrays. *Cancer Res.* 2007;67(6):2632-42.

[239] Sjoblom T, Jones S, Wood LD, Parsons DW, Lin J, Barber TD, et al. The consensus coding sequences of human breast and colorectal cancers. *Science.* 2006;314(5797):268-74.

[240] Milde T, Oehme I, Korshunov A, Kopp-Schneider A, Remke M, Northcott P, et al. HDAC5 and HDAC9 in medulloblastoma: novel markers for risk stratification and role in tumor cell growth. *Clin Cancer Res.* 2010;16(12):3240-52.

[241] Lachenmayer A, Toffanin S, Cabellos L, Alsinet C, Hoshida Y, Villanueva A, et al. Combination therapy for hepatocellular carcinoma: additive preclinical efficacy of the HDAC inhibitor panobinostat with sorafenib. *J Hepatol.* 2012;56(6):1343-50.

[242] Ouaissi M, Sielezneff I, Silvestre R, Sastre B, Bernard JP, Lafontaine JS, et al. High histone deacetylase 7 (HDAC7) expression is significantly associated with adenocarcinomas of the pancreas. *Ann Surg Oncol.* 2008;15(8):2318-28.

[243] Weichert W. HDAC expression and clinical prognosis in human malignancies. *Cancer Lett.* 2009;280(2):168-76.

[244] Moreno DA, Scrideli CA, Cortez MA, de Paula Queiroz R, Valera ET, da Silva Silveira V, et al. Differential expression of HDAC3, HDAC7 and HDAC9 is associated with prognosis and survival in childhood acute lymphoblastic leukaemia. *Br J Haematol.* 2010;150(6):665-73.

[245] Skov V, Larsen TS, Thomassen M, Riley CH, Jensen MK, Bjerrum OW, et al. Increased gene expression of histone deacetylases in patients with Philadelphia-negative chronic myeloproliferative neoplasms. *Leuk Lymphoma.* 2012;53(1):123-9.

[246] Choi YW, Bae SM, Kim YW, Lee HN, Kim YW, Park TC, et al. Gene expression profiles in squamous cell cervical carcinoma using array-based comparative genomic hybridization analysis. *Int J Gynecol Cancer.* 2007;17(3):687-96.

[247] Lucio-Eterovic AK, Cortez MA, Valera ET, Motta FJ, Queiroz RG, Machado HR, et al. Differential expression of 12 histone deacetylase (HDAC) genes in astrocytomas and normal brain tissue: class II and IV are hypoexpressed in glioblastomas. *BMC Cancer.* 2008;8:243.

[248] Rajendran R, Garva R, Krstic-Demonacos M, Demonacos C. Sirtuins: molecular traffic lights in the crossroad of oxidative stress, chromatin remodeling, and transcription. *J Biomed Biotechnol.* 2011;2011:368276.

[249] Batta K, Das C, Gadad S, Shandilya J, Kundu TK. Reversible acetylation of non histone proteins: role in cellular function and disease. *Subcell Biochem.* 2007;41:193-212.

[250] Finkel T, Deng CX, Mostoslavsky R. Recent progress in the biology and physiology of sirtuins. *Nature.* 2009;460(7255):587-91.

[251] Shore D, Squire M, Nasmyth KA. Characterization of two genes required for the position-effect control of yeast mating-type genes. *Embo J.* 1984;3(12):2817-23.

[252] Yuan H, Su L, Chen WY. The emerging and diverse roles of sirtuins in cancer: a clinical perspective. *Onco Targets Ther.* 2013;6:1399-416.

[253] de Oliveira MV, Andrade JM, Paraiso AF, Santos SH. Sirtuins and cancer: new insights and cell signaling. *Cancer Invest.* 2013;31(10):645-53.

[254] Deubzer HE, Schier MC, Oehme I, Lodrini M, Haendler B, Sommer A, et al. HDAC11 is a novel drug target in carcinomas. *Int J Cancer.* 2013;132(9):2200-8.

[255] Sawa H, Murakami H, Ohshima Y, Sugino T, Nakajyo T, Kisanuki T, et al. Histone deacetylase inhibitors such as sodium butyrate and trichostatin A induce apoptosis through an increase of the bcl-2-related protein Bad. *Brain Tumor Pathol.* 2001;18(2):109-14.

[256] Jenuwein T, Allis CD. Translating the histone code. *Science.* 2001;293(5532):1074-80.

[257] Manuyakorn A, Paulus R, Farrell J, Dawson NA, Tze S, Cheung-Lau G, et al. Cellular histone modification patterns predict prognosis and treatment response in resectable pancreatic adenocarcinoma: results from RTOG 9704. *J Clin Oncol.* 2010;28(8):1358-65.

[258] Seligson DB, Horvath S, McBrian MA, Mah V, Yu H, Tze S, et al. Global levels of histone modifications predict prognosis in different cancers. *Am J Pathol.* 2009;174(5):1619-28.

[259] Van Den Broeck A, Brambilla E, Moro-Sibilot D, Lantuejoul S, Brambilla C, Eymin B, et al. Loss of histone H4K20 trimethylation occurs in preneoplasia and influences prognosis of non-small cell lung cancer. *Clin Cancer Res.* 2008;14(22):7237-45.

[260] Barlesi F, Giaccone G, Gallegos-Ruiz MI, Loundou A, Span SW, Lefesvre P, et al. Global histone modifications predict prognosis of resected non small-cell lung cancer. *J Clin Oncol.* 2007;25(28):4358-64.

[261] Li GM. Decoding the histone code: Role of H3K36me3 in mismatch repair and implications for cancer susceptibility and therapy. *Cancer Res.* 2013;73(21):6379-83.

[262] Croce CM. Oncogenes and cancer. *N Engl J Med.* 2008;358(5):502-11.

[263] Meyer N, Penn LZ. Reflecting on 25 years with MYC. *Nat Rev Cancer.* 2008;8(12):976-90.

[264] Shchors K, Shchors E, Rostker F, Lawlor ER, Brown-Swigart L, Evan GI. The Myc-dependent angiogenic switch in tumors is mediated by interleukin 1beta. *Genes Dev.* 2006;20(18):2527-38.

[265] Borrello MG, Degl'Innocenti D, Pierotti MA. Inflammation and cancer: the oncogene-driven connection. *Cancer Lett.* 2008;267(2):262-70.

[266] Grivennikov SI, Karin M. Inflammation and oncogenesis: a vicious connection. *Curr Opin Genet Dev.* 2010;20(1):65-71.

[267] Huang J, Sengupta R, Espejo AB, Lee MG, Dorsey JA, Richter M, et al. p53 is regulated by the lysine demethylase LSD1. *Nature.* 2007;449(7158):105-8.

[268] Rak J, Mitsuhashi Y, Bayko L, Filmus J, Shirasawa S, Sasazuki T, et al. Mutant ras oncogenes upregulate VEGF/VPF expression: implications for induction and inhibition of tumor angiogenesis. *Cancer Res.* 1995;55(20):4575-80.

[269] Rak J, Mitsuhashi Y, Sheehan C, Tamir A, Viloria-Petit A, Filmus J, et al. Oncogenes and tumor angiogenesis: differential modes of vascular endothelial growth factor up-regulation in ras-transformed epithelial cells and fibroblasts. *Cancer Res.* 2000;60(2):490-8.

[270] Futscher BW, O'Meara MM, Kim CJ, Rennels MA, Lu D, Gruman LM, et al. Aberrant methylation of the maspin promoter is an early event in human breast cancer. *Neoplasia.* 2004;6(4):380-9.

[271] Futscher BW, Oshiro MM, Wozniak RJ, Holtan N, Hanigan CL, Duan H, et al. Role for DNA methylation in the control of cell type specific maspin expression. *Nat Genet.* 2002;31(2):175-9.

[272] Bettstetter M, Woenckhaus M, Wild PJ, Rummele P, Blaszyk H, Hartmann A, et al. Elevated nuclear maspin expression is associated with microsatellite instability and high tumour grade in colorectal cancer. *J Pathol.* 2005;205(5):606-14.

[273] Wasylishen AR, Kalkat M, Kim SS, Pandyra A, Chan PK, Oliveri S, et al. MYC activity is negatively regulated by a C-terminal lysine cluster. *Oncogene.* 2014;33(8):1066-72.

[274] Patel JH, Du Y, Ard PG, Phillips C, Carella B, Chen CJ, et al. The c-MYC oncoprotein is a substrate of the acetyltransferases hGCN5/PCAF and TIP60. *Mol Cell Biol.* 2004;24(24):10826-34.

[275] Zhang K, Faiola F, Martinez E. Six lysine residues on c-Myc are direct substrates for acetylation by p300. *Biochem Biophys Res Commun.* 2005;336(1):274-80.

[276] Yang MH, Nickerson S, Kim ET, Liot C, Laurent G, Spang R, et al. Regulation of RAS oncogenicity by acetylation. *Proc Natl Acad Sci USA.* 2012;109(27):10843-8.

[277] Diab A, Zickl L, Abdel-Wahab O, Jhanwar S, Gulam MA, Panageas KS, et al. Acute myeloid leukemia with translocation t(8;16) presents with features which mimic acute promyelocytic leukemia and is associated with poor prognosis. *Leuk Res.* 2013;37(1):32-6.

[278] Krivtsov AV, Armstrong SA. MLL translocations, histone modifications and leukaemia stem-cell development. *Nat Rev Cancer.* 2007;7(11):823-33.

[279] Chaffanet M, Gressin L, Preudhomme C, Soenen-Cornu V, Birnbaum D, Pebusque MJ. MOZ is fused to p300 in an acute monocytic leukemia with t(8;22). *Genes Chromosomes Cancer.* 2000;28(2):138-44.

[280] Rozman M, Camos M, Colomer D, Villamor N, Esteve J, Costa D, et al. Type I MOZ/CBP (MYST3/CREBBP) is the most common chimeric transcript in acute myeloid leukemia with t(8;16)(p11;p13) translocation. *Genes Chromosomes Cancer.* 2004;40(2):140-5.

[281] Wang L, Gural A, Sun XJ, Zhao X, Perna F, Huang G, et al. The leukemogenicity of AML1-ETO is dependent on site-specific lysine acetylation. *Science.* 2011;333(6043):765-9.

[282] Deguchi K, Ayton PM, Carapeti M, Kutok JL, Snyder CS, Williams IR, et al. MOZ-TIF2-induced acute myeloid leukemia requires the MOZ nucleosome binding motif and TIF2-mediated recruitment of CBP. *Cancer Cell.* 2003;3(3):259-71.

[283] Kitabayashi I, Aikawa Y, Nguyen LA, Yokoyama A, Ohki M. Activation of AML1-mediated transcription by MOZ and inhibition by the MOZ-CBP fusion protein. *Embo J.* 2001;20(24):7184-96.

[284] Li M, Luo RZ, Chen JW, Cao Y, Lu JB, He JH, et al. High expression of transcriptional coactivator p300 correlates with aggressive features and poor prognosis of hepatocellular carcinoma. *J Transl Med.* 2011;9:5.

[285] Xiao XS, Cai MY, Chen JW, Guan XY, Kung HF, Zeng YX, et al. High Expression of p300 in Human Breast Cancer Correlates with Tumor Recurrence and Predicts Adverse Prognosis. *Chin J Cancer Res.* 2011;23(3):201-7.

[286] Hou X, Li Y, Luo RZ, Fu JH, He JH, Zhang LJ, et al. High expression of the transcriptional co-activator p300 predicts poor survival in resectable non-small cell lung cancers. *Eur J Surg Oncol.* 2012;38(6):523-30.

[287] Hobbs CA, Wei G, DeFeo K, Paul B, Hayes CS, Gilmour SK. Tip60 protein isoforms and altered function in skin and tumors that overexpress ornithine decarboxylase. *Cancer Res.* 2006;66(16):8116-22.

[288] Knights CD, Catania J, Di Giovanni S, Muratoglu S, Perez R, Swartzbeck A, et al. Distinct p53 acetylation cassettes differentially influence gene-expression patterns and cell fate. *J Cell Biol.* 2006;173(4):533-44.

[289] Vousden KH, Lane DP. p53 in health and disease. *Nat Rev Mol Cell Biol.* 2007;8(4):275-83.

[290] Soussi T, Ishioka C, Claustres M, Beroud C. Locus-specific mutation databases: pitfalls and good practice based on the p53 experience. *Nat Rev Cancer.* 2006;6(1):83-90.

[291] Gu W, Roeder RG. Activation of p53 sequence-specific DNA binding by acetylation of the p53 C-terminal domain. *Cell.* 1997;90(4):595-606.

[292] Sakaguchi K, Herrera JE, Saito S, Miki T, Bustin M, Vassilev A, et al. DNA damage activates p53 through a phosphorylation-acetylation cascade. *Genes Dev.* 1998;12(18):2831-41.

[293] Ito A, Lai CH, Zhao X, Saito S, Hamilton MH, Appella E, et al. p300/CBP-mediated p53 acetylation is commonly induced by p53-activating agents and inhibited by MDM2. *Embo J.* 2001;20(6):1331-40.

[294] Sykes SM, Mellert HS, Holbert MA, Li K, Marmorstein R, Lane WS, et al. Acetylation of the p53 DNA-binding domain regulates apoptosis induction. *Mol Cell.* 2006;24(6):841-51.

[295] Tang Y, Luo J, Zhang W, Gu W. Tip60-dependent acetylation of p53 modulates the decision between cell-cycle arrest and apoptosis. *Mol Cell.* 2006;24(6):827-39.

[296] Brooks CL, Gu W. Ubiquitination, phosphorylation and acetylation: the molecular basis for p53 regulation. *Curr Opin Cell Biol.* 2003;15(2):164-71.

[297] Friedler A, Veprintsev DB, Freund SM, von Glos KI, Fersht AR. Modulation of binding of DNA to the C-terminal domain of p53 by acetylation. *Structure.* 2005;13(4):629-36.

[298] Mujtaba S, He Y, Zeng L, Yan S, Plotnikova O, Sachchidanand, et al. Structural mechanism of the bromodomain of the coactivator CBP in p53 transcriptional activation. *Mol Cell.* 2004;13(2):251-63.

[299] Li AG, Piluso LG, Cai X, Gadd BJ, Ladurner AG, Liu X. An acetylation switch in p53 mediates holo-TFIID recruitment. *Mol Cell.* 2007;28(3):408-21.

[300] Le Cam L, Linares LK, Paul C, Julien E, Lacroix M, Hatchi E, et al. E4F1 is an atypical ubiquitin ligase that modulates p53 effector functions independently of degradation. *Cell.* 2006;127(4):775-88.

[301] Toledo F, Wahl GM. Regulating the p53 pathway: in vitro hypotheses, in vivo veritas. *Nat Rev Cancer.* 2006;6(12):909-23.

[302] Banerjee S, Kumar BR, Kundu TK. General transcriptional coactivator PC4 activates p53 function. *Mol Cell Biol.* 2004;24(5):2052-62.

[303] Batta K, Kundu TK. Activation of p53 function by human transcriptional coactivator PC4: role of protein-protein interaction, DNA bending, and posttranslational modifications. *Mol Cell Biol.* 2007;27(21):7603-14.

[304] Nakamura Y. Isolation of p53-target genes and their functional analysis. *Cancer Sci.* 2004;95(1):7-11.
[305] Chuikov S, Kurash JK, Wilson JR, Xiao B, Justin N, Ivanov GS, et al. Regulation of p53 activity through lysine methylation. *Nature.* 2004;432(7015):353-60.
[306] Huang J, Perez-Burgos L, Placek BJ, Sengupta R, Richter M, Dorsey JA, et al. Repression of p53 activity by Smyd2-mediated methylation. *Nature.* 2006;444(7119):629-32.
[307] Shi X, Kachirskaia I, Yamaguchi H, West LE, Wen H, Wang EW, et al. Modulation of p53 function by SET8-mediated methylation at lysine 382. *Mol Cell.* 2007;27(4):636-46.
[308] Chen HZ, Tsai SY, Leone G. Emerging roles of E2Fs in cancer: an exit from cell cycle control. *Nat Rev Cancer.* 2009;9(11):785-97.
[309] Goodrich DW. The retinoblastoma tumor-suppressor gene, the exception that proves the rule. *Oncogene.* 2006;25(38):5233-43.
[310] Saddic LA, West LE, Aslanian A, Yates JR, 3rd, Rubin SM, Gozani O, et al. Methylation of the retinoblastoma tumor suppressor by SMYD2. *J Biol Chem.* 2010;285(48):37733-40.
[311] Carr SM, Munro S, Kessler B, Oppermann U, La Thangue NB. Interplay between lysine methylation and Cdk phosphorylation in growth control by the retinoblastoma protein. *Embo J.* 2011;30(2):317-27.
[312] Cho HS, Hayami S, Toyokawa G, Maejima K, Yamane Y, Suzuki T, et al. RB1 methylation by SMYD2 enhances cell cycle progression through an increase of RB1 phosphorylation. *Neoplasia.* 2012;14(6):476-86.
[313] Cho HS, Suzuki T, Dohmae N, Hayami S, Unoki M, Yoshimatsu M, et al. Demethylation of RB regulator MYPT1 by histone demethylase LSD1 promotes cell cycle progression in cancer cells. *Cancer Res.* 2011;71(3):655-60.
[314] Kontaki H, Talianidis I. Lysine methylation regulates E2F1-induced cell death. *Mol Cell.* 2010;39(1):152-60.
[315] Sen R, Baltimore D. Multiple nuclear factors interact with the immunoglobulin enhancer sequences. *Cell* 1986. 46: 705-716. *J Immunol.* 2006;177(11):7485-96.
[316] 31Silverman N, Maniatis T. NF-kappaB signaling pathways in mammalian and insect innate immunity. *Genes Dev.* 2001;15(18):2321-42.
[317] Sethi G, Tergaonkar V. Potential pharmacological control of the NF-kappaB pathway. *Trends Pharmacol Sci.* 2009;30(6):313-21.
[318] Karin M. Nuclear factor-kappaB in cancer development and progression. *Nature.* 2006;441(7092):431-6.
[319] Vallabhapurapu S, Karin M. Regulation and function of NF-kappaB transcription factors in the immune system. *Annu Rev Immunol.* 2009;27:693-733.
[320] Li F., Sethi G. Targeting transcription factor NF-kappaB to overcome chemoresistance and radioresistance in cancer therapy. *Biochim Biophys Acta.* 2010;1805(2):167-80.
[321] Mantovani A. Molecular pathways linking inflammation and cancer. *Curr Mol Med.* 2010;10(4):369-73.
[322] Wong ET, Tergaonkar V. Roles of NF-kappaB in health and disease: mechanisms and therapeutic potential. *Clin Sci* (Lond). 2009;116(6):451-65.
[323] Cheung N, Chan LC, Thompson A, Cleary ML, So CW. Protein arginine-methyltransferase-dependent oncogenesis. *Nat Cell Biol.* 2007;9(10):1208-15.

[324] Medzhitov R, Horng T. Transcriptional control of the inflammatory response. *Nat Rev Immunol.* 2009;9(10):692-703.

[325] Lal G, Zhang N, van der Touw W, Ding Y, Ju W, Bottinger EP, et al. Epigenetic regulation of Foxp3 expression in regulatory T cells by DNA methylation. *J Immunol.* 2009;182(1):259-73.

[326] Wells AD. New insights into the molecular basis of T cell anergy: anergy factors, avoidance sensors, and epigenetic imprinting. *J Immunol.* 2009;182(12):7331-41.

[327] Wierda RJ, Geutskens SB, Jukema JW, Quax PH, van den Elsen PJ. Epigenetics in atherosclerosis and inflammation. *J Cell Mol Med.* 2010;14(6A):1225-40.

[328] Chen L, Fischle W, Verdin E, Greene WC. Duration of nuclear NF-kappaB action regulated by reversible acetylation. *Science.* 2001;293(5535):1653-7.

[329] Chen LF, Greene WC. Regulation of distinct biological activities of the NF-kappaB transcription factor complex by acetylation. *J Mol Med.* 2003;81(9):549-57.

[330] Yang SR, Valvo S, Yao H, Kode A, Rajendrasozhan S, Edirisinghe I, et al. IKK alpha causes chromatin modification on pro-inflammatory genes by cigarette smoke in mouse lung. *Am J Respir Cell Mol Biol.* 2008;38(6):689-98.

[331] Anest V, Hanson JL, Cogswell PC, Steinbrecher KA, Strahl BD, Baldwin AS. A nucleosomal function for IkappaB kinase-alpha in NF-kappaB-dependent gene expression. *Nature.* 2003;423(6940):659-63.

[332] Yamamoto Y, Verma UN, Prajapati S, Kwak YT, Gaynor RB. Histone H3 phosphorylation by IKK-alpha is critical for cytokine-induced gene expression. *Nature.* 2003;423(6940):655-9.

[333] Barnes PJ. Role of HDAC2 in the pathophysiology of COPD. *Annu Rev Physiol.* 2009;71:451-64.

[334] Ito K, Barnes PJ, Adcock IM. Glucocorticoid receptor recruitment of histone deacetylase 2 inhibits interleukin-1beta-induced histone H4 acetylation on lysines 8 and 12. *Mol Cell Biol.* 2000;20(18):6891-903.

[335] Aguilar-Quesada R, Munoz-Gamez JA, Martin-Oliva D, Peralta-Leal A, Quiles-Perez R, Rodriguez-Vargas JM, et al. Modulation of transcription by PARP-1: consequences in carcinogenesis and inflammation. *Curr Med Chem.* 2007;14(11):1179-87.

[336] Hassa PO, Buerki C, Lombardi C, Imhof R, Hottiger MO. Transcriptional coactivation of nuclear factor-kappaB-dependent gene expression by p300 is regulated by poly(ADP)-ribose polymerase-1. *J Biol Chem.* 2003;278(46):45145-53.

[337] Hassa PO, Haenni SS, Buerki C, Meier NI, Lane WS, Owen H, et al. Acetylation of poly(ADP-ribose) polymerase-1 by p300/CREB-binding protein regulates coactivation of NF-kappaB-dependent transcription. *J Biol Chem.* 2005;280(49):40450-64.

[338] Aggarwal BB. Nuclear factor-kappaB: the enemy within. *Cancer Cell.* 2004;6(3):203-8.

[339] Li Y, Reddy MA, Miao F, Shanmugam N, Yee JK, Hawkins D, et al. Role of the histone H3 lysine 4 methyltransferase, SET7/9, in the regulation of NF-kappaB-dependent inflammatory genes. Relevance to diabetes and inflammation. *J Biol Chem.* 2008;283(39):26771-81.

[340] Covic M, Hassa PO, Saccani S, Buerki C, Meier NI, Lombardi C, et al. Arginine methyltransferase CARM1 is a promoter-specific regulator of NF-kappaB-dependent gene expression. *Embo J.* 2005;24(1):85-96.

[341] Hong H, Kao C, Jeng MH, Eble JN, Koch MO, Gardner TA, et al. Aberrant expression of CARM1, a transcriptional coactivator of androgen receptor, in the development of prostate carcinoma and androgen-independent status. *Cancer.* 2004;101(1):83-9.
[342] Meyer R, Wolf SS, Obendorf M. PRMT2, a member of the protein arginine methyltransferase family, is a coactivator of the androgen receptor. *J Steroid Biochem Mol Biol.* 2007;107(1-2):1-14.
[343] Brasacchio D, Okabe J, Tikellis C, Balcerczyk A, George P, Baker EK, et al. Hyperglycemia induces a dynamic cooperativity of histone methylase and demethylase enzymes associated with gene-activating epigenetic marks that coexist on the lysine tail. *Diabetes.* 2009;58(5):1229-36.
[344] El-Osta A, Brasacchio D, Yao D, Pocai A, Jones PL, Roeder RG, et al. Transient high glucose causes persistent epigenetic changes and altered gene expression during subsequent normoglycemia. *J Exp Med.* 2008;205(10):2409-17.
[345] Taylor CT. Interdependent roles for hypoxia inducible factor and nuclear factor-kappaB in hypoxic inflammation. *J Physiol.* 2008;586(Pt 17):4055-9.
[346] Villeneuve LM, Reddy MA, Lanting LL, Wang M, Meng L, Natarajan R. Epigenetic histone H3 lysine 9 methylation in metabolic memory and inflammatory phenotype of vascular smooth muscle cells in diabetes. *Proc Natl Acad Sci U S A.* 2008;105(26):9047-52.
[347] Huang B, Yang XD, Zhou MM, Ozato K, Chen LF. Brd4 coactivates transcriptional activation of NF-kappaB via specific binding to acetylated RelA. *Mol Cell Biol.* 2009;29(5):1375-87.
[348] Greene WC, Chen LF. Regulation of NF-kappaB action by reversible acetylation. Novartis Found Symp. 2004;259:208-17; discussion 18-25.
[349] Wu X, Qi J, Bradner JE, Xiao G, Chen LF. Bromodomain and extraterminal (BET) protein inhibition suppresses human T cell leukemia virus 1 (HTLV-1) Tax protein-mediated tumorigenesis by inhibiting nuclear factor kappaB (NF-kappaB) signaling. *J Biol Chem.* 2013;288(50):36094-105.
[350] Kiernan R, Bres V, Ng RW, Coudart MP, El Messaoudi S, Sardet C, et al. Post-activation turn-off of NF-kappa B-dependent transcription is regulated by acetylation of p65. *J Biol Chem.* 2003;278(4):2758-66.
[351] Deng WG, Wu KK. Regulation of inducible nitric oxide synthase expression by p300 and p50 acetylation. *J Immunol.* 2003;171(12):6581-8.
[352] Pradhan M, Baumgarten SC, Bembinster LA, Frasor J. CBP mediates NF-kappaB-dependent histone acetylation and estrogen receptor recruitment to an estrogen response element in the BIRC3 promoter. *Mol Cell Biol.* 2012;32(2):569-75.
[353] Hamamoto R, Saloura V, Nakamura Y. Critical roles of non-histone protein lysine methylation in human tumorigenesis. *Nat Rev Cancer.* 2015;15(2):110-24.
[354] Stark GR, Wang Y, Lu T. Lysine methylation of promoter-bound transcription factors and relevance to cancer. *Cell Res.* 2011;21(3):375-80.
[355] Lu T, Yang M, Huang DB, Wei H, Ozer GH, Ghosh G, et al. Role of lysine methylation of NF-kappaB in differential gene regulation. *Proc Natl Acad Sci U S A.* 2013;110(33):13510-5.
[356] Levy D, Kuo AJ, Chang Y, Schaefer U, Kitson C, Cheung P, et al. Lysine methylation of the NF-kappaB subunit RelA by SETD6 couples activity of the histone

methyltransferase GLP at chromatin to tonic repression of NF-kappaB signaling. *Nat Immunol.* 2011;12(1):29-36.

[357] Yang P, Guo L, Duan ZJ, Tepper CG, Xue L, Chen X, et al. Histone methyltransferase NSD2/MMSET mediates constitutive NF-kappaB signaling for cancer cell proliferation, survival, and tumor growth via a feed-forward loop. *Mol Cell Biol.* 2012;32(15):3121-31.

[358] Levy DE, Darnell JE, Jr. Stats: transcriptional control and biological impact. *Nat Rev Mol Cell Biol.* 2002;3(9):651-62.

[359] Icardi L, De Bosscher K, Tavernier J. The HAT/HDAC interplay: multilevel control of STAT signaling. *Cytokine Growth Factor Rev.* 2012;23(6):283-91.

[360] Zhuang S. Regulation of STAT signaling by acetylation. *Cell Signal.* 2013;25(9):1924-31.

[361] Ray S, Boldogh I, Brasier AR. STAT3 NH2-terminal acetylation is activated by the hepatic acute-phase response and required for IL-6 induction of angiotensinogen. *Gastroenterology.* 2005;129(5):1616-32.

[362] Nie Y, Erion DM, Yuan Z, Dietrich M, Shulman GI, Horvath TL, et al. STAT3 inhibition of gluconeogenesis is downregulated by SirT1. *Nat Cell Biol.* 2009;11(4):492-500.

[363] Ohbayashi N, Ikeda O, Taira N, Yamamoto Y, Muromoto R, Sekine Y, et al. LIF- and IL-6-induced acetylation of STAT3 at Lys-685 through PI3K/Akt activation. *Biol Pharm Bull.* 2007;30(10):1860-4.

[364] Kramer OH, Baus D, Knauer SK, Stein S, Jager E, Stauber RH, et al. Acetylation of Stat1 modulates NF-kappaB activity. *Genes Dev.* 2006;20(4):473-85.

[365] Gupta M, Han JJ, Stenson M, Wellik L, Witzig TE. Regulation of STAT3 by histone deacetylase-3 in diffuse large B-cell lymphoma: implications for therapy. *Leukemia.* 2012;26(6):1356-64.

[366] Martinez-Balbas MA, Bauer UM, Nielsen SJ, Brehm A, Kouzarides T. Regulation of E2F1 activity by acetylation. *Embo J.* 2000;19(4):662-71.

[367] Kumar BR, Swaminathan V, Banerjee S, Kundu TK. p300-mediated acetylation of human transcriptional coactivator PC4 is inhibited by phosphorylation. *J Biol Chem.* 2001;276(20):16804-9.

[368] Matsuzaki H, Daitoku H, Hatta M, Aoyama H, Yoshimochi K, Fukamizu A. Acetylation of Foxo1 alters its DNA-binding ability and sensitivity to phosphorylation. *Proc Natl Acad Sci U S A.* 2005;102(32):11278-83.

[369] Chi H, Flavell RA. Acetylation of MKP-1 and the control of inflammation. *Sci Signal.* 2008;1(41):pe44.

[370] Chen Y, Wang J, Wang X, Liu X, Li H, Lv Q, et al. STAT3, a Poor Survival Predicator, Is Associated with Lymph Node Metastasis from Breast Cancer. *J Breast Cancer.* 2013;16(1):40-9.

[371] Yang J, Huang J, Dasgupta M, Sears N, Miyagi M, Wang B, et al. Reversible methylation of promoter-bound STAT3 by histone-modifying enzymes. *Proc Natl Acad Sci U S A.* 2010;107(50):21499-504.

[372] Kim E, Kim M, Woo DH, Shin Y, Shin J, Chang N, et al. Phosphorylation of EZH2 activates STAT3 signaling via STAT3 methylation and promotes tumorigenicity of glioblastoma stem-like cells. *Cancer Cell.* 2013;23(6):839-52.

[373] Serman L, Nikuseva Martic T, Serman A, Vranic S. Epigenetic alterations of the Wnt signaling pathway in cancer: a mini review. Bosn *J Basic Med Sci*. 2014;14(4):191-4.
[374] Surana R, Sikka S, Cai W, Shin EM, Warrier SR, Tan HJ, et al. Secreted frizzled related proteins: Implications in cancers. *Biochim Biophys Acta*. 2014;1845(1):53-65.
[375] Staal FJ, Clevers HC. WNT signalling and haematopoiesis: a WNT-WNT situation. *Nat Rev Immunol*. 2005;5(1):21-30.
[376] Gregorieff A, Clevers H. Wnt signaling in the intestinal epithelium: from endoderm to cancer. *Genes Dev*. 2005;19(8):877-90.
[377] Ying Y, Tao Q. Epigenetic disruption of the WNT/beta-catenin signaling pathway in human cancers. *Epigenetics*. 2009;4(5):307-12.
[378] Batra S, Shi Y, Kuchenbecker KM, He B, Reguart N, Mikami I, et al. Wnt inhibitory factor-1, a Wnt antagonist, is silenced by promoter hypermethylation in malignant pleural mesothelioma. *Biochem Biophys Res Commun*. 2006;342(4):1228-32.
[379] Liu TH, Raval A, Chen SS, Matkovic JJ, Byrd JC, Plass C. CpG island methylation and expression of the secreted frizzled-related protein gene family in chronic lymphocytic leukemia. *Cancer Res*. 2006;66(2):653-8.
[380] Suzuki H, Watkins DN, Jair KW, Schuebel KE, Markowitz SD, Chen WD, et al. Epigenetic inactivation of SFRP genes allows constitutive WNT signaling in colorectal cancer. *Nat Genet*. 2004;36(4):417-22.
[381] Schiefer L, Visweswaran M, Perumal V, Arfuso F, Groth D, Newsholme P, et al. Epigenetic regulation of the secreted frizzled-related protein family in human glioblastoma multiforme. *Cancer Gene Ther*. 2014;21(7):297-303.
[382] Katoh M. Expression and regulation of WNT1 in human cancer: up-regulation of WNT1 by beta-estradiol in MCF-7 cells. *Int J Oncol*. 2003;22(1):209-12.
[383] Huang CL, Liu D, Nakano J, Ishikawa S, Kontani K, Yokomise H, et al. Wnt5a expression is associated with the tumor proliferation and the stromal vascular endothelial growth factor--an expression in non-small-cell lung cancer. *J Clin Oncol*. 2005;23(34):8765-73.
[384] Sato N, Fukushima N, Maitra A, Matsubayashi H, Yeo CJ, Cameron JL, et al. Discovery of novel targets for aberrant methylation in pancreatic carcinoma using high-throughput microarrays. *Cancer Res*. 2003;63(13):3735-42.
[385] Shu J, Jelinek J, Chang H, Shen L, Qin T, Chung W, et al. Silencing of bidirectional promoters by DNA methylation in tumorigenesis. *Cancer Res*. 2006;66(10):5077-84.
[386] Veeck J, Niederacher D, An H, Klopocki E, Wiesmann F, Betz B, et al. Aberrant methylation of the Wnt antagonist SFRP1 in breast cancer is associated with unfavourable prognosis. *Oncogene*. 2006;25(24):3479-88.
[387] Veeck J, Noetzel E, Bektas N, Jost E, Hartmann A, Knuchel R, et al. Promoter hypermethylation of the SFRP2 gene is a high-frequent alteration and tumor-specific epigenetic marker in human breast cancer. *Mol Cancer*. 2008;7:83.
[388] Chim CS, Chan WW, Pang A, Kwong YL. Preferential methylation of Wnt inhibitory factor-1 in acute promyelocytic leukemia: an independent poor prognostic factor. *Leukemia*. 2006;20(5):907-9.
[389] Yu J, Tao Q, Cheng YY, Lee KY, Ng SS, Cheung KF, et al. Promoter methylation of the Wnt/beta-catenin signaling antagonist Dkk-3 is associated with poor survival in gastric cancer. Cancer. 2009;115(1):49-60.

[390] Prasad CP, Mirza S, Sharma G, Prashad R, DattaGupta S, Rath G, et al. Epigenetic alterations of CDH1 and APC genes: relationship with activation of Wnt/beta-catenin pathway in invasive ductal carcinoma of breast. *Life Sci.* 2008;83(9-10):318-25.

[391] Toyokawa G, Cho HS, Masuda K, Yamane Y, Yoshimatsu M, Hayami S, et al. Histone lysine methyltransferase Wolf-Hirschhorn syndrome candidate 1 is involved in human carcinogenesis through regulation of the Wnt pathway. *Neoplasia.* 2011;13(10):887-98.

[392] Ge X, Jin Q, Zhang F, Yan T, Zhai Q. PCAF acetylates {beta}-catenin and improves its stability. *Mol Biol Cell.* 2009;20(1):419-27.

[393] Chu CS, Lo PW, Yeh YH, Hsu PH, Peng SH, Teng YC, et al. O-GlcNAcylation regulates EZH2 protein stability and function. *Proc Natl Acad Sci U S A.* 2014;111(4):1355-60.

[394] Crea F, Danesi R, Farrar WL. Cancer stem cell epigenetics and chemoresistance. *Epigenomics.* 2009;1(1):63-79.

[395] Kanno R, Janakiraman H, Kanno M. Epigenetic regulator polycomb group protein complexes control cell fate and cancer. *Cancer Sci.* 2008;99(6):1077-84.

[396] Ohm JE, McGarvey KM, Yu X, Cheng L, Schuebel KE, Cope L, et al. A stem cell-like chromatin pattern may predispose tumor suppressor genes to DNA hypermethylation and heritable silencing. *Nat Genet.* 2007;39(2):237-42.

[397] Widschwendter M, Fiegl H, Egle D, Mueller-Holzner E, Spizzo G, Marth C, et al. Epigenetic stem cell signature in cancer. *Nat Genet.* 2007;39(2):157-8.

[398] Sparmann A, van Lohuizen M. Polycomb silencers control cell fate, development and cancer. *Nat Rev Cancer.* 2006;6(11):846-56.

[399] Hurt EM, Farrar WL. Cancer stem cells: the seeds of metastasis? *Mol Interv.* 2008;8(3):140-2.

[400] Chou RH, Yu YL, Hung MC. The roles of EZH2 in cell lineage commitment. *Am J Transl Res.* 2011;3(3):243-50.

[401] Economopoulou P, Kaklamani VG, Siziopikou K. The role of cancer stem cells in breast cancer initiation and progression: potential cancer stem cell-directed therapies. *Oncologist.* 2012;17(11):1394-401.

[402] Lee J, Son MJ, Woolard K, Donin NM, Li A, Cheng CH, et al. Epigenetic-mediated dysfunction of the bone morphogenetic protein pathway inhibits differentiation of glioblastoma-initiating cells. *Cancer Cell.* 2008;13(1):69-80.

[403] Suva ML, Riggi N, Janiszewska M, Radovanovic I, Provero P, Stehle JC, et al. EZH2 is essential for glioblastoma cancer stem cell maintenance. *Cancer Res.* 2009;69(24):9211-8.

[404] Pei L, Choi JH, Liu J, Lee EJ, McCarthy B, Wilson JM, et al. Genome-wide DNA methylation analysis reveals novel epigenetic changes in chronic lymphocytic leukemia. *Epigenetics.* 2012;7(6):567-78.

[405] Lu D, Zhao Y, Tawatao R, Cottam HB, Sen M, Leoni LM, et al. Activation of the Wnt signaling pathway in chronic lymphocytic leukemia. *Proc Natl Acad Sci U S A.* 2004;101(9):3118-23.

[406] Martin V, Agirre X, Jimenez-Velasco A, Jose-Eneriz ES, Cordeu L, Garate L, et al. Methylation status of Wnt signaling pathway genes affects the clinical outcome of Philadelphia-positive acute lymphoblastic leukemia. *Cancer Sci.* 2008;99(9):1865-8.

[407] Hou HA, Kuo YY, Liu CY, Chou WC, Lee MC, Chen CY, et al. DNMT3A mutations in acute myeloid leukemia: stability during disease evolution and clinical implications. *Blood.* 2012;119(2):559-68.

[408] Yamazaki J, Estecio MR, Lu Y, Long H, Malouf GG, Graber D, et al. The epigenome of AML stem and progenitor cells. *Epigenetics.* 2013;8(1):92-104.

[409] Rheinbay E, Suva ML, Gillespie SM, Wakimoto H, Patel AP, Shahid M, et al. An aberrant transcription factor network essential for Wnt signaling and stem cell maintenance in glioblastoma. *Cell Rep.* 2013;3(5):1567-79.

[410] Shen H, Laird PW. Interplay between the cancer genome and epigenome. *Cell.* 2013;153(1):38-55.

[411] Lewis PW, Muller MM, Koletsky MS, Cordero F, Lin S, Banaszynski LA, et al. Inhibition of PRC2 activity by a gain-of-function H3 mutation found in pediatric glioblastoma. *Science.* 2013;340(6134):857-61.

[412] Juo YY, Gong XJ, Mishra A, Cui X, Baylin SB, Azad NS, et al. Epigenetic therapy for solid tumors: from bench science to clinical trials. *Epigenomics.* 2015;7(2):215-35.

[413] Baylin SB, Jones PA. A decade of exploring the cancer epigenome - biological and translational implications. *Nat Rev Cancer.* 2011;11(10):726-34.

[414] Redmond KM, Wilson TR, Johnston PG, Longley DB. Resistance mechanisms to cancer chemotherapy. *Front Biosci.* 2008;13:5138-54.

[415] Sierra JR, Cepero V, Giordano S. Molecular mechanisms of acquired resistance to tyrosine kinase targeted therapy. *Mol Cancer.* 2010;9:75.

[416] Turke AB, Zejnullahu K, Wu YL, Song Y, Dias-Santagata D, Lifshits E, et al. Preexistence and clonal selection of MET amplification in EGFR mutant NSCLC. *Cancer Cell.* 2010;17(1):77-88.

[417] Takeuchi A, Shiota M, Tatsugami K, Yokomizo A, Tanaka S, Kuroiwa K, et al. p300 mediates cellular resistance to doxorubicin in bladder cancer. *Mol Med Rep.* 2012;5(1):173-6.

[418] Lee SO, Chun JY, Nadiminty N, Lou W, Feng S, Gao AC. Interleukin-4 activates androgen receptor through CBP/p300. *Prostate.* 2009;69(2):126-32.

[419] Wei SH, Brown R, Huang TH. Aberrant DNA methylation in ovarian cancer: is there an epigenetic predisposition to drug response? *Ann N Y Acad Sci.* 2003;983:243-50.

[420] Gifford G, Paul J, Vasey PA, Kaye SB, Brown R. The acquisition of hMLH1 methylation in plasma DNA after chemotherapy predicts poor survival for ovarian cancer patients. *Clin Cancer Res.* 2004;10(13):4420-6.

[421] Plumb JA, Strathdee G, Sludden J, Kaye SB, Brown R. Reversal of drug resistance in human tumor xenografts by 2'-deoxy-5-azacytidine-induced demethylation of the hMLH1 gene promoter. *Cancer Res.* 2000;60(21):6039-44.

[422] Strathdee G, MacKean MJ, Illand M, Brown R. A role for methylation of the hMLH1 promoter in loss of hMLH1 expression and drug resistance in ovarian cancer. *Oncogene.* 1999;18(14):2335-41.

[423] Zeller C, Dai W, Steele NL, Siddiq A, Walley AJ, Wilhelm-Benartzi CS, et al. Candidate DNA methylation drivers of acquired cisplatin resistance in ovarian cancer identified by methylome and expression profiling. *Oncogene.* 2012;31(42):4567-76.

[424] Glasspool RM, Teodoridis JM, Brown R. Epigenetics as a mechanism driving polygenic clinical drug resistance. *Br J Cancer.* 2006;94(8):1087-92.

[425] Balch C, Huang TH, Brown R, Nephew KP. The epigenetics of ovarian cancer drug resistance and resensitization. *Am J Obstet Gynecol.* 2004;191(5):1552-72.

[426] Miyamoto N, Izumi H, Noguchi T, Nakajima Y, Ohmiya Y, Shiota M, et al. Tip60 is regulated by circadian transcription factor clock and is involved in cisplatin resistance. *J Biol Chem.* 2008;283(26):18218-26.

[427] Tada Y, Wada M, Kuroiwa K, Kinugawa N, Harada T, Nagayama J, et al. MDR1 gene overexpression and altered degree of methylation at the promoter region in bladder cancer during chemotherapeutic treatment. *Clin Cancer Res.* 2000;6(12):4618-27.

[428] Nakayama M, Wada M, Harada T, Nagayama J, Kusaba H, Ohshima K, et al. Hypomethylation status of CpG sites at the promoter region and overexpression of the human MDR1 gene in acute myeloid leukemias. *Blood.* 1998;92(11):4296-307.

[429] Stronach EA, Alfraidi A, Rama N, Datler C, Studd JB, Agarwal R, et al. HDAC4-regulated STAT1 activation mediates platinum resistance in ovarian cancer. *Cancer Res.* 2011;71(13):4412-22.

[430] Hegi ME, Diserens AC, Gorlia T, Hamou MF, de Tribolet N, Weller M, et al. MGMT gene silencing and benefit from temozolomide in glioblastoma. *N Engl J Med.* 2005;352(10):997-1003.

[431] Hegi ME, Liu L, Herman JG, Stupp R, Wick W, Weller M, et al. Correlation of O6-methylguanine methyltransferase (MGMT) promoter methylation with clinical outcomes in glioblastoma and clinical strategies to modulate MGMT activity. *J Clin Oncol.* 2008;26(25):4189-99.

[432] Esteller M, Hamilton SR, Burger PC, Baylin SB, Herman JG. Inactivation of the DNA repair gene O6-methylguanine-DNA methyltransferase by promoter hypermethylation is a common event in primary human neoplasia. *Cancer Res.* 1999;59(4):793-7.

[433] Esteller M, Garcia-Foncillas J, Andion E, Goodman SN, Hidalgo OF, Vanaclocha V, et al. Inactivation of the DNA-repair gene MGMT and the clinical response of gliomas to alkylating agents. *N Engl J Med.* 2000;343(19):1350-4.

[434] Zhang W, Peyton M, Xie Y, Soh J, Minna JD, Gazdar AF, et al. Histone deacetylase inhibitor romidepsin enhances anti-tumor effect of erlotinib in non-small cell lung cancer (NSCLC) cell lines. *J Thorac Oncol.* 2009;4(2):161-6.

[435] Witta S. Histone deacetylase inhibitors in non-small-cell lung cancer. *J Thorac Oncol.* 2012;7(16 Suppl 5):S404-6.

[436] Schenk T, Chen WC, Gollner S, Howell L, Jin L, Hebestreit K, et al. Inhibition of the LSD1 (KDM1A) demethylase reactivates the all-trans-retinoic acid differentiation pathway in acute myeloid leukemia. *Nat Med.* 2012;18(4):605-11.

[437] Lokken AA, Zeleznik-Le NJ. Breaking the LSD1/KDM1A addiction: therapeutic targeting of the epigenetic modifier in AML. *Cancer Cell.* 2012;21(4):451-3.

[438] West AC, Johnstone RW. New and emerging HDAC inhibitors for cancer treatment. *J Clin Invest.* 2014;124(1):30-9.

[439] Bojang P, Jr., Ramos KS. The promise and failures of epigenetic therapies for cancer treatment. *Cancer Treat Rev.* 2014;40(1):153-69.

[440] Fenichel MP. FDA approves new agent for multiple myeloma. *J Natl Cancer Inst.* 2015;107(6):djv165.

[441] Hamm CA, Costa FF. Epigenomes as therapeutic targets. *Pharmacol Ther.* 2015;151:72-86.

[442] Rodriguez-Paredes M, Esteller M. Cancer epigenetics reaches mainstream oncology. *Nat Med.* 2011;17(3):330-9.

[443] Shanmugam MK, Sethi G. Role of Epigenetics in Inflammation-Associated Diseases. In: Kundu TK, editor. Epigenetics: Development and Disease. *Subcellular Biochemistry.* Heidelberg New York London: Springer Dordrecht; 2012. p. 627-58.

In: Epigenetics: Mechanisms and Clinical Perspectives
Editor: Yildiz Dincer
ISBN: 978-1-63484-490-1

Chapter 4

ALZHEIMER'S DISEASE: AN INSIGHT FROM EPIGENETIC PERSPECTIVE

***Yıldız Dincer*[1,*] *and Zeynep Sezgin*[1,2]**
[1]Istanbul University, Cerrahpasa Medical Faculty,
Department of Medical Biochemistry, Istanbul, Turkey
[2]Yeni Yuzyıl University Medical Faculty,
Department of Medical Biochemistry, Istanbul, Turkey

ABSTRACT

Alzheimer's Disease is one of the most common neurodegenerative disorders. Many efforts have been directed to prevent AD due to its rising prevalence and the lack of an effective curative treatment. Epigenetic changes are involved in regulation of gene expression, and may mediate various pathologies. Epigenetic changes are reversible so that can be easily modulated. Modulation of dysregulated epigenetic mechanisms is a promising therapeutic approach for many diseases. There is some evidence for epigenetic dysregulation at various levels contributing to AD pathogenesis. Despite the recent rapid accumulation of knowledge about AD pathogenesis, the role of epigenetic modifications has not been understood exactly. This chapter provides a brief overview about the role of epigenetic changes that are linked to AD pathogenesis and emerging targets for new therapeutic strategies in this field.

Keywords: Alzheimer's disease, epigenetic modifications, DNA methylation, histone acetylation, sirtuins, microRNAs

ABBREVIATIONS

ABCA7	ATP-binding cassette subfamily A, member 7
ACE	angiotensin-converting enzyme

[*] Corresponding author: yldz.dincer@gmail.com.

AD	Alzheimer's disease
AGE	advanced glycation endproducts
AICD	APP intracellular domain
anti-TNF-α	antitumor necrosis factor-α
APP	amyloid precursor protein
Aβ	amyloid β-peptid
BACE1/2	β-site APP cleaving enzyme 1/2
BBB	blood-brain barrier
BIN1	bridging integrator 1
CAT	catalase
CD2AP	CD2-associated protein
COX-1/2	cyclooxygenase-1
CR	calorie restriction
CR-1	complement receptor 1
DNMTs	DNA methyltransferases
ECE-1	endothelin-converting enzyme 1
EPHA1	EPH receptor A1
FOXO	forkhead box protein O
GSK-	3β Glycogen synthase kinase 3β
HATs	histone acetyltransferases
HDACs	histone deacetylases
IDE	insulin-degrading enzyme
LEARn	latent early-life associated regulation
LRP1 IL-1β	interleukin-1β lipoprotein receptor-related protein 1
MCI	Mild Cognitive Impairment
MDA	malondialdehyde
miRNAs	microRNAs
NAD^+	nicotinamide adenine dinucleotide
NCT	nicastrin
NEP	neprilysin
NF-κB	nuclear factor κ-light chain enhancer of activated B cells
NSAIDs	nonsteroidal anti-inflammatory drugs
PBMCs	peripheral blood mononuclear cells
PICALM	phosphatidylinositol binding clathrin assembly protein
PGC-1α	peroxisome proliferator-activated reseptor-γ coactivator-1α
PPAR- γ	peroxisome proliferator-activated reseptor-γ
PS1/2	presenilin 1/2
ROS	reactive oxygen species
SAH	S-adenosylhomocysteine
SAM	S-adenosyl methionine
Sir2	silent information regulator factor 2
SIRTs	sirtuins
SOD	superoxide dismutase
3xTg-AD	triple transgenic animal model of AD
8-OHdG	8-hydroxydeoxyguanosine

1. INTRODUCTION

Alzheimer's disease (AD), originally described by the German neurologist Alois Alzheimer in 1907, is the most common form of dementia in elderly people. AD begins with mild memory loss and then progress with severe cognitive impairment and functional decline. The major pathological hallmarks of AD are accumulation of extracellular amyloid plaques and formation of intracellular neurofibrillary tangles. Amyloid plaques are mainly constituted by amyloid β-peptid (Aβ) that occurs primarily in 40mer and 42mer forms. Intracellular neurofibrillary tangles contain hyperphoshphorylated tau that is a microtubule-associated protein. Extensive loss of synapses and neurons in the temporal and frontal cortices and hippocampi are also hallmarks of the disease. The pathogenesis of AD is highly complex and involves genetic, epigenetic and physiological alterations in the brain that are still not completely understood. Accumulation of Aβ is neurotoxic and leads to neural, synaptic and cognitive dysfunction. In general, AD can be categorized into two types: early-onset familial AD that is seen in patients younger than 65 years, and late-onset sporadic AD that is seen in patients older than 65 years. Both genetic and environmental factors can contribute to development of these AD forms. Early-onset familial AD constitutes 5% of all AD cases and can be explained genetically by mutations in the amyloid precursor protein (APP) gene or in the genes encoding APP processing enzymes. The late-onset sporadic AD (95% of all AD cases) is likely caused by Aβ overproduction and/or dysfunctions in Aβ solubility or aggregation, endocytosis, degradation, transcytosis and removal [1]. With the increase in life expectancy, the prevalence of AD is going up rapidly worldwide. The number of affected individuals has been estimated as more than 100 million by the year 2050 due to the aging of the population (http://www.alz.co.uk/research/statistics,http://www.alz.org/downloads/facts_figures_2012.pdf). Progressive loss of memory and other cognitive functions in AD and the lack of effective treatment options to prevent its progression result in a high burden on social and economic resources. Aging is the greatest risk factor for sporadic forms of AD. However, epidemiological studies indicate that genetic alterations and polymorphisms, abnormal immune or inflammatory responses, low educational level, smoking, physical inactivity, depression, midlife hypertension, high cholesterol level, diabetes, APOε4 allele positivity, traumatic head injury, oxidative stress, certain drugs and hormone replacement are contributing factors for AD. The distruption of epigenetic mechanisms which control expression of AD-related genes is recently discovered another contributory factor. Despite the tremendous progress within the last a few decades in our understanding of AD pathogenesis, the accurate molecular and epigenetic mechanisms of AD have not been revealed completely.

2. PATHOGENESIS OF ALZHEIMER'S DISEASE

Cellular mechanisms provoking anomalies in AD are still under clarification, but Aβ deposition, hyperphosphorylated tau protein, neuroinflammation, oxidative stress, metal dishomeostasis, mitochondrial dysfunction, certain genetic mutations and presence of APOε4 allele have been shown to play key role in AD pathogenesis. Two main proteins take play a major role in the pathogenesis of AD: Aβ and tau.

2.1. Aβ Hypothesis

According to this hypothesis, Aβ and its aggregates are responsible for neurodegeneration and dementia in AD. Aβ peptides are natural products of brain metabolism, but the balance between Aβ production and clearance is disrupted in AD. Aβ is generated by the proteolytic cleavage of APP which is synthesized in ribosomes on endoplasmic reticulum and then transported to the golgi apparatus. APP is a transmembrane protein which is cleaved by a series of secretases: α then γ, or β then γ for processing. Aberrant cleavage of APP leads to Aβ overproduction and neuritic plaque formation. Proteolytic processing of APP occurs by one of two pathways, the non-amyloidogenic or the amyloidogenic (Figure 1). For *non-amyloidogenic cleavage* of APP, α-secretase cleaves within the Aβ encoding region, releasing a soluble fragment, sAPPα, into the extracellular space. sAPPα was shown to have neurotrophic and neuroprotective properties [2]. After this first cleavage the remaining C-terminal fragment which includes 83 amino acid residues (C-83) is subsequently cleaved by γ-secretase within the transmembrane region, generating p3 (Aβ17-40) and APP intracellular domain (AICD). AICD and p3 (3 kDa) peptide are released into cytosol and extracellular space, respectively. AICD potentially translocates to the nucleus and functions as a transcription factor [3, 4]. As for *amyloidogenic cleavage*, APP is cleavaged by β-secretase and then γ-secretase. β-secretases include β-site APP cleaving enzyme 1 (BACE1) and 2 (BACE2). After the cleavage of APP by β-secretases, $sAPP_{\beta}$ frament is released whereas remaining C-terminal fragment of 99 amino acid residues (C-99) is still membrane bound. C-99 is further cleaved by γ-secretase, generating AICD and an insoluble fragment, Aβ, which contains 38–43 amino acid residues. Several peptides in varying lengths can be generated from the cleavages by β and γ-secretases. Majority of these Aβ fragments is Aβ40. The another fragment, Aβ42, is less common but more fibrillogenic. It forms insoluble aggregates. Aβ accumulation in the brain induces oxidative stress and inflammatory response thus leads to neurotoxicity which contributes to impairment of cognitive functions.

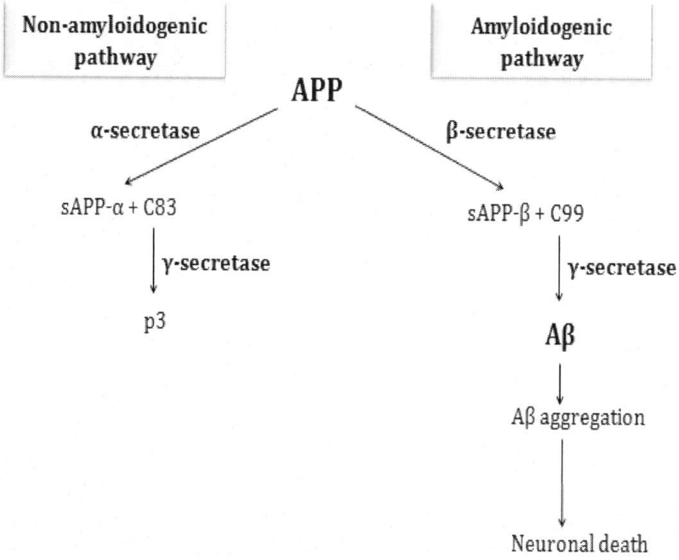

Figure 1. Processing of beta amyloid precursor protein.

The balance between different secretase activities is very important in the maintenance of the physiologic Aβ levels. Decreased processing of APP into Aβ has been found to be correlated with decreased activity of BACE1 [5]. In AD, accumulation of Aβ has been suggested to occur prior to tau pathology, and Aβ aggregates may be one of a cascade of pathological events ranging from synaptic dysfunction, oxidative stress, mitochondrial dysfunction, loss of calcium regulation and inflammation leading to tau hyperphosphorylation [6]. It has been thought that increased production of Aβ and/or decreased degradation by Aβ-degrading enzymes, or reduced clearance across the blood-brain barrier may cause aggregation and accumulation of Aβ in the brain. Impairment of synaptic functions and related signaling pathways by Aβ oligomers result in changes in neuronal activities and cause to release of neurotoxic mediators from glial cells.

The degradation and clearance of Aβ from the brain has been reported to be impaired in patients with AD. Aβ clearance happens via transporting across the blood-brain barrier (BBB) and metabolic degradation. Aβ can be transported across the BBB into the blood or cerebrospinal fluid by low-density lipoprotein receptor-related protein 1 (LRP1) [7], AGE advanced glycation endproducts) receptor, ApoE, and β-2-macroglobulin [8, 9]. ApoE binds extracellular Aβ and forms ApoE/Aβ complex which is cleared by low-density lipoprotein receptor-related protein (LRP) via the endosomal/lysosomal pathway [10]. In vitro and in vivo studies have shown that Aβ undergoes proteolytic degradation by Aβ-degrading enzymes such as neprilysin (NEP), insulin-degrading enzyme (IDE), endothelin-converting enzyme (ECE)-1, angiotensin-converting enzyme (ACE), uPA/tPA-plasmin system, cathepsin D, gelatinase-A and -B, matrix metalloproteinase-9, coagulation factor XIa, antibody light chain c23.5 and hk14, and α2-macroglobulin complexes [11]. Thus, both distruption of BBB or decreased activities of these Aβ-degrading enzymes may contribute to AD pathogenesis.

2.2. Tau Hypothesis

Tau is an intracellular protein and is belong to a family of microtubule-associated proteins that normally promotes microtubule assembly and stabilization. Tau protein exerts neurotoxic effects when it is hyperphosphorylated. Hyperphosphorylation causes an impairment in its binding ability with tubulin and subsequently results in loss of its normal function. Hyperphosphorylated tau have reduced capasity to promote microtubule assembly. In addition, hyperphosphorylation promotes its polymerization. It aggregates into filaments itself forming neurofibrillary tangles and causes the disorders known as tauopathies [12]. Abnormal hyperphosphorylation of tau is the key player of AD neurodegeneration as the major component of neurofibrillary tangles in AD. Abnormally hyperphosphorylated tau has been isolated from AD brain in 1990s [13]. In vitro studies have shown that tau phosphorylation at different sites may cause different effects in its biological function as well in pathogenic role. Phosphorylation of tau at Ser262, Thr231, and Ser235 has been shown to inhibit its binding to microtubules by ~35%, ~25%, and ~10%, respectively [14].

Synaptic Loss

Synaptic loss is emerged as an early event in the pathogenesis of AD that is confirmed in rodent brain after acute injection of soluble Aβ oligomers as well in APP transgenic mouse models [15]. Recent gene array investigations have consistently demonstrated that expression levels of neurotransmitter receptor genes and the genes involved in synaptic vesicle trafficking/release, receptor trafficking, postsynaptic density scaffolding, cell adhesion regulating synaptic stability and neuromodulatory systems are altered in early stage of AD [16-19]. Axonal transport defects, oxidative stress, mitochondrial dysfunction, neuroinflammation and possibly Aβ accumulation in the brain might result in synaptic loss that is observed in AD patients [20].

2.3. Oxidative Stress in Alzheimer's Disease

Oxidative stress is a state that occurs when there is an imbalance between production and clearance of reactive oxygen species (ROS) such as superoxide anion, hydrogen peroxide and hydroxyl radical. ROS have unpaired electrons in seperate orbitals in *theirs* outer *shell* and exhibit a high reactivity. ROS are produced as a byproduct of the normal oxygen metabolism and they have important role in cell signaling and homeostasis. However, in the case of environmental stress such as UV, ionizing radiation, certain drugs and exposure of certain chemicals ROS levels can increase dramatically causing oxidative stress. Interacting with nucleic acids, proteins and lipids, ROS cause structural and functional defects in these molecules. Transition metal ions such as iron and cupper act as pro-oxidant. In the abundant presence of these metal ions, ROS production increases via Fenton and Haber-Weiss reactions. There is consistent evidence for presence of extensive oxidative stress in AD brains [21, 22]. Different studies have revealed that the levels of protein carbonyls and 3-nitrotyrosine (protein oxidation markers), 8-hydroxydeoxyguanosine (8-OHdG) and 8-hydroxyguanosine (DNA and RNA oxidation markers), malondialdehyde (MDA), 4-hydroxynonenal, and F2-isoprostanes (lipid peroxidation markers) are elevated in AD brains [23, 24]. Increased production of hydrogen peroxide and nitric oxide, elevated levels of oxidatively modified proteins and lipids have been shown in transgenic mouse models carrying mutations in APP and presenilin genes (PS1 and PS2) [25].

Increased oxidative stress determined in AD is attributed the pro-oxidant role of Aβ. Aβ possesses pro-oxidant properties that has an influence on oxidative processes. At high micromolar and nanomolar concentrations, Aβ aggregates to form fibrils and induces ROS formation, lipid peroxidation, protein oxidation and DNA damage, and subsequently neuronal death. Transition metals including iron and copper are required for its pro-oxidant activity. Direct interaction between Aβ and transition metal ions may result in increased ROS formation [26]. In an early study, Aβ treatment has been shown to increase the levels of hydrogen peroxide and lipid peroxides in cell models [27]. Similar results have also been obtained in animal studies. It has been reported that levels of 4-hydroxy-2-noneal and isoprostanoids, carbonylic proteins are increased in animal models of AD [28]. On the other hand, as a closed circuit, oxidative stress also promotes Aβ production by decreasing α-secretase activity while promoting β- and γ-secretase expression and activity [29].

On the other hand, human organism is equipped with efficient antioxidant defence mechanisms. Harmful effects of ROS can be prevented by antioxidant defence system that is constituted by small antioxidant molecules and antioxidant enzymes such as superoxide dismutase (SOD), catalase (CAT) and glutathione peroxidase. Impaired antioxidant defence is the most important factor for oxidative stress. Defective antioxidant defence, elevated oxidative stress and increased Aβ deposition have been shown in APP overexpressing transgenic mice [30, 31]. Dumont et al. have demonstrated that facilitation of the mitochondrial antioxidant defence improves resistance to Aβ, decreases plaque formation or increases plaque degradation in a transgenic AD mouse model [31]. Recently, in accordance with Dumont et al. [31], possible role of SOD in AD has been demonstrated in Tg2576 APP-overexpressing AD mouse model. In this model cytoplasmic copper/zinc superoxide dismutase deletion has been found to be associated with increased Aβ oligomerization, accelerated loss of memory and spatial learning [32]. Aging is a natural and inevitable process and is associated with an prolonged ROS production. Due to the decline in the normal antioxidant defence mechanisms with aging, human body is insufficient in response to ROS-mediated damage [33]. Free-radical theory by Harman (1956) [34] proposes that ROS production via aerobic respiration increases with age and causes cumulative damage on cell components during the life time. As respect with the fact that AD is seen in elderly individuals, increased oxidative stress in AD patients may be partially attributed to advanced age.

Oxidative stress may also influence hyperphosphorylation and polymerization of tau. It has been suggested that the elevated amounts of fatty acid oxidation that is demonstrated in AD patients, can facilitate the polymerization of tau [35].

Inflammation is a ROS producing state. Neuroinflammation is a contributory factor for AD that will be described below. Production of ROS by glial cells due to inflammation may exacerbate the oxidative stress in AD patients.

2.4. Metal Dyshomeostasis in Alzheimer's Disease

Metal homeostasis is altered in AD brain. Metal ions such as copper (Cu), zinc (Zn) and iron (Fe) are essential for neuronal function. As respect with the fact that metal dyshomeostasis contributes to oxidative stress, disrupted homeostasis of these metal ions in brain are thought to underlie the pathogenesis of AD [36]. Metal dyshomeostais in AD has attracted the interest of researchers, and reliable results have been obtained. Copper, zinc and iron trigger signaling cascades that mediate cellular physiology by binding to Aβ. APP and its Aβ fragments have specific binding sites for Cu^{2+} ions [37]. In vitro studies have demonstrated that Aβ-Cu complexes catalytically produce hydrogen peroxide and hydroxyl radical and eventually induce oxidative damage [38]. There is growing evidence that Cu can facilitate Aβ aggregation by binding to Aβ which results in Aβ plaque burden [39].

Iron homeostasis is controlled by transferrin and ferritin. Fe, ferritin and transferrin levels have been found to be altered in the hippocampus and cerebral cortex of AD brains [40]. Iron accumulates within senile plaques and neurofibrillary tangles of AD brains. Excess iron leads to increased production of ROS and finally causes oxidative stress. Oxidation of cellular lipids and proteins results in synaptic dysfunction whereas oxidative damage to DNA induces apoptotic pathways and neuronal death.

Zinc may accelerate the precipitation of Aβ and produce protease-resistant "non-structured" aggregates [41]. It has been suggested that compounds affecting zinc homeostasis can decrease Aβ deposition in the brain [42]. Deregulation of neuronal Zn^{2+} homeostasisis is believed to be strictly connected to oxidative stress. Zn^{2+} can trigger ROS production by interfering with the activity of the electron transport chain (inhibits complex I and III) [43, 44].

On the other hand, zinc and iron promote tau phosphorylation and aggregation by binding to tau [45]. In consistent with these data, treatment with metal chelators such as clioquinol and desferrioxamine have provided some success in altering the progression of AD [46, 47]. The development of therapeutic interventions targeted to restore metal balance is an active area of current investigations in the field of AD.

2.5. Mitochondrial Dysfunction in Alzheimer's Disease

Mitochondrial dysfunction plays important role in both brain aging and AD. Structural and metabolic changes have been observed in mitochondria of neurons in AD brain. Mitochondrial dysfunction is an early event in AD. Mitochondrial abnormalities, increased mitochondrial degradation by autophagy, and decreased cytocrome oxidase (complex IV) activity that is associated with increased ROS production have been determined in AD brains [48]. Deficient energy production and impaired metabolic activity have been shown in neurons of AD brain and this situation has been explained by reduction in the number and size of mitochondria [49]. The mitochondria is the major source of ROS in the central nervous system, and is also particularly vulnerable to oxidative stress. Oxidation of mitochondrial DNA renders it vulnarable to somatic mutations which leads to initiation of erroneous APP processing [50]. Moreover, Aβ deposits can contribute to more mitochondrial damage by distrupting lipid polarity and protein mobility in mitochondrial membranes, and inhibiting key enzymes of the mitochondrial respiratory chain [51, 52].

2.6. Neuroinflammation in Alzheimer's Disease

Neuroinflammation plays a prominent and early role in AD pathogenesis. Neuroinflammation is a response to an inflammatory stimuli comprising over activation of microglia and subsequent over expression of proinflammatory cytokines. Although inflammatory response in AD is useful for restoration of tissue integrity, it becomes harmful when chronically induced. Production of pro-inflammatory cytokines, prostoglandins and ROS increases in chronic inflammatory state which in turn leads to aggravation of Aβ deposition and induction of neuronal dysfuntion. The another link between neuroinflammation and Aβ deposition is BACE1 that is involved in amyloidogenic processing of APP. BACE1 gene has been shown to have four putative NFκB binding elements in its promoter region, thus its expression is upregulated upon inflammatory stimuli leading to increased production of Aβ [53]. Furthermore, the accumulation of Aβ in plaques may produce sequential inflammatory events and excitotoxicity which in turn cause neurodegeneration and cognitive impairment. Microglia activation has been suggested to occur in mild and early stage of AD pathogenesis. Vom Berg et al. [54] have demonstrated that suppression of expression and signaling of interleukin-12 and interleukin-23 leads to decreases in microglial activity, Aβ level and Aβ plaque burden in a mouse model of AD.

It has been shown that suppression of neuroinflammation is concordant with reduction in Aβ plaques and hyperphosphorylated tau in brain, and is associated with improvements in cognitive and behavioral deficts in AD mouse models [55, 56]. Piro et al. [57] have reported that genetic inactivation of monoacyl-glycerol lipase, an enzyme that hydrolyzes endocannabinoids to generate the primary arachidonic acid pool for neuroinflammatory prostaglandins, controls arachidonic acid release, reduces neuroinflammation and lowers Aβ levels and plaques in an AD mouse model of Aβ deposition.

Due to the role of neuroinflammatory process in AD pathogenesis, reducing neuroinflammation by inflammation-based therapeutic strategies can be used in the prevention and treatment of AD. Nonsteroidal anti-inflammatory drugs (NSAIDs) and peroxisome proliferator-activated reseptor-γ (PPAR-γ) agonists have been shown to reduce the expression and activity of β-secretase and lower Aβ secretion [58]. NSAIDs inhibit the enzymatic activity of cyclooxygenase-1 (COX-1) and inducible COX-2 which catalyze the first committed step in the synthesis of prostaglandins. COX inhibitors can decrease levels of highly amyloidogenic Aβ1-42 peptide. Suppression of neuroinflammation with NSAIDs has been suggested to rescue memory and cognitive decline. However, retrospective epidemiological studies have demonstrated that prolonged treatment with NSAIDs delays onset of AD when initiated early stage or before disease initiation [59], but its efficiency has not been shown in neither mild nor moderate forms of AD [60]. On the other hand, treatment with antitumor necrosis factor-α (anti-TNF-α) or interleukin-1β (IL-1β) antibodies is efficient in reducing the pathology in animal models [61, 62].

2.7. Mutations and Polymorphisms in Alzheimer's Disease

Mutations in several disease-associated genes such as APP, PS1and PS2 all of which appear to be involved in APP processing have been shown to cause early-onset autosomal dominant AD [63]. Although it seems unlikely to be significant in late-onset AD susceptibility [64] some investigators have suggested the possibility that the rare variants in these three genes may contribute to the risk for late-onset AD [65]. Heterozygous missense mutations in or near Aβ-coding exons 16 and 17 are the most frequently detected mutations in APP [66]. Whole-gene duplications [67, 68], rare recessive small deletions [69] and recessive missense mutations [70] have also been identified in early onset AD. These mutations result in altered Aβ production, changed $Aβ_{42}/Aβ_{40}$ ratio and increased fibril formation. Mutations in PS-1 and PS-2 also cause to increased production of more aggregative $Aβ_{42}$ by impairing the γ-secretase-mediated cleavage of APP [71, 72]. Recently large-scale genome-wide association studies have changed the face of the complex genetics of AD, and these data have disclosed the novel findings about the candidate risk genes for the development of AD. According to findings of these studies, single nucleotid polymorphisms in or near complement receptor 1 (CR1), phosphatidylinositol binding clathrin assembly protein (PICALM), bridging integrator 1 (BIN1), CD2-associated protein (CD2AP), CD33, EPH receptor A 1 (EPHA1) and ATP-binding cassette subfamily A, member 7 (ABCA7) may increase the risk for AD by affecting APP processing, Aβ pathway, synaptic cell functioning, and immune system. [73-75].

Apolipoprotein E (APOE) is a polymorphic glycoprotein which contains 299-amino acid. It is highly expressed in the brain and mainly secreted by astrocytes. APOE exists in three isoforms (E2, E3, and E4) in humans encoded by the APOE epsilon (ε)2, ε3 and ε4 related genes [76]. Aside from its well-known role as a lipid-transporting entity, APOE is involved in neurogenesis, synaptic integrity, plasticity and repair [77]. Although the APOε4 allele is represented in approximately %25 of population, it has been postulated to be a risk factor for sporadic AD in older adulthood [78]. The influence of APOε4-allele on development, progression, clinical presentation of AD and rate of cognitive decline have not been fully described. Cross-sectional studies in patients with AD have shown that the APOε4 allele is associated with impaired memory function, more severe atrophic changes and hypometabolism in the medial temporal lobe structures [79-82]. Autopsy studies on AD have indicated the APOE ε4 allele to be associated with greater densites of Aβ deposition, senile plaques and neurofibrillary tangles, especially in the hippocampus [83, 84]. Besides this, the ε2 allele may have a protective effect against AD, and carriers of ε2 allele are associated with less amyloid-β plaques and neurofibrillary tangles than both ε4 carriers and ε3 homozygotes [85, 86].

Although presence of APOε4 allele has been suggested to be associated with faster cognitive decline by some studies, there are also opposing data. Moreover, some studies have reported a converse effect of APOε4 on the rate of cognitive decline [87-92]. These contradictory data for the influence of APOε4 allele on AD development might provide clear support for an antagonistic pleiotrophy hypothesis that the APOε4 allele confers some cognitive advantage in early life despite adverse consequences in old age [93].

3. Epigenetic Changes in Alzheimer's Disease

Epigenetic modifications regulate gene expression without altering the DNA sequence. Epigenetic changes are dynamic and unlike genetic mutations, they can be reversed by environmental factors and therapeutics. These modifications that can occur at specific gene loci in specific cells result in specific cellular phenotypes. Environmental factors can induce epigenetic changes in individuals in each period of life and these factors are becoming increasingly relevant with respect to the shift from healthy to diseased state [94]. Physical and behavioral factors, nutrition, lifestyle, physical and mental exercise, environmental pollutants, pesticides, chemicals and certain metals affect the epigenetic mechanisms. Especially, nutrition is one of the main epigenetic regulators. It readily modifies epigenetic pathways. Dietary methyl donors involved in one-carbon metabolism such as folate, vitamin B12, betain, choline and methionine play a pivotal role in DNA methylation as well in maintanence of genomic stability. In addition to their antioxidant properties, dietary polyphenols have been shown to modify the epigenome. There is an overwhelming scientific consensus that epigenetic changes contribute to aging and aging-associated diseases. Recently, the diet containing nutrients that are able to modify epigenetic mechanisms has been suggested as a preventive approach for AD [95].

Studies have indicated that genetic mutations are involved in the pathogenesis of early-onset (familial) AD (around 3% to 5% of cases) but epigenetic mechanisms are likely to be important in the etiology of sporadic form of AD. Epigenetic modulations mediated by DNA methylation, histone modification and non-coding RNAs may explain the complex pathogenic mechanisms in AD. These mechanisms mediate genome-environment interactions by transforming variable combinations of genetic and environmental factors into long-term adaptive changes in gene expression.

In a recent study, expression of BACE1 gene has been examined in both triple transgenic animal model of AD (3xTg-AD mice) and peripheral blood mononuclear cells (PBMCs) from patients with AD or Mild Cognitive Impairment (MCI). According to findings of this study; BACE1 mRNA level is increased in brain of aged 3xTg-AD mice, both in the cortex and hippocampus, the promoter region of the BACE1 is significantly more acetylated without significant alterations in the promoter accessibility, *BACE1* mRNA level is significantly increased in AD patients whereas it is slightly elevated in MCI patients in comparison to controls subjects. Furthermore, as consistent with a higher transcription rate of the gene, BACE1 promoter is significantly more accessible in AD patients but chromatin in the *BACE1* promoter is closed in MCI patients [96].

γ-Secretase is a multiprotein complex that is an intramembrane-cleaving protease involved in APP processing. Nicastrin (NCT) is one of the integral components of γ-Secretase and is essential for stabilization of the enzyme [97]. Overexpression of NCT is related with increased Aβ production [98]. Downregulation of NCT gene has been reported in the brain of aged transgenic mice probably due to nucleosome remodelling which may lead to a decreased accessibility of transcription machinery or hypermethylation of CpG island of the gene promotor [96].

3.1. DNA Methylation

DNA methylation is the first epigenetic modification identified on DNA that regulates gene expression in all known vertebrates. DNA methyltransferases (DNMTs), DNMT1, DNMT2, DNMT3a, DNMT3b, catalyze the transfer of a methyl group to the carbon atom in position 5 of a cytosine moiety of single-stranded DNA in the presence of the methyl donor S-adenosyl methionine (SAM) [99, 100]. DNMT3a and 3b fall in the group of de novo methyltransferases (they methylate previously unmethylated CpG sequences), while DNMT1 copies the preexisting methylation marks onto the new strand during replication [101]. SAM is derived from one-carbon metabolism which is regulated by availability of folate, vitamin B12 and vitamin B6. In promoter regions, DNA methylation occurs on the 5' position of cytosine residue in CpG rich regions called CpG islands. Methylation at this site disrupts the binding of transcription factors and recruitment of methyl-CpG-binding domain proteins that are associated with chromatin compaction and gene silencing. On the contrary, methylation within the body of the gene usually marks active transcription [102]. Recently it has been discovered that methylated cytosine (5mC) can be furher hydroxylated, predominantly in brain tissue. This reaction is mediated by the TET proteins family. 5-hydroxymethylcytosine (5hmC) formation is suggested as a potential mechanism leading to DNA demethylation [103].

Studies with monozygotic twin siblings who had spent a long period of their lives apart have demonstrated that extrinsic factors derived from environmental exposure affect DNA methylation patterns [104]. DNA methylation may be affected by aging, deficiencies of vitamin B12, B6 and folic acid, oxidative stress, pesticides and heavy metal exposure. Some of these factors such as arsenic, cadmium, lead and pesticides interfere with the regulation of DNMTs [105]. Others such as deficiencies of vitamin B12, B6 and folic acid influence DNA methylation patterns altering SAM availability via metabolites of the SAM cycle (Figure 2).

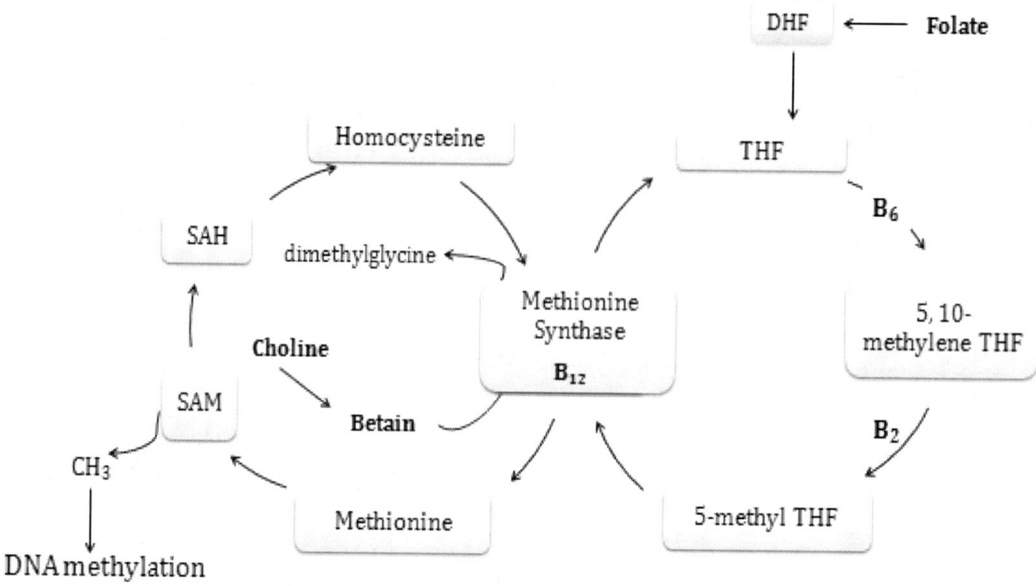

Figure 2. SAM cycle.

5-methyltetrahydrofolate donates its methyl group to homocysteine (HCY) and converts it to methionine in a reaction which requires vitamin B12. Then methionine is converted to SAM. SAM is thereby converted to S-adenosylhomocysteine (SAH), which in turn is converted to HCY in a reversible reaction. SAM level has been demonstrated to be decreased in post mortem AD brain, and increased HCY has been suggested as a risk factor for AD [106]. The lower availability of SAM can be related to the altered expression of genes involved in APP metabolism and Aβ accumulation.

Normal aging involves the gradual decline in memory and cognitive functions that are correlated with a profound changes (global DNA hypomethylation and hypermethylation of CpG islands) in DNA methylation [107]. DNA methylation modifications can be divided into two groups in AD: global DNA methylation modifications and gene-specific DNA methylation modifications [108]. Genomic and sequence-specific DNA methylation patterns are distinct between different brain areas. In 2010 Mastroeni et al. [109] have demonstrated that global DNA and RNA methylation status in entorhinal cortex layer II neurons are significantly diminished in AD. In accordance with this data, 5mC and 5hmC levels have been found to be decreased in hippocampus of AD patients [110]. Recently, DNA hypomethylation has been demonstrated to be correlated with a greater amyloid plaque burden, enhanced APP production, and increased activities of BACE-1 and PS1 [110, 111]. On the other hand, DNA hypermethylation is also present in AD. A postmortem human frontal cortex study has shown the presence of DNA hypermethylation in AD cases [112]. Evidence from genetic and immunohistochemical studies on AD cases supports the notion that aberrant DNA methylation modification of a number of specific genes, which are believed to be related to AD, can contribute to onset and progression of the disease. In this context, APP gene has been investigated as a primary target.

APP Gene

In an early study, Toghi et al. (1999) [113] have reported the hypomethylation in APP promoter in aged brain, which would be consistent with an increase in APP expression with aging. Afterwards, methylation status of different regions of APP promoter has been examined in AD by Barrachina and Ferrer, (2009) [114], and no significant difference has been found between control and disease samples obtained from frontal cortex and hippocampi of postmortem brains in various stages of AD. They have concluded that small changes in methylation of DNA promoters in vulnerable cells might have not been detected in total homogenates [114].

BACE, PS1 and DNMT Genes

According to findings of in vitro studies, BACE and PS1 genes are regulated by methylation. The reduction of folate and vitamin B12 in culture medium can cause a decrease in SAM concentration with subsequent increase in PS1 and BACE levels and an increase in Aβ formation [115-117]. Most recently, the methylation status of PS-1, BACE1, DNMT1, DNMT3A and DNMT3B promoters in blood DNA obtained from AD patients and controls have been analysed. It has been found that none of the studied regions is differently methylated between AD and controls [118].

Neprilysin (NEP)

NEP is one of the enzymes involved in Aβ degradation. Its expression is decreased in both AD and aged brains. Promoter region for NEP has been found to be hypermethylated in murine cerebral endothelial cells model. It has been concluded that hypermethylation of NEP promoter supresses NEP's expression at mRNA and protein levels, which in turn causes to increased Aβ accumulation [119].

Some other genes that are thought to be involved in AD pathogenesis have also been investigated in respect with the promoter methylation status. The promoters for *cAMP-responsive element* that is associated with synaptic plasticity have been found to be hypermethylated in AD [112]. *Glycogen synthase kinase 3β (GSK3β)* is the major kinase that phophorylates tau protein in brain. It has been reported that GSK3β activity and expression levels are increased in AD brains which may be due to hypomethylation of its promoter region [120]. Promoter methylation of *Neuronal protein phospatase 2A* has been found to be downregulated in affected brain regions of AD patients, causing the accumulation of both phosphorylated tau and APP isoforms, and increased secretion of Aβ peptides [121, 122]. Promoters of *cyclooxygenase-2, brain-derived neurotrophic factor and NF-κB* have been reported to be hypomethylated, whereas promoters of *cAMP resonse element-binding protein* and *synaptophysin* have been reported to be hypermethylated in frontal cortex of postmortem AD brain [112].

3.2. Histone Modifications

Reversible post-translational modifications such as phosphorylation, acetylation, methylation, glycosylation and ubiquitination modulate biological activity of proteins [123]. In recent years, protein acetylation at the lysine ε-amino group has gained considerable interest due to its ability to regulate important cellular processes. Modifications in histone proteins may alter the access of the transcriptional machinery to genes in DNA. Acetylation of histon tails is the most frequently studied and best defined histone modification. Addition of an acetyl group from acetyl-coenzyme A to lysine residues on the N-terminus that protrudes from the surface of the nucleoseome which is named as "histone tail" is accomplished by histone acetyltransferases (HATs). Removal of these acetyl groups from histone tails is catalyzed by histone deacetylases (HDACs) [124]. Acetylation of histone tails neutralizes their positive charge and decreases the interaction between histone tails and negatively charged phosphate groups of DNA backbone. This conformational relaxation enhances accessibility of the gene for transcription machinery. On the contrary, deacetylation of histone tails produces a more condensed chromatin and results in gene silencing. HDACs are classified into four classes as I, II, III and IV according to their sequence homology. Class I, II and IV HDACs are Zn-dependent whereas class III HDACs that are called as sirtuins are nicotinamide adenine dinucleotide (NAD^+)-dependent.

HDAC2 is a class I HDAC that is suggested to negatively regulate memory and synaptic plasticity in healthy mouse brain. The protein level of HDAC2 in AD mouse models and also in postmortem brain tissues of AD have been found to be increased along with cognitive impairment [125]. After the discovery of the fact that histone acetylation vigorously participates to activity-dependent neural plasticity via regulation of critical gene transcription, inhibition of histone acetylation has been investigated in mouse models of AD. In the study conducted by Francis et al. acetylation level of hippocampal histone 4 (H4) in APP/PS1 double mutant transgenic mice has been found to be 50% lower than in wild-type littermates, and this has been linked with impairment in associative learning that is observed in the transgenic mice. Acute administration of HDAC inhibitor, trichostatin A, has rescued the deficiency in H4 acetylation and increased associative memory formation in APP/PS1 mice [126]. Ricobaraza et al. (2009) [127] have demonstrated that intra-perionatal injection of phenylbutyrate (a HDAC inhibitor) improves spatial memory consolidation in 16 month-old Tg2576 mice. This study has not reported a change in amyloid pathology, but rather determined a reduction in pathological phospho-tau levels. In another study, phenylbutyrate treatment has been shown to rescue decreased hippocampal H4 acetylation and impaired dendritic spine density [127, 128]. Notably, these studies have demonstrated that HDAC inhibitors are effective just after the onset of disease. In the case of advanced stage of disease progression, prolonged treatment with sodium butyrate has shown to improve associative memory in APPPS1-21 mice even when administered at a very advanced stage of pathology. The recovery of memory function has been found to be associated with elevated hippocampal histone acetylation and increased expression of genes implicated in associative learning [129]. Recently, orally administrated MS-275 (Entinostat), a class I HDAC inhibitor, has been shown to improve behavioral impairment, reduce neuroinflammation and amyloid plaque deposition in hippocampus and cortical regions of APPPS1-21 mice [130]. Taken together, HDAC inhibitors reduce amyloid plaque deposition, possibly by regulating expression of enzymes that control Aβ processing and clearance [131].

There are some evidence for the potential role of Aβ on histone acetylation. Lithner et al., 2013 [132] have reported that treatment with Aβ oligomers results in increased levels of acetylated H3K14 and loss of dendritic spines in cortical/hippocampal cultures. Further studies are needed to reveal causative interaction between Aβ and histone modifications.

HDAC6 is a member of class II HDAC and is primarily localized in the cytoplasm. By modifying acetylation levels of α-tubulin and cortactin, an F-actin-binding protein, HDAC6 regulates microtubule-dependent and actin-dependent cell motility. HDAC6 is suggested as a potential modulator of tau phosphorylation and accumulation. Ding et al. [133] have demonstrated *in vitro* and *in vivo* that HDAC6 co-localizes with tau in hippocampus of AD brain; selective inhibition of tubulin deacetylation activity of HDAC6 by tubacin does not disrupt HDAC6-tau interaction but decreases site-specific tau phosphorylation (at Thr231); protein level of HDAC6 is elevated in cortex and hippocampus of postmortem AD brain as compared to those in young normal brain.

First identified in saccharomyces cerevisiae, sirtuins which stands for silent information regulator factor 2 (Sir2), is grouped as class III HDACs. NAD^+ is required for sirtuin activity [134]. Not only histones but also non-histone proteins such as transcription factors, metabolic and apoptotic modulators are the target for sirtuins. Sirtuins are highly conserved from prokaryotes to humans, the sirtuin family comprises seven homologs (SIRT1-7) in mammals. They differ from each other according to their tissue specificity, cellular localization, expression patterns, enzymatic activity and substrates [135]. SIRT1, SIRT6 and SIRT7 are nuclear proteins. SIRT1 shuttles between nucleus and cytoplasm. SIRT2 is predominantly localized in cytoplasm [136]. SIRT3, SIRT4 and SIRT5 are mitochondrial proteins. Recent findings suggest that localization of SIRT3 changes from the mitochondria to the nucleus, when SIRT3 and SIRT5 are co-expressed [137]. Certain sirtuins have also mono-ADP-ribosyl transferase activity. SIRT4 and SIRT6 have weak deacetylase activity but they are able to transfer ADP-ribosyl group to their substrates [138]. SIRT5 was identified as a lysine demalonylase and desuccinylase [139].

In early 2000s, epidemiological studies have revealed that iindividuals who maintain a low calorie diet have a reduced risk of developing AD. The influence of calorie restriction (CR) on AD pathology has been examined in AD mouse models, and it has been shown that CR reduces amyloid plaques by activating SIRT1 [140, 141]. Neuroprotective effect of CR has been attributed to increase in NAD^+/NADH ratio and subsequent activation of SIRT1 during the CR. SIRT1 is the most frequently studied sirtuin in AD. It is highly expressed in certain brain regions including cortex, hippocampus, cerebellum and hypothalamus. Non-histone substrates of SIRT1 are core RNA polymerase I transcriptional machinery, HAT p300/CBP [142], tumor supressors p53 and p73 [143, 144], nuclear factor κ-light chain enhancer of activated B cells (NF-κB) [145], forkhead box protein O (FOXO) family of transcription factors [146] and peroxisome proliferator-activated receptor γ (PPARγ) coactivator-1α (PGC-1α) [147]. All of these molecules are involved in regulation of neurogenesis, mitochondrial biogenesis, apoptosis and antioxidant defence in the brain. SIRT1 inhibits p53 and NF-κB and consequently reduces their pro-apoptotic and pro-inflammatory effects, respectively. SIRT1 may be involved in neuroprotection due to its inhibitor effect on p53-mediated apoptosis and upregulating role in α-secretase production [148]. According to results of studies with cell culture models and transgenic mouse models, SIRT1 overexpression increases α-secretase activity, reduces Aβ formation, promotes neuronal survival, and is neuroprotective against Aβ toxicity [141, 149, 150]. SIRT1 deacetylates retinoic acid receptor and activates transcription of ADAM10 which encodes α-secretase. Since α-secretase is responsible for non-amyloidogenic cleavage of APP, upregulation of α-secretase shifts APP processing to reduce the pathological accumulation of toxic Aβ peptides [149].

There is some evidence for benefical effect of SIRT1 activation on taupathy. It has been demonstrated that breakdown of tau is inhibited when tau is acetylated by histone acetyltransferase p300. SIRT1 deacetylates the acetylated tau and promotes its degredation and clearance. SIRT1 downregulation has been shown to inhibit tau polyubiquitination and tau turnover, thus results in accumulation of phospho-tau [151]. Recently Du et al., 2014 [152] have demonstrated that activation of SIRT1 attenuates cerebral ventricular streptozotocin-induced tau hyperphosphorylation and cognitive injuries in rat hippocampi.

Metabolic regulator role of SIRT1 is well documented. SIRT1 controls metabolic processes by activating PGC-1α, a master regulator of mitochondrial biogenesis. PGC-1α activation results in increased insulin sensitivity and inhibition of mitochondrial dysfunction. In this context, SIRT1 can mediate neuroprotection [153]. As an another neuroprotective mechanism, SIRT1 induces basal rate of autophagy which is an important way for the removal of toxic misfolded protein aggregates such as Aβ [154].

Julien et al. [155] have reported that cortical SIRT1 expression is decreased in autopsy specimens of AD patients but not in the individuals with MCI; SIRT1 mRNA and protein levels are negatively correlated with the duration of symptoms and the accumulation of tau, but are weakly correlated with the Aβ42. In accordance with these data, a significant decline in serum SIRT1 level has been determined in patients with both AD and MCI in comparison to healthy elderly individuals [156].

Taken together, SIRT1 activation seems as an useful approach to prevent onset and progression of AD, thus has gained great interest. Resveratrol is a naturally occurring polyphenol found in red grapes and red wine. It is known with its antioxidant, anti-carcinogenic and anti-inflammatory activities. In addition to these properties it is a powerful activator of SIRT1 [157]. SIRT1 activation by resveratrol has been shown to protect against polyglutamine toxicity [158] and plaque formation in AD models. Also, it may prevent apoptotic death of neurons by deacetylating and subsequently repressing p53 activity [159]. Therapeutic role of resveratrol in AD has been extensively investigated. In transgenic mouse model of AD, resveratrol has been shown to prevent learning impairment by increasing deaceylation of PGC-1α and p53 [160]. In addition, resveratrol has been demonstrated to inhibit amyloid burden by increasing mitochondrial complex IV protein levels in mouse model of AD [161]. Resveratrol is undergoing evaluation in clinical trials (www. clinicaltrials.gov).

SIRT3 is a mitochondrial protein. SIRT3 decreases mitochondrial membrane potential and production of ROS [162]. SIRT3 has been thought to be involved in neuroprotection. Because deacetylation activity of SIRT3 may protect neurons via decreasing ROS levels or increasing O_2 consumption in mitochondria [163].

In addition to histon acetylation, dysregulation of histone methylation has also been suggested to be related to synaptic plasticty and cognition [164, 165].

3.3. MicroRNAs

MicroRNAs (miRNAs) are small non-coding RNAs (21-25 nucleotides in length) and regulate gene expression post-transcriptionally. miRNAs bind to target mRNAs at their 3'-untranslated regions and supress their translation and/or promote their degradation [166]. More than 30,000 miRNAs have been identified in humans. As indicated by a large number of studies in the last decade, miRNAs play an important role in various processes such as proliferation, differentiation and apoptosis in both healthy and pathological states [167]. miRNAs can be used as potential diagnostic tools for many diseases due to their stability and convenience of detection in serum. The finding that specific circulating miRNAs are altered in plasma and serum of cancer patients has opened the door to novel diagnostic and possibly therapeutic tools for AD.

Orchestrating the changes in gene expression and protein production, miRNAs take place in memory consolidation and contribute to pathogenesis of neurodegenerative diseases. miRNAs are highly expressed in different neuronal compartments of the brain, and many of them are brain-specific or brain-enriched [167]. miRNAs have selective distribution within neurons and synapses. They have been found to regulate translation of proteins needed for neurite outgrowth, synaptogenesis and activity-driven synaptic plasticity [168-170]. They may also orchestrate fine-tuning of gene expression required for learning, memory and cognitive performance in general.

A number of miRNAs have been strongly implicated in AD, particularly brain-enriched or brain-specific miRNAs. Identifying deregulated miRNAs, miRNA profiling studies have opened a new area for researches in the field of AD. A number of specific miRNAs have been determined to be dysregulated in patients with AD. These specific miRNAs may affect the AD-related pathology either directly (via APP and BACE1), or indirectly by affecting genes related with neurogenesis and immune response [171-173]. In vitro studies have revealed that miR-106a, -520c [174], members of the miR-20a family, such as -20a, -106a/b and -17 [175], miR-16 and -101 [176, 177] and most recently miR-147, -655, -323-3p and -153 [178] directly regulate APP.

On the other hand, Aβ is a powerful regulator of miRNA levels. Schonrock et al. [179] have found that Aβ leads to a severe change in miRNA profiles; incubation of primary hippocampal neurons with Aβ42 preparation causes to down-regulation of substantial proportion of miRNAs (miR-9, -181c, -148b, -30c, -20b, -361, -21, 409-3p and Let-71).

The expression of certain miRNAs changes over time, supporting the transient effect on miRNA expression during AD development. Loss of miRNA cluster containing miR-29a, -29b1 and -9 in sporadic forms of AD has been found to corralate with increased BACE1 expression and increased Aβ levels [172]. In an in vitro study, the level of BACE1 has been found to be down-regulated in mice overexpressing miR-29c and it has been concluded that miR-29c might be an endogenous BACE1 regulator [180]. Animal studies have indicated an inverse correlation between BACE1 protein level and miR-298, -328 [181], -103 and -107 [182].

Previous evidence of post-mortem AD brains and serum samples have demonstrated a dysregulation in a number of miRNAs, including miR-9, miR-29a, miR-29b, miR-101, miR-106, miR-125b, miR-181c, miR-137 and miR-125b that are potentially contibute to increased APP expression and Aβ production [171, 172, 183-187].

4. Future Perspectives

There is strong evidence that epigenetic alterations are inextricably linked with several major pathological processes in AD. Pathological alterations finally leading the AD are established very early in life. This can be explained by latent early-life associated regulation (LEARn) model. According to this model various environmental factors including diet, are associated genes in a long-term fashion, beginning at early developmental stages. These modifications that do not have pathological consequences until later in life are mediated by epigenetic alterations in the promoter regions of genes [188]. Regardless a cause or consequence, epigenetic dysregulations are an important factor contributing to development of AD. Unlike DNA mutations, epigenetic disruptions are potentially reversible and therefore provide an important target for pharmacological intervention. The possibility of transgenic animal models of AD has facilitated to search the role of epigenetic mechanisms on disease-related pathologies such as Aβ and tau. Some epigenetic changes occur long before the appearance of AD symptoms. If such changes are identified and detected early, it could be possible to prevent onset of AD via epigenetic therapeutic approaches. Despite the intense investigation in this field previous findings are not enough for approval of epigenetic drugs in AD. DNA methylation alterations in promoter of AD-related genes should be better understood. Further researches are needed to reveal DNA methylation changes affecting the transcription of AD-related genes, particularly effect of demethylation on hypermethylated gene promoters should be examined in AD models. SIRT1 is frequently studied but not much is known about the role of other sirtuins in AD. Other sirtuins, particularly SIRT2 and SIRT3 also seem to play a role in AD. It would be interesting to see whether other sirtuins have an effect in AD pathogenesis. Recently, SIRT1 activators have become a novel target for AD management and are undergoing evaluation in clinical trials. miRNAs are promising molecular tools to identify main pathways implicated in different aspects of the disease such as accumulation of Aβ, hyperphosphorylation of tau, neuroinflammation and neuronal cell death. Accumulating evidence suggests that alterations in the miRNA profiles of brain contribute to AD pathogenesis. Thus, understanding of miRNA-mediated gene silencing which contributes to AD pathogenesis provides a new perspective in AD management.

REFERENCES

[1] Wang, YJ; Zhou, HD; Zhou, XF. Clearance of amyloid-beta in Alzheimer's disease: progress, problems and perspectives. *Drug Discov Today*, 2006, 11, 931-938.

[2] Anantharaman, M; Tangpong, J; Keller, JN; Murphy, MP; Markesbery, WR; Kiningham, KK. St Clair DK. β-Amyloid mediated nitration of manganese superoxide dismutase: implication for oxidative stress in a APPNLh/NLh X PS-1P264L/P264L double knock-in mouse model of Alzheimer's disease. *Am J Pathol.*, 2006, 168, 1608–1618.

[3] Cao, X; Sudhof, TC. A transcriptionally [correction of transcriptively] active complex of APP with Fe65 and histone acetyltransferase Tip60. *Science*, 2001, 293, 115–120.

[4] Cao, X; Sudhof, TC. Dissection of amyloid-beta precursor protein-dependent transcriptional transactivation. *J Biol Chem.*, 2004, 279, 24601–24611.

[5] Boddapati, S; Levites, Y; Sierks, MR. Inhibiting β-secretase activity in Alzheimer's disease cell models with single-chain antibodies specifically targeting APP. *J Mol Biol.*, 2011, 405, 436–447.

[6] Zheng, WH; Bastianetto, S; Mennicken, F; Ma, W; Kar, S. Amyloid β peptide induces tau phosphorylation and loss of cholinergic neurons in rat primary septal cultures. *Neuroscience*, 2002, 115, 201–211.

[7] Castellano, JM; Deane, R; Gottesdiener, AJ; Verghese, PB; Stewart, FR; West, T; Paoletti, AC; Kasper, TR; DeMattos, RB; Zlokovic, BV; Holtzman, DM. Low-density lipoprotein receptor overexpression enhances the rate of brain-to-blood Ab eta clearance in a mouse model of beta-amyloidosis. *Proc Natl Acad Sci U S A*, 2012, 109, 15502-15507.

[8] Deane, R; Du Yan, S; Submamaryan, RK; LaRue, B; Jovanovic, S; Hogg, E; Welch, D; Manness, L; Lin, C; Yu, J; Zhu, H; Ghiso, J; Frangione, B; Stern, A; Schmidt, AM; Armstrong, DL; Arnold, B; Liliensiek, B; Nawroth, P; Hofman, F; Kindy, M; Stern, D; Zlokovic, B. RAGE mediates amyloid-beta peptide transport across the blood-brain barrier and accumulation in brain. *Nature Medicine*, 2003, 9, 907–913.

[9] Deane, R; Wu, Z; Zlokovic, BV. RAGE (yin) versus LRP (yang) balance regulates alzheimer amyloid beta-peptide clearance through transport across the blood-brain barrier. *Stroke*, 2004, 35, 2628–2631.

[10] Holtzman, DM. Role of apoe/Abeta interactions in the pathogenesis of Alzheimer's disease and cerebral amyloid angiopathy. *J Mol Neurosci.*, 2001, 17, 147–155.

[11] Wang, DS; Dickson, DW; Malter, JS. β-Amyloid Degradation and Alzheimer's Disease. *J Biomed Biotechnol*, 2006, 2006, 1–12.

[12] Lee, VM; Goedert, M; Trojanowski, JQ. Neurodegenerative tauopathies. *Annu Rev Neurosci.*, 2001, 24, 1121–1159.

[13] Alonso, AD; Zaidi, T; Grundke-Iqbal, I; Iqbal, K. Role of abnormally phosphorylated tau in the breakdown of microtubules in Alzheimer disease. *Proc Natl Acad Sci USA*, 1994, 91, 5562–5566.

[14] Sengupta, A; Kabat, J; Novak, M; Wu, Q; Grundke-Iqbal, I; Iqbal, K. Maximal inhibition of tau binding to microtubules requires the phosphorylation of tau at both Thr 231 and Ser 262. *Neurobiol Aging*, 1998, 19S, S124– S524.

[15] Freir, DB; Fedriani, R; Scully, D; Smith, IM; Selkoe, DJ; Walsh, DM; Regan, CM. Abeta oligomers inhibit synapse remodelling necessary for memory consolidation. *Neurobiol Aging*, 2011, 32, 2211–2218.

[16] Mufson, EJ; Binder, L; Counts, SE; DeKosky, ST; de Toledo-Morrell, L; Ginsberg, SD; Ikonomovic, MD; Perez, SE; Scheff, SW. Mild cognitive impairment: pathology and mechanisms. *Acta Neuropathol.*, 2012, 123, 13–30.

[17] Scheff, SW; Price, DA; Schmitt, FA; DeKosky, ST; Mufson, EJ. Synaptic alterations in CA1 in mild Alzheimer disease and mild cognitive impairment. *Neurology*, 2007, 68, 1501–1508.

[18] Berchtold, NC; Coleman, PD; Cribbs, DH; Rogers, J; Gillen, DL; Cotman, CW. Synaptic genes are extensively downregulated across multiple brain regions in normal human aging and Alzheimer's disease. *Neurobiol Aging*, 2013, 34, 1653–1661.

[19] Chang, RY; Nouwens, AS; Dodd, PR; Etheridge, N. The synaptic proteome in Alzheimer's disease. *Alzheimers Dement.*, 2013, 9, 499–511.

[20] Krstic, D; Knuesel, I. Deciphering the mechanism underlying late-onset Alzheimer disease. *Nat Rev Neurol.*, 2013, 9, 25–34.

[21] Praticò, D. Oxidative stress hypothesis in Alzheimer's disease: a reappraisal. *Trends Pharmacol Sci.*, 2008, 29, 609–615.

[22] Ferreiro, E; Baldeiras, I; Ferreira, IL; Costa, RO; Rego, AC; Pereira, CF; Oliveira, CR. Mitochondrial- and endoplasmic reticulum-associated oxidative stress in Alzheimer's disease: from pathogenesis to biomarkers. *Int J Cell Biol.*, 2012, 735206.

[23] Beal, MF. Oxidatively modified proteins in aging and disease. *Free Radical Bio Med.*, 2002, 32, 797–803.

[24] Praticò, D; Sung, S. Lipid peroxidation and oxidative imbalance: early functional events in Alzheimer's disease. *J Alzheimers Dis.*, 2004, 6, 171–175.

[25] Matsuoka, Y; Picciano, M; La Francois, J; Duff, K. Fibrillar β-amyloid evokes oxidative damage in a transgenic mouse model of Alzheimer's disease. *Neuroscience*, 2001, 104, 609–613.

[26] Kontush, A. Amyloid-beta: an antioxidant that becomes a pro-oxidant and critically contributes to Alzheimer's disease. *Free Radic Biol Med.*, 2001, 31(9), 1120-1131.

[27] Behl, C; Davis, JB; Lesley, R; Schubert, D. Hydrogen peroxide mediates amyloid β protein toxicity. *Cell*, 1994, 77, 817–827.

[28] Butterfield, DA. Amyloid b-peptide (1–42)-induced Oxidative Stress and Neurotoxicity: Implications for Neurodegeneration in Alzheimer's Disease Brain. *Free Radic Res.*, 2002, 36, 1307–1313.

[29] Quiroz-Baez, R; Rojas, E; Arias, C. Oxidative stress promotes JNK-dependent amyloidogenic processing of normally expressed human APP by differential modification of α-, β- and γ-secretase expression. *Neurochem Int.*, 2009, 55, 662–670.

[30] Li, F; Calingasan, NY; Yu, F; Mauck, WM; Toidze, M; Almeida, CG; Takahashi, RH; Carlson, GA; Flint Beal, M; Lin, MT; Gouras, GK. Increased plaque burden in brains of APP mutant MnSOD heterozygous knockout mice. *J Neurochem.*, 2004, 89, 1308–1312.

[31] Dumont, M; Wille, E; Stack, C; Calingasan, NY; Beal, MF; Lin, MT. Reduction of oxidative stress, amyloid deposition, and memory deficit by manganese superoxide dismutase overexpression in a transgenic mousemodel of Alzheimer's disease. *The FASEB Journal*, 2009, 23, 2459–2466.

[32] Murakami, K; Murata, N; Noda, Y; Tahara, S; Kaneko, T; Kinoshita, N; Hatsuta, H; Murayama, S; Barnham, KJ; Irie, K; Shirasawa, T; Shimizu, T. SOD1 (copper/zinc superoxide dismutase) deficiency drives amyloid β protein oligomerization andmemory loss in mouse model of Alzheimer disease. *J Biol Chem.*, 2011, 286, 44557–44568.

[33] Poon, HF; Calabrese, V; Scapagnini, G; Butterfield, DA. Free radicals and brain aging. *Clin Geriatr Med.*, 2004, 20, 329.

[34] Harman, D. Aging: a theory based on free radical and radiation chemistry. *J Geront.*, 1956, 11, 289.

[35] Gamblin, TC; King, ME; Kuret, J; Berry, RW; Binder, LI. Oxidative regulation of fatty acid-induced tau polymerization. *Biochemistry*, 2000, 39, 14203–14210.

[36] Bush, AI. The Metal Theory of Alzheimer's Disease. *J Alzheimers Dis.*, 2012, 30, 1–5.

[37] Kozłowski, H; Łuczkowski, M; Valensin, D; Valensin, G. Metal ion binding properties of proteins related to neurodegeneration. Sigel A, Sigel H and Sigel RKO.Eds., Wiley & Sons. *Metal Ions in Life Sciences*, 2006, 1, 61-87.

[38] Nunomura, A; Perry, G; Aliev, G; Hirai, K; Takeda, A; Balraj, EK; Jones, PK; Ghanbari, H; Wataya, T; Shimohama, S; Chiba, S; Atwood, CS; Petersen, RB; Smith, MA. Oxidative damage is the earliest event in Alzheimer disease. *J Neuropathol Exp Neurol.*, 2001, 60, 759–767.

[39] Greenough, MA; Camakaris, J; Bush, AI. Metal dyshomeostasis and oxidative stress in Alzheimer's disease. *Neurochem Int.*, 2013, 62, 540-555.

[40] Sipe, JC; Lee, P; Beutler, E. Brain iron metabolism and neurodegenerative disorders. *Dev Neurosci.*, 2002, 24, 188–196.

[41] Bush, AI. The metallobiology of Alzheimer's disease. *Trends Neurosci.*, 2003, 26, 207–214.

[42] Cherny, RA; Atwood, CS; Xilinas, ME; Gray, DN; Jones, WD; McLean, CA; Barnham, KJ; Volitakis, I; Fraser, FW; Kim, Y; Huang, X; Goldstein, LE; Moir, RD; Lim, JT; Beyreuther, K; Zheng, H; Tanzi, RE; Masters, CL; Bush, AI. Treatment with a copper-zinc chelator markedly and rapidly inhibits b-amyloid accumulation in Alzheimer's disease transgenic mice. *Neuron*, 2001, 30, 665–676.

[43] Frazzini, V; Rockabrand, E; Mocchegiani, E; Sensi, SL. Oxidative stress and brain aging, is zinc the link? *Biogerontology*, 2006, 7, 307-314.

[44] Maret, W; Sandstead, HH. Zinc requirements and the risks and benefits of zinc supplementation. *J Trace Elem Med Biol.*, 2006, 20, 3-18.

[45] Yamamoto, A; Shin, RW; Hasegawa, K; Naiki, H; Sato, H; Yoshimasu, F; Kitamoto, T. Iron (III) induces aggregation of hyperphosphorylated tau and its reduction to iron (II) reverses the aggregation: implications in the formation of neurofibrillary tangles of Alzheimer's disease. *J Neurochem.*, 2002, 82, 1137-1147.

[46] Smith, DG; Cappai, R; Barnham, KJ. The redox chemistry of the Alzheimer's disease amyloid beta peptide. *Biochim Biophys Acta*, 2007, 1768, 1976-1990.

[47] Lannfelt, L; Blennow, K; Zetterberg, H; Batsman, S; Ames, D; Harrison, J; Masters, CL; Targum, S; Bush, AI; Murdoch, R; Wilson, J; Ritchie, CW. PBT2-201-EURO study group. Safety; efficacy; and biomarker findings of PBT2 in targeting Abeta as a modifying therapy for Alzheimer's disease: a phase IIa, double-blind, randomized, placebo-controlled trial. *Lancet Neurol.*, 2008, 7, 779-786.

[48] Zhu, X; Perry, G; Moreira, PI; Aliev, G; Cash, AD; Hirai, K; Smith, MA. Mitochondrial abnormalities and oxidative imbalance in Alzheimer disease, *J Alzheimers Dis.*, 2006, 9, 147–153.

[49] Webster, R. Neurotransmitters, Drugs and Brain Function. *John Wiley & Sons*, 2003.

[50] Swerdlow, RH. Brain aging, Alzheimer's disease, and mitochondria. *Biochim Biophys Acta.*, 2011, 1812, 1630-1639.

[51] Rodrigues, CMP; Sol´a, S; Brito, MA; Brondino, CD; Brites, D; Moura, JJG. Amyloid β-peptide disrupts mitochondrial membrane lipid and protein structure: protective role of tauroursodeoxycholate. *Biochem Biophys Res Commun.*, 2001, 281, 468–474.

[52] Casley, CS; Canevari, L; Land, JM; Clark, JB; Sharpe, MA. β-Amyloid inhibits integrated mitochondrial respiration and key enzyme activities. *J Neurochem.*, 2002, 80, 91–100.

[53] Chen, CH; Zhou, W; Liu, S; Deng, Y; Cai, F; Tone, M; Tone, Y; Tong, Y; Song, W. Increased NF-κB signalling up-regulates BACE1 expression and its therapeutic potential in Alzheimer's disease. *Int J Neuropsychopharmacol.*, 2012, 15(1), 77-90.

[54] Vom Berg, J; Prokop, S; Miller, KR; Obst, J; Kälin, RE; Lopategui-Cabezas, I; Wegner, A; Mair, F; Schipke, CG; Peters, O; Winter, Y; Becher, B; Heppner, FL. Inhibition of IL-12/IL-23 signaling reduces Alzheimer's disease-like pathology and cognitive decline. *Nat Med.*, 2012, 18, 1812-9.

[55] Choi, SH; Bosetti, F. Cyclooxygenase-1 null mice show reduced neuroinflammation in response to beta-amyloid. *Aging (Albany NY)*, 2009, 1, 234–244.

[56] Liu, S; Liu, Y; Hao, W; Wolf, L; Kiliaan, AJ; Penke, B; Rübe, CE; Walter, J; Heneka, MT; Hartmann, T; Menger, MD; Fassbender, K. TLR2 is a primary receptor for Alzheimer's amyloid b peptide to trigger neuroinflammatory activation. *J Immunol.*, 2012, 188, 1098–1107.

[57] Piro, JR; Benjamin, DI; Duerr, JM; Pi, Y; Gonzales, C; Wood, KM; Schwartz, JW; Nomura, DK; Samad, TA. A Dysregulated Endocannabinoid-Eicosanoid Network Supports Pathogenesis in a Mouse Model of Alzheimer's Disease. *Cell Reports*, 2012, 1, 617–623.

[58] Sastre, M; Dewachter, I; Landreth, GE; Willson, TM; Klockgether, T; van Leuven, F; Heneka, MT. Nonsteroidal anti-inflammatory drugs and peroxisome proliferator-activated receptor-gamma agonists modulate immunostimulated processing of amyloid precursor protein through regulation of beta-secretase. *J Neurosci.*, 2003, 23, 9796–9804.

[59] Szekely, CA; Breitner, JCS; Fitzpatrick, AL; Rea, TD; Psaty, BM; Kuller, LH; Zandi, PP. NSAID use and dementia risk in the Cardiovascular Health Study: role of APOE and NSAID type. *Neurology*, 2008, 70, 17–24.

[60] Imbimbo, BP; Solfrizzi, V; Panza, F. Are NSAIDs useful to treat Alzheimer's disease or mild cognitive impairment? *Front Aging Neurosci.*, 2010, 2, 19.

[61] Kitazawa, M; Cheng, D; Tsukamoto, MR; Koike, MA; Wes, PD; Vasilevko, V; Cribbs, DH; LaFerla, FM. Blocking IL-1 signaling rescues cognition, attenuates tau pathology, and restores neuronal b-catenin pathway function in an Alzheimer's disease model. *J Immunol.*, 2011, 187, 6539–6549.

[62] Shi, JQ; Shen, W; Chen, J; Wang, BR; Zhong, LL; Zhu, YW; Zhu, HQ; Zhang, QQ; Zhang, YD; Xu, J. Anti-TNF-a reduces amyloid plaques and tau phosphorylation and induces CD11c-positive dendritic-like cell in the APP/PS1 transgenic mouse brains. *Brain Res.*, 2011, 1368, 239–247.

[63] Bertram, L; Lill, CM; Tanzi, RE. The genetics of Alzheimer disease: back to the future. *Neuron*, 2010, 68, 270–281.

[64] Gerrish, A; Russo, G; Richards, A; Moskvina, V; Ivanov, D; Harold, D; et al. The role of variation at AbPP, PSEN1 PSEN2, and MAPT in late onset Alzheimer's disease. *J Alzheimers Dis*, 2012, 28, 377–87.

[65] Cruchaga, C; Chakraverty, S; Mayo, K; Vallania, FLM; Mitra, RD; Faber, K; et al. Rare variants in APP PSEN1 and PSEN2 increase risk for AD in late-onset Alzheimer's disease families. *PLoS ONE*, 2012, 7(2), e31039.

[66] Goate, A; Chartier-Harlin, MC; Mullan, M; Brown, J; Crawford, F; Fidani, L; Giuffra, L; Haynes, A; Irving, N; James, L; et al. Segregation of a missense mutation in the amyloid precursor protein gene with familial Alzheimer's disease. *Nature*, 1991, 349, 704–06.

[67] Sleegers, K; Brouwers, N; Gijselinck, I; Theuns, J; Goossens, D; Wauters, J; Del-Favero, J; Cruts, M; van Duijn, CM; Van Broeckhoven, C. APP duplication is suffi cient to cause early onset Alzheimer's dementia with cerebral amyloid angiopathy. *Brain*, 2006, 129, 2977–83.

[68] Kasuga, K; Shimohata, T; Nishimura, A; Shiga, A; Mizuguchi, T; Tokunaga, J; Ohno, T; Miyashita, A; Kuwano, R; Matsumoto, N; Onodera, O; Nishizawa, M; Ikeuchi, T.

Identification of independent APP locus duplication in Japanese patients with early-onset Alzheimer disease. *J Neurol Neurosurg Psychiatry*, 2009, 80, 1050–1052.

[69] Rogaev, EI; Sherrington, R; Rogaeva, EA; Levesque, G; Ikeda, M; Liang, Y; Chi, H; Lin, C; Holman, K; Tsuda, T; et al. Familial Alzheimer's disease in kindreds with missense mutations in a gene on chromosome 1 related to the Alzheimer's disease type 3 gene. *Nature*, 1995, 376, 775–778.

[70] Di Fede, G; Catania, M; Morbin, M; Rossi, G; Suardi, S; Mazzoleni, G; Merlin, M; Giovagnoli, AR; Prioni, S; Erbetta, A; Falcone, C; Gobbi, M; Colombo, L; Bastone, A; Beeg, M; Manzoni, C; Francescucci, B; Spagnoli, A; Cantù, L; Del Favero, E; Levy, E; Salmona, M; Tagliavini, F. A recessive mutation in the APP gene with dominant-negative effect on amyloidogenesis. *Science*, 2009, 323, 1473–1477.

[71] Bentahir, M; Nyabi, O; Verhamme, J; Tolia, A; Horré, K; Wiltfang, J; Esselmann, H; De Strooper, B. Presenilin clinical mutations can aff ect gamma-secretase activity by different mechanisms. *J Neurochem*., 2006, 96, 732–742.

[72] Kumar-Singh, S; Theuns, J; Van Broeck, B; Pirici, D; Vennekens, K; Corsmit, E; Cruts, M; Dermaut, B; Wang, R; Van Broeckhoven, C. Mean age-of-onset of familial alzheimer disease caused by presenilin mutations correlates with both increased Abeta42 and decreased Abeta40. *Hum Mutat*., 2006, 27, 686–695.

[73] Harold, D; Abraham, R; Hollingworth, P; et al. Genome-wide association study identifies variants at CLU and PICALM associated with Alzheimer's disease. *Nat Genet*., 2009, 41, 1088–1093.

[74] Hollingworth, P; Harold, D; Sims, R; et al. Common variants at ABCA7, MS4A6A/MS4A4E, EPHA1, CD33 and CD2AP are associated with Alzheimer's disease. *Nat Genet*., 2011, 43, 429–435.

[75] Seshadri, S; Fitzpatrick, AL; Ikram, MA; et al. Genome-wide analysis of genetic loci associated with Alzheimer disease. *JAMA*, 2010, 303, 1832–1840.

[76] Raygani, AV; Zahrai, M; Raygani, AV; Doosti, M; Javadi, E; Rezaei, M; Pourmotabbed, T. Association between apolipoprotein E polymorphism and Alzheimer disease in Tehran, Iran. *Neurosci Lett*., 2005, 375, 1-6.

[77] Mahley, RW. Apolipoprotein E: cholesterol transport protein with expanding role in cell biology. *Science*, 1988, 240, 622–630.

[78] Rocchi, A; Pellegrini, S; Siciliano, G; Murri, L. Causative and susceptibility genes for Alzheimer's disease: a review. *Brain Res Bull*., 2003, 61, 1–24.

[79] Lehtovirta, M; Laakso, MP; Soininen, H; Helisalmi, S; Mannermaa, A; Helkala, EL; Partanen, K; Ryynänen, M; Vainio, P; Hartikainen, P. Volumes of hippocampus, amygdala and frontal lobe in Alzheimer's patients with different apolipoprotein E genotypes. *Neuroscience*, 1995, 67, 65–72.

[80] Lehtovirta, M; Soininen, H; Helisalmi, S; Mannermaa, A; Helkala, EL; Hartikainen, P; Hänninen, T; Ryynänen, M; Riekkinen, PJ. Clinical and neuropsychological characteristics in familial and sporadic Alzheimer's disease: relation to apolipoprotein E polymorphism. *Neurology*, 1996, 46, 413–419.

[81] Bigler, ED; Lowry, CM; Anderson, CV; Johnson, SC; Terry, J; Steed, M. Dementia, quantitative neuroimaging, and apolipoprotein E genotype. *AJNR Am J Neuroradiol*., 2000, 21, 1857–1868.

[82] Hashimoto, M; Yasuda, M; Tanimukai, S; Matsui, M; Hirono, N; Kazui, H; Mori, E. Apolipoprotein E e4 and the pattern of regional brain atrophy in Alzheimer's disease. *Neurology*, 2001, 57, 1461–1466.

[83] Schmechel, DE; Saunders, AM; Strittmatter, WJ; Crain, BJ; Hulette, CM; Joo, SH; Pericak-Vance, MA; Goldgaber, D; Roses, AD. Increased amyloid beta-peptide deposition in cerebral cortex as a consequence of apolipoprotein E genotype in late-onset Alzheimer disease. *Proc Natl Acad Sci USA*, 1993, 90, 9649–9653.

[84] Berg, L; McKeel, DW: Jr. Miller, JP; Storandt, M; Rubin, EH; Morris, JC; Baty, J; Coats, M; Norton, J; Goate, AM; Price, JL; Gearing, M; Mirra, SS; Saunders, AM. Clinicopathologic studies in cognitively healthy aging and Alzheimer's disease, relation of histologic markers to dementia severity, age, sex, and apolipoprotein E genotype. *Arch Neurol.*, 1998, 55, 326–335.

[85] Lippa, C; Smith, T; Saunders, A; Hulette, C; Pulaski-Salo, D; Roses, A. Apolipoprotein E-epsilon 2 and Alzheimer's disease: Genotype influences pathologic phenotype. *Neurology*, 1997, 48, 515–519.

[86] Nagy, Z; Esiri, M; Jobst, K; Johnston, C; Litchfield, S; Sim, E; Smith, A. Influence of the apolipoprotein-e genotype on amyloid deposition and neurofibrillary tangle formation in Alzheimer's disease. *Neuroscience*, 1995, 69, 757–761.

[87] Frisoni, GB; Govoni, S; Geroldi, C; Bianchetti, A; Calabresi, L; Franceschini, G; Trabucchi, M. Gene dose of the epsilon 4 allele of apolipoprotein E and disease progression in sporadic late-onset Alzheimer's disease. *Ann Neurol.*, 1995, 37, 596–604.

[88] Hirono, N; Hashimoto, M; Yasuda, M; Kazui, H; Mori, E. Accelerated memory decline in Alzheimer's disease with apolipoprotein epsilon 4 allele. *J Neuropsychiatry Clin Neurosci.*, 2003, 15, 354–358.

[89] Hoyt, BD; Massman, PJ; Schatschneider, C; Cooke, N; Doody, RS. Individual growth curve analysis of APOE epsilon 4-associated cognitive decline in Alzheimer disease. *Arch Neurol.*, 2005, 62, 454–459.

[90] Kleiman, T; Zdanys, K; Black, B; Rightmer, T; Grey, M; Garman, K; Macavoy, M; Gelernter, J; van Dyck, C. Apolipoprotein E epsilon 4 allele is unrelated to cognitive or functional decline in Alzheimer's disease: retrospective and prospective analysis. *Dement Geriatr Cogn Disord.*, 2006, 22, 73–82.

[91] Martins, CA; Oulhaj, A; deJager, CA; Williams, JH. APOE alleles predict the rate of cognitive decline in Alzheimer disease: a nonlinear model. *Neurology*, 2005, 65, 1888–1893.

[92] Tschanz, JT; Corcoran, CD; Schwartz, S; Treiber, K; Green, RC; Norton, MC; Mielke, MM; Piercy, K; Steinberg, M; Rabins, PV; Leoutsakos, JM; Welsh-Bohmer, KA; Breitner, JC; Lyketsos, CG. Progression of cognitive, functional, and neuropsychiatric symptom domains in a population cohort with Alzheimer dementia: the Cache County Dementia Progression study. *Am J Geriatr Psychiatry*, 2011, 19, 532–542.

[93] Han, SD; Bondi, MW. Revision of the apolipoprotein E compensatory mechanism recruitment hypothesis. *Alzheimers Dement.*, 2008, 4, 251–254.

[94] Jang, H; Serra, C. Nutrition, epigenetics, and diseases. *Clin Nutr Res.*, 2014, 3, 1–8.

[95] Sezgin, Z; Dincer, Y. Alzheimer's disease and epigenetic diet. *Neurochem Int.*, 2014, 78, 105-116.

[96] Marques, SCF; Lemos, R; Ferreiro, E; Martins, M; de Mendonça, A; Santana, I; Outeiro, TF; Pereira, CMF. Epigenetic regulation of BACE1 in Alzheimer's disease patients and in transgenic mice. *Neuroscience*, 2012, 220, 256-66.
[97] Kong, LN; Zuo, PP; Mu, L; Liu, YY; Yang, N. Gene expression profile of amyloid beta protein-injected mouse model for Alzheimer disease. *Acta Pharmacol Sin.*, 2005, 26, 666–672.
[98] Papassotiropoulos, A; Fountoulakis, M; Dunckley, T; Stephan, DA; Reiman, EM. Genetics; transcriptomics; and proteomics of Alzheimer's disease. *J Clin Psychiatry*, 2006, 67, 652–670.
[99] Yen, RW; Vertino, PM; Nelkin, BD; Yu, JJ; el-Deiry, W; Cumaraswamy, A; Lennon, GG; Trask, BJ; Celano, P; Baylin, SB. Isolation and characterization of the cDNA encoding human DNA methyltransferase. *Nucleic Acids Res.*, 1992, 20, 2287–2291.
[100] Okano, M; Bell, DW; Haber, DA; Li, E. DNA methyltransferases Dnmt3a and Dnmt3b are essential for de novo methylation and mammalian development. *Cell*, 1999, 99, 247–257.
[101] Jeltsch, A. Molecular enzymology of mammalian DNA methyltransferases. *Curr Top Microbiol Immunol.*, 2006, 301, 203–225.
[102] Suzuki, MM; Bird, A. DNA methylation landscapes: provocative insights from epigenomics. *Nat Rev Genet.*, 2008, 9, 465–476.
[103] Tan, L; Shi, YG. Tet family proteins and 5-hydroxymethylcytosine in development and disease. *Development*, 2012, 139, 1895-1902.
[104] Feil, R; Fraga, MF. Epigenetics and the environment: emerging patterns andimplications. *Nat Rev.*, 2012, 13, 97–109.
[105] Fang, M; Chen, D; Yang, CS. Dietary polyphenols may affect DNAmethylation. *J Nutr.*, 2007, 137, 223S–228S.
[106] Seshadri, S; Beiser, A; Selhub, J; Jacques, PF; Rosenberg, IH; D'Agostino, RB; Wilson, PW; Wolf, PA. Plasma homocysteine as a risk factor for dementia and Alzheimer's disease. *N Engl J Med.*, 2002, 346(7), 476-83.
[107] Hernandez, DG; Nalls, MA; Gibbs, JR; Arepalli, S; van der Brug, M; Chong, S; Moore, M; Longo, DL; Cookson, MR; Traynor, BJ; Singleton, AB. Distinct DNA methylation changes highly correlated with chronological age in the human brain. *Hum Mol Genet.*, 2011, 20(6), 1164-72.
[108] Lu, H; Liu, X; Deng, Y; Qing, H. DNA methylation, a hand behind neurodegenerative diseases. *Front Aging Neurosci.*, 2013, 5, 85.
[109] Mastroeni, D; Grover, A; Delvaux, E; Whiteside, C; Coleman, PD; Rogers, J. Epigenetic changes in Alzheimer's disease: decrements in DNA methylation. *Neurobiol Aging*, 2010, 31, 2025–2037.
[110] Chouliaras, L; Mastroeni, D; Delvaux, E; Grover, A; Kenis, G; Hof, PR; Steinbusch, HW; Coleman, PD; Rutten, BP; van den Hove, DL. Consistent decrease in global DNA methylation and hydroxymethylation in the hippocampus of Alzheimer's disease patients. *Neurobiol Aging*, 2013, 34, 2091–2099.
[111] Coppieters, N; Dieriks, BV; Lill, C; Faull, RL; Curtis, MA; Dragunow, M. Global changes in DNA methylation and hydroxymethylation in Alzheimer's disease human brain. *Neurobiol Aging*, 2014, 35, 1334-1344.

[112] Rao, JS; Keleshian, VL; Klein, S; Rapoport, SI. Epigenetic modifications in frontal cortex from Alzheimer's disease and bipolar disorder patients. *Transl Psychiatry*, 2012, 2, e132.

[113] Tohgi, H; Utsugisawa, K; Nagane, Y; Yoshimura, M; Genda, Y; Ukitsu, M. Reduction with age in methylcytosine in the promoter region -224 approximately -101 of the amyloid precursor protein gene in autopsy human cortex. *Brain Res Mol Brain Res.*, 1999, 70(2), 288–292.

[114] Barrachina, M; Ferrer, I. DNA methylation of Alzheimer disease and tauopathy-related genes in postmortem brain. *J Neuropathol Exp Neurol.*, 2009, 68, 880–891.

[115] Scarpa, S; Fuso, A; D'anselmi, F; Cavallaro, RA. Presenilin 1 gene silencing by S-adenosylmethionine: a treatment for Alzheimer disease? *FEBS Lett.*, 2003, 541, 145–148.

[116] Fuso, A; Seminara, L; Cavallaro, RA; D'Anselmi, F; Scarpa, S. S-adenosylmethionine/homocysteine cycle alterations modify DNA methylation status with consequent deregulation of PS1 and BACE and beta-amyloid production. *Mol Cell Neurosci.*, 2005, 28, 195–204.

[117] Fuso, A; Nicolia, V; Cavallaro, RA; Ricceri, L; D'Anselmi, F; Coluccia, P; Calamandrei, G; Scarpa, S. B-vitamin deprivation induces hyperhomocysteinemia and brain S-adenosylhomocysteine, depletes brain S-adenosylmethionine, and enhances PS1 and BACE expression and amyloid-beta deposition in mice. *Mol Cell Neurosci.*, 2008, 37, 731–746.

[118] Tannorella, P; Stoccoro, A; Tognoni, G; Petrozzi, L; Salluzzo, MG; Ragalmuto, A; Siciliano, G; Haslberger, A; Bosco, P; Bonuccelli, U; Migliore, L; Coppedè, F. Methylation analysis of multiple genes in blood DNA of Alzheimer's disease and healthy individuals. *Neurosci Lett.*, 2015, 600, 143-147.

[119] Chen, KL; Wang, SS; Yang, YY; Yuan, RY; Chen, RM; Hu, CJ. The epigenetic effects of amyloid-beta (1-40) on global DNA and neprilysin genes in murine cerebral endothelial cells. *Biochem Biophys Res Commun.*, 2009, 378, 57–61.

[120] Nicolia, V; Fuso, A; Cavallaro, RA; DiLuzio, A; Scarpa, S. B vitamin deficiency promotes tau phosphorylation through regulation of GSK3beta and PP2A. *J Alzheimers Dis.*, 2010, 19, 895–907.

[121] Sontag, E; Nunbhakdi-Craig, V; Sontag, JM; Diaz-Arrastia, R; Ogris, E; Dayal, S; Lentz, SR; Arning, E; Bottiglieri, T. Protein phosphatase 2A methyltransferase links homocysteine metabolism with tau and amyloid precursor protein regulation. *J Neurosci.*, 2007, 27, 2751–2759.

[122] Sontag, JM; Nunbhakdi-Craig, V; Montgomery, L; Arning, E; Bottiglieri, T; Sontag, E. Folate deficiency induces in vitro and mouse brain region-specific downregulation of leucine carboxyl methyltransferase-1 and protein phosphatase 2A B(alpha) subunit expression that correlate with enhanced tau phosphorylation. *J Neurosci.*, 2008, 28, 11477–11487.

[123] Choudhary, C; Weinert, BT; Nishida, Y; Verdin, E; Mann, M. The growing landscape of lysine acetylation links metabolism and cell signalling. *Nat Rev Mol Cell Biol.*, 2014, 15, 536–550.

[124] Yang, XJ; Seto, E. HATs and HDACs: from structure, function and regulation to novel strategies for therapy and prevention. *Oncogene*, 2007, 26, 5310–5318.

[125] Graff, J; Rei, D; Guan, JS; Wang, WY; Seo, J; Hennig, KM; Nieland, TJ; Fass, DM; Kao, PF; Kahn, M; Su, SC; Samiei, A; Joseph, N; Haggarty, SJ; Delalle, I; Tsai, LH. An epigenetic blockade of cognitive functions in the neurodegenerating brain. *Nature*, 2012, 483, 222–226.
[126] Francis, YI; Fa, M; Ashraf, H; Zhang, H; Staniszewski, A; Latchman, DS; Arancio, O. Dysregulation of histone acetylation in the APP/PS1 mouse model of Alzheimer's disease. *J Alzheimers Dis.*, 2009, 18(1), 131–139.
[127] Ricobaraza, A; Cuadrado-Tejedor, M; Pérez-Mediavilla, A; Frechilla, D; Del Río, J; García-Osta, A. Phenylbutyrate ameliorates cognitive deficit and reduces tau pathology in an Alzheimer's disease mouse model. *Neuropsychopharmacology*, 2009, 34, 1721-1732.
[128] Ricobaraza, A; Cuadrado-Tejedor, M; Marco, S; Pérez-Otaño, I; García-Osta, A. Phenylbutyrate rescues dendritic spine loss associated with memory deficits in a mouse model of Alzheimer disease. *Hippocampus*, 2010, 22, 1040-1050.
[129] Govindarajan, N; Agis-Balboa, C; Walter, J; Sananbenesi, F; Fischer, A. Sodium butyrate improves memory function in an Alzheimer's disease mouse model when administered at an advanced stage of disease progression. *J Alzheimers Dis.*, 2011, 24, 1-11.
[130] Zhang, Z; Schluesener, YHJ. Oral administration of histone deacetylase inhibitor MS-275 ameliorates neuroinflammation and cerebral amyloidosis and improves behavior in a mouse model. *J Neuropathol Exp Neurol.*, 2013, 72, 178-185.
[131] Sung, YM; Lee, T; Yoon, H; Di Battista, AM; Song, J; Sohn, Y; Moffat, E; Turner, RS; Jung, M; Kim, J; Hoe, HS. Mercaptoacetamide-based class II HDAC inhibitor lowers Ab levels and improves learning and memory in a mouse model of Alzheimer's disease. *Exp Neurol.*, 2013, 239, 192-201.
[132] Lithner, CU; Lacor, PN; Zhao, WQ; Mustafiz, T; Klein, WL; Sweatt, JD; Hernandez, CM. Disruption of neocortical histone H3 homeostasis by soluble Abeta: implications for Alzheimer's disease. *Neurobiol Aging*, 2013, 34(9), 2081–2090.
[133] Ding, H; Dolan, PJ; Johnson, GV. Histone deacetylase 6 interacts with the microtubule-associated protein tau. *J Neurochem.*, 2008, 106, 2119–2130.
[134] Imai, SI; Armstrong, CM; Kaeberlein, M; Guarente, L. Transcriptional silencing and longevity protein Sir2 is an NAD-dependent histone deacetylase. *Nature*, 2000, 403, 795-799.
[135] Haigis, MC; Guarente, LP. Mammalian sirtuins—emerging roles in physiology, aging, and calorie restriction. *Genes Dev.*, 2006, 20, 2913–2921.
[136] Michishita, E; Park, JY; Burneskis, JM; Barrett, JC; Horikawa, I. Evolutionarily conserved and nonconserved cellular localizations and functions of human SIRT proteins. *Mol Biol Cell*, 2005, 16, 4623–4635.
[137] Nakamura, Y; Ogura, M; Tanaka, D; Inagaki, N. Localization of mouse mitochondrial SIRT proteins: shift of SIRT3 to nucleus by co-expression with SIRT5. *Biochem Biophys Res Commun.*, 2008, 366, 174–179.
[138] Imai, SI; Guarente, L. Ten years of NAD-dependent SIR2 family deacetylases – implications for metabolic diseases. *Trends Pharm Sci.*, 2010, 31, 212–220.
[139] Du, J; Zhou, Y; Su, X; Yu, JJ; Khan, S; Jiang, H; Kim, J; Woo, J; Kim, JH; Choi, BH; He, B; Chen, W; Zhang, S; Cerione, RA; Auwerx, J; Hao, Q; Lin, H. Sirt5 Is an NAD-

Dependent Protein Lysine Demalonylase and Desuccinylase. *Science*, 2011, 334, 806–809.

[140] Patel, NV; Gordon, MN; Connor, KE; Good, RA; Engelman, RW; Mason, J; Morgan, DG; Morgan, TE; Finch, CE. Caloric restriction attenuates Abeta-deposition in Alzheimer transgenic models. *Neurobiol Aging*, 2005, 26, 995-1000.

[141] Qin, W; Yang, T; Ho, L; Zhao, Z; Wang, J; Chen, L; Zhao, W; Thiyagarajan, M; MacGrogan, D; Rodgers, JT; Puigserver, P; Sadoshima, J; Deng, H; Pedrini, S; Gandy, S; Sauve, AA; Pasinetti, GM. Neuronal SIRT1 activation as a novel mechanism underlying the prevention of Alzheimer disease amyloid neuropathology by calorie restriction. *J Biol Chem.*, 2006, 281, 21745-21754.

[142] Bouras, T; Fu, M; Sauve, AA; Wang, F; Quong, AA; Perkins, ND; Hay, RT; Gu, W; Pestell, RG. SIRT1 deacetylation and repression of p300 involves lysine residues 1020/1024 within the cell cycle regulatory domain 1. *J Biol Chem.*, 2005, 280, 10264–10276.

[143] Luo, J; Nikolaev, AY; Imai, S; Chen, D; Su, F; Shiloh, A; Guarente, L; Gu, W. Negative control of p53 by Sir2alpha promotes cell survival under stress. *Cell*, 2001, 107, 137–148.

[144] Michan, S; Sinclair, D. Sirtuins in mammals: insights into their biological function. *Biochem J.*, 2007, 404, 1–13.

[145] Yeung, F; Hoberg, JE; Ramsey, CS; Keller, MD; Jones, DR; Frye, RA; Mayo, MW. Modulation of NF-kappaB-dependent transcription and cell survival by the SIRT1 deacetylase. *EMBO J.*, 2004, 23, 2369–2380.

[146] Motta, MC; Divecha, N; Lemieux, M; Kamel, C; Chen, D; Gu, W; Bultsma, Y; McBurney, M; Guarente, L. Mammalian SIRT1 represses forkhead transcription factors. *Cell*, 2004, 116, 551–563.

[147] Nemoto, S; Fergusson, MM; Finkel, T. SIRT1 functionally interacts with the metabolic regulator and transcriptional coactivator PGC-1{alpha}. *J Biol Chem.*, 2005, 280, 16456–16460.

[148] Cohen, TJ; Guo, JL; Hurtado, DE; Kwong, LK; Mills, IP; Trojanowski, JQ; Lee, VM. The acetylation of tau inhibits its function and promotes pathological tau aggregation. *Nat Commun.*, 2011, 2, 252.

[149] Donmez, G; Wang, D; Cohen, DE; Guarente, L. SIRT1 suppresses beta-amyloid production by activating the alpha-secretase gene ADAM10. *Cell*, 2010, 142, 320–332.

[150] Theendakara, V; Patent, A; Peters Libeu, CA; Philpot, B; Flores, S; Descamps, O; Poksay, KS; Zhang, Q; Cailing, G; Hart, M; John, V; Rao, RV; Bredesen, DE. Neuroprotective Sirtuin ratio reversed by ApoE4. *Proc Natl Acad Sci USA*, 2013, 110, 18303-18308.

[151] Min, SW; Cho, SH; Zhou, Y; Schroeder, S; Haroutunian, V; Seeley, WW; Huang, EJ; Shen, Y; Masliah, E; Mukherjee, C; Meyers, D; Cole, PA; Ott, M; Gan, L. Acetylation of tau inhibits its degradation and contributes to tauopathy. *Neuron*, 2010, 67, 953-966.

[152] Du, LL; Xie, JZ; Cheng, XS; Li, XH; Kong, FL; Jiang, X; Ma, ZW; Wang, JZ; Chen, C; Zhou, XW. Activation of sirtuin 1 attenuates cerebral ventricular streptozotocin-induced tau hyperphosphorylation and cognitive injuries in rat hippocampi. *Age (Dordr)*, 2014, 36, 613-623.

[153] Rodgers, JT; Lerin, C; Haas, W; Gygi, SP; Spiegelman, BM; Puigserver, P. Nutrient control of glucose homeostasis through a complex of PGC-1 alpha and SIRT1. *Nature*, 2005, 434, 113–118.

[154] Lee, IH; Cao, L; Mostoslavsky, R; Lombard, DB; Liu, J; Bruns, NE; Tsokos, M; Alt, FW; Finkel, T. A role for the NAD-dependent deacetylase Sirt1 in the regulation of autophagy. *Proc Natl Acad Sci USA*, 2008, 105(9), 3374–3379.

[155] Julien, C; Tremblay, C; Émond, V; Lebbadi, M; Salem, N; Bennett, DA; Calon, F. SIRT1 Decrease Parallels the Accumulation of tau in Alzheimer Disease. *J Neuropathol Exp Neurol.*, 2009, 68(1), 48.

[156] Kumar, R; Chaterjee, P; Sharma, PK; Singh, AK; Gupta, A; Gill, K; Tripathi, M; Dey, AB; Dey, S. Sirtuin1: A Promising Serum Protein Marker for Early Detection of Alzheimer's Disease. *PLoS ONE*, 2013, 8(4), e61560.

[157] Grubisha, O; Smith, BC; Denu, JM. Small molecule regulation of Sir2 protein deacetylases. *FEBS J.*, 2005, 272, 4607–4616.

[158] Pallos, J; Bodai, L; Lukacsovich, T; Purcell, JM; Steffan, JS; Thompson, LM; Marsh, JL. Inhibition of specific HDACs and sirtuins suppresses pathogenesis in a Drosophila model of Huntington's disease. *Hum Mol Genet.*, 2008, 17, 3767–3775.

[159] Anekonda, TS. Resveratrol–a boon for treating Alzheimer's disease? *Brain Res Rev.*, 2006, 52, 316–326.

[160] Kim, D; Nguyen, MD; Dobbin, MM; Fischer, A; Sananbenesi, F; Rodgers, JT; Delalle, I; Baur, JA; Sui, G; Armour, SM; Puigserver, P; Sinclair, DA; Tsai, LH. SIRT1 deacetylase protects against neurodegeneration in models for Alzheimer's disease and amyotrophic lateral sclerosis. *The EMBO J.*, 2007, 26(13), 3169–3179.

[161] Porquet, D; Griñán-Ferré, C2; Ferrer, I3; Camins, A1; Sanfeliu, C4; Del Valle, J5; Pallàs, M1. Neuroprotective role of trans-resveratrol in a murine model of familial Alzheimer's disease. *J Alzheimers Dis.*, 2014, 42(4), 1209-1220.

[162] Shi, T; Wang, F; Stieren, E; Tong, Q. SIRT3, a mitochondrial sirtuin deacetylase, regulates mitochondrial function and thermogenesis in brown adipocytes. *J Biol Chem.*, 2005, 280, 13560–13567.

[163] Anekonda, TS; Reddy, PH. Neuronal protection by sirtuins in Alzheimer's disease. *J Neurochem.*, 2006, 96, 305–313.

[164] Millan, MJ. An epigenetic framework for neurodevelopmental disorders: from pathogenesis to potential therapy. *Neuropharmacology.*, 2013, 68, 2-82.

[165] Jarome, TJ; Lubin, FD. Histone lysine methylation: critical regulator of memory and behavior. *Rev Neurosci.*, 2013, 24, 375-387.

[166] Filipowicz, W; Jaskiewicz, L; Kolb, FA; Pillai, RS. Post- transcriptional gene silencing by siRNAs and miRNAs. *Curr Opin Struct Biol.*, 2005, 15, 331–341.

[167] Landgraf, P; Rusu, M; Sheridan, R; Sewer, A; Iovino, N; Aravin, A; Pfeffer, S; Rice, A; Kamphorst, AO; Landthaler, M; Lin, C; Socci, ND; Hermida, L; Fulci, V; Chiaretti, S; Foà, R; Schliwka, J; Fuchs, U; Novosel, A; Müller, RU; Schermer, B; Bissels, U; Inman, J; Phan, Q; Chien, M; Weir, DB; Choksi, R; De Vita, G; Frezzetti, D; Trompeter, HI; Hornung, V; Teng, G; Hartmann, G; Palkovits, M; Di Lauro, R; Wernet, P; Macino, G; Rogler, CE; Nagle, JW; Ju, J; Papavasiliou, FN; Benzing, T; Lichter, P; Tam, W; Brownstein, MJ; Bosio, A; Borkhardt, A; Russo, JJ; Sander, C; Zavolan, M; Tuschl, T. A mammalian microRNA expression atlas based on small RNA library sequencing. *Cell*, 2007, 129, 1401–1414.

[168] Schratt, GM; Tuebing, F; Nigh, EA; Kane, CG; Sabatini, ME; Kiebler, M; Greenberg, ME. A brain-specific microRNA regulates dendritic spine development. *Nature*, 2006, 439, 283–289.

[169] Fiore, R; Khudayberdiev, S; Christensen, M; Siegel, G; Flavell, SW; Kim, TK; Greenberg, ME; Schratt, G. Mef2- mediated transcription of the miR379-410 cluster regulates activity-dependent dendritogenesis by fine-tuning Pumilio2 protein levels. *EMBO J.*, 2009, 28, 697–710.

[170] Gao, J; Wang, WY; Mao, YW; Graff, J; Guan, JS; Pan, L; Mak, G; Kim, D; Su, SC; Tsai, LH. A novel pathway regulates memory and plasticity via SIRT1 and miR-134. *Nature*, 2010, 466, 1105–1109.

[171] Cogswell, JP; Ward, J; Taylor, IA; Waters, M; Shi, Y; Cannon, B; Kelnar, K; Kemppainen, J; Brown, D; Chen, C; Prinjha, RK; Richardson, JC; Saunders, AM; Roses, AD; Richards, CA. Identification of miRNA changes in Alzheimer's disease brain and CSF yields putative biomarkers and insights into disease pathways. *J Alzheimers Dis.*, 2008, 14, 27–41.

[172] Hebert, SS; Horre, K; Nicolai, L; Papadopoulou, AS; Mandemakers, W; Silahtaroglu, AN; Kauppinen, S; Delacourte, A; De Strooper, B. Loss of microRNA cluster miR-29a/b-1 in sporadic Alzheimer's disease correlates with increased BACE1/beta-secretase expression. *Proc Natl Acad Sci. USA.*, 2008, 105, 6415–6420.

[173] Soreq, H; Wolf, Y. NeurimmiRs: microRNAs in the neuroimmune interface. *Trends Mol Med.*, 2011, 17, 548-555.

[174] Patel, N; Hoang, D; Miller, N; Ansaloni, S; Huang, Q; Rogers, JT; Lee, JC; Saunders, AJ. MicroRNAs can regulate human APP levels. *Mol Neurodegener.*, 2008, 3, 10.

[175] Hebert, SS; Horre, K; Nicolai, L; Bergmans, B; Papadopoulou, AS; Delacourte, A; De Strooper, B. MicroRNA regulation of Alzheimer's Amyloid precursor protein expression. *Neurobiol Dis.*, 2009, 33, 422–428.

[176] Long, JM; Lahiri, DK. MicroRNA-101 downregulates Alzheimer's amyloid-beta precursor protein levels in human cell cultures and is differentially expressed. *Biochem Biophys Res Commun.*, 2011, 404, 889–895.

[177] Vilardo, E; Barbato, C; Ciotti, M; Cogoni, C; Ruberti, F. MicroRNA-101 regulates amyloid precursor protein expression in hippocampal neurons. *J Biol Chem.*, 2010, 285, 18344–18351.

[178] Delay, C; Calon, F; Mathews, P; Hebert, SS. Alzheimer-specific variants in the 3′ UTR of Amyloid precursor protein affect microRNA function. *Mol Neurodegener.*, 2011, 6, 70.

[179] Schonrock, N; Ke, YD; Humphreys, D; Staufenbiel, M; Ittner, LM; Preiss, T; Gotz, J. Neuronal microRNA deregulation in response to Alzheimer's disease amyloid-beta. *PLoS One*, 2010, 5, e11070.

[180] Zong, Y; Wang, H; Dong, W; Quan, X; Zhu, H; Xu, Y; Huang, L; Ma, C; Qin, C. miR-29c regulates BACE1 protein expression. *Brain Res.*, 2011, 1395, 108–115.

[181] Boissonneault, V; Plante, I; Rivest, S; Provost, P. MicroRNA-298 and microRNA-328 regulate expression of mouse beta-amyloid precursor protein-converting enzyme 1. *J Biol Chem.*, 2009, 284, 1971–1981.

[182] Faghihi, MA; Modarresi, F; Khalil, AM; Wood, DE; Sahagan, BG; Morgan, TE; Finch, CE; St Laurent III, G; Kenny, PJ; Wahlestedt, C. Expression of a noncoding RNA is

elevated in Alzheimer's disease and drives rapid feed-forward regulation of beta-secretase. *Nat Med.*, 2008, 14, 723–730.

[183] Hebert, SS; Sergeant, N; Buee, L. MicroRNAs and the regulation of tau metabolism. *Int J Alzheimers Dis.*, 2012, 2012(9), 406561.

[184] Lukiw, WJ; Alexandrov, PN. Regulation of complement factor H (CFH) by multiple miRNAs in Alzheimer's disease (AD) brain. *Mol Neurobiol.*, 2012, 46, 11–19.

[185] Nunez-Iglesias, J; Liu, CC; Morgan, TE; Finch, CE; Zhou, XJ. Joint genome-wide profiling of miRNA and mRNA expression in Alzheimer's disease cortex reveals altered miRNA regulation. *PLoS One*, 2010, 5, e8898.

[186] Geekiyanage, H; Jicha, GA; Nelson, PT; Chan, C. Blood serum miRNA: non-invasive biomarkers for Alzheimer's disease. *Exp Neurol.*, 2012, 235, 491–496.

[187] Tan, L; Yu, JT; Liu, QY; Tan, MS; Zhang, W; Hu, N; Wang, YL; Sun, L; Jiang, T; Tan, L. Circulating miR-125b as a biomarker of Alzheimer's disease. *J Neurol Sci.*, 2014, 336, 52–56.

[188] Lahiri, DK; Maloney, B; Zawia, NH. The LEARn model: an epigenetic explanation for idiopathic neurobiological diseases. *Mol Psychiatry*, 2009, 14, 992– 1003.

In: Epigenetics: Mechanisms and Clinical Perspectives
Editor: Yildiz Dincer
ISBN: 978-1-63484-490-1

Chapter 5

EPIGENETIC MECHANISMS IN CARDIOVASCULAR DISEASES

Mehmet Güven[1,*], PhD, and Bahadır Batar[1,2], PhD

[1]Department of Medical Biology, Cerrahpasa Medical Faculty, Istanbul University, Istanbul, Turkey

[2]Department of Molecular Virology, Immunology and Medical Genetics, Comprehensive Cancer Center, Ohio State University Wexner Medical Center, Columbus, Ohio, US

ABSTRACT

Cardiovascular disease (CVD) is not a single condition, but an umbrella term used to describe a range of common diseases affecting the heart and the circulatory system. The term commonly includes diseases of the cardiac muscle and of the vascular system supplying the heart, brain, and other vital organs. Many of these conditions can be life-threatening. CVD is the leading cause of death in the world, affecting all populations, irrespective of demographic or socioeconomic differences and is responsible for one third of all deaths.

In the present chapter, we focus on the epigenetic control of embryonic cardiac development and the role of epigenetic mechanisms in CVD observed from results in human, animal and cell culture studies' approaches. We discuss the main epigenetic mechanisms involved in heart development and major CVDs such as coronary heart disease (CHD), heart failure (HF), myocardial infarction (MI), hypertension, stroke, arrhythmias, cardiomyopathy and cardiac hypertrophy (CH). Additionally, this chapter also focuses on the epigenetic modifiers that are involved in the development of CVD, and the potential utility of epigenetics-based therapeutic strategies in CVD.

[*] Correspondence author: mgguven@istanbul.edu.tr.

Keywords: epigenetic, cardiovascular disease, coronary heart disease, heart, post-translational modification

Abbreviations

11βHSD2	11 β-hydroxysteroid dehydrogenase 2
5-azaC	5-azacytidine
5hmC	5-hydroxy-methylcytosine
5mC	Methylation of cytosine
AMOTL2	Angiomotin like protein-2
ANF	*Atrial natriuretic factor*
ANRIL	Antisense non-coding RNA in the *INK4* locus
ARH	
GAP24	Rho-GTPase activating protein-24
AT1bR	Angiotensin type 1b receptor
BAF	BRG1-associated factor
BNP	Brain Natriuretic Peptide
BRG1	*Brahma-related gene-1*
BRM	Brahma
Bvht	Braveheart
CARD8	Caspase recruitment domain 8
Cbp	Creb-binding protein
CH	Cardiac hypertrophy
CHD	Coronary heart disease
CHD7	Chromodomain helicase DNA-binding 7
COL15A1	Collagen type XV A1
CRB	CREB binding protein
CVD	Cardiovascular disease
CYP1A1	Cytochrome P450A1
DAC	5-aza-2-deoxycytidine
DGCR8	DiGeorge syndrome critical region gene 8
DNMT	DNA methyltransferase
DOT1L	DOT1-Like Histone H3K79 Methyltransferase
DOT1L	Telomeric silencing-1-like
EGFR	Epidermal growth factor receptor
ENaC	Epithelial sodium channel
ER	Estrogen receptor
ESC	Embryonic stem cell
F2RL3	Coagulation factor II (thrombin) receptor-like 3
Fendrr	Foxf1 adjacent non-coding developmental regulatory RNA
Gata4	GATA binding protein 4
GR	Glucocorticoid
H3K4	Histone H3 lysine 4
HAND1	*Heart and neural crest derivatives expressed 1*

HAT	Histone acetyltransferase
Hcy	Homocysteine
HDAC	Histone deacetylase
HF	Heart failure
HHcy	Hyperhomocysteinemia
IL	Interleukin
IncRNA	Long non-coding RNA
Irx4	Iroquois homeobox 4
IUGR	Intrauterine growth restriction
Jarid2	Jumonji AT rich interactive domain 2
KDM1A	Lysine-specific histone demethylase 1A
LINE-1	Long interspersed nucleotide elements-1
LIPCAR	Long intergenic noncoding RNA predicting cardiac remodeling
LSD-1	Lysine (K)-specific demethylase 1A
MBD	Methyl-CpG binding domain protein
MeCP2	Methyl-CpG–binding protein 2
MED13	Mediator complex subunit 13
MEF2C	Myocyte enhancer factor 2C
MHC	Myosin heavy chain
MI	Myocardial infarction
MIAT	Myocardial infarction-associated IncRNA transcript
MiRNA	MicroRNA
MLC2V	Myosin light chain 2 ventricular
Msx2	Msh homeobox 2
MyoD	Myogenic differentiation
NAD	Nicotinamide adenine dinucleotide
NF-κB	Nuclear factor-κB
NKCC1	Na+/K+/2Cl− cotransporter 1
PBAF	Polybromo-associated BAF
PBMC	Peripheral blood mononuclear cell
PPAR	Peroxisome proliferator-activated receptor
PRC2	Polycomb silencing complex 2
RNAPII	RNA polymerase II
ROS	Reactive oxygen species
RXRa	Retinoid X receptor-alpha
sACE	Somatic angiogensin-converting enzyme
SAM	S-adenosylmethionine
SMC	Smooth muscle cell
SMYD1	SET and mynd domain containing 1
SRF	Serum response factor
SWI/SNF	SWItch/Sucrose non-fermentable
TET	Ten-eleven translocation
TNF	Tumor necrosis factor
VANGL2	Van Gogh-like 2
VDR	Vitamin D Receptor
VSMC	Vascular smooth muscle cell

1. Epigenetic Mechanisms

Epigenetics refers to genetic factors that change the phenotype of an organism or the biological functions without changing the DNA sequence. DNA base modifications, post-translational modifications of histone proteins, and RNA-based mechanisms are major processes involved in epigenetics.

Among the epigenetic factors, methylation of cytosine (5mC) bases within the DNA is the essential pathway by which DNA is modified that is relevant for epigenetic changes. DNA methylation usually occurs within the context of CpG dinucleotide sequences (CpG-sites) [1]. The main function of DNA methylation is to modulate the expression of the genetic information by modifying the accessibility of DNA to the transcription binding factors. DNA methylation plays a key role in X-chromosome inactivation, genomic imprinting, and tissue-restricted gene expression during development and differentiation [2]. The process of DNA methylation is conducted by enzymes called DNA methyltransferases (DNMTs) [1]. Methylation in DNA can be actively removed through oxidative demethylation by the Ten-eleven translocation (TET) family (TET1, TET2 and TET3) [3]. TET protein family is involved in oxidizing of 5mC to 5-hydroxymethylcytosine (5hmC). 5hmC is another important epigenetic modification on DNA in mammalian cells.

Although histones are the proteins that make up nucleosomes, the post-translational modifications of these proteins are another epigenetic mechanism affecting gene expression. "Histone Code Hypothesis" proposes that various combinations of histone modifications may control chromatin structure and transcriptional status. Among histone modifications defined by the histone code, the acetylation, methylation and phosphorylation are important mechanisms of epigenetic regulation [4]. Histone acetylation is catalyzed by proteins knows as histone acetyltransferases (HATs) and histone deacetylation is carried out by histone deacetylases (HDACs) [5]. Similarly, histone methylation is regulated by histone methyltransferases and histone demethylases. Histone phosphorylation is also catalyzed by several distinct kinases that are mostly specific for individual histone residues [6].

RNA-based mechanisms are the most recently defined of the three major types of epigenetic regulation. RNA-based mechanisms exclusively focus on microRNAs (miRNAs), short non-coding RNAs, which modulates post-transcriptional repression of epigenetic regulators of gene expression. MiRNAs are endogenously produced RNA molecules that are approximately 22 nucleotides in length. At least 30% of human genes are thought to be regulated by miRNAs [7]. MiRNAs are transcribed by miRNA genes in a primary transcript (pri-miRNA). Drosha-DGCR8 complex processes pri-miRNA as precursor-miRNAs (pre-miRNAs) in the nucleus. Pre-miRNAs are transported into the cytosol where Dicer cleaves the pre-miRNAs to generate a short double-stranded miRNA-miRNA duplex [8, 9]. Mature miRNA targets to complementary mRNA sequence to downregulate the translation of mRNA. Long non-coding RNAs (IncRNAs) which are transcripts 200 nucleotides in length could promote gene activation and inhibition by interacting with epigenetic regulators. IncRNAs are transcribed by Polymerase II enzyme, capped of 5′ end, alternatively spliced and polyadenylated.

2. Epigenetic Mechanisms in Heart Development

Heart being one of the first organs to develop during embryogenesis has a complex development process which involves coordinated cellular proliferation, migration, differentiation, programmed cell death and structural remodeling including looping and septation. Nucleosome remodeling, DNA methylation, and histone modification are processes controlled by enzymes that are dependent on the correct activation of a gene program which is necessary for the accurate growth of the mammalian heart during embryonic period and the terminal differentiation in the adult. Defects in the aforementioned processes result in deleterious phenotypes like septal defects, atrioventricular malformations, or developmental absence of right or the left heart and may end with mortality anytime during embryogenesis [10]. Epigenetic mechanisms have been implicated during the development of the heart in both directly by regulation of gene expression patterns as well as indirectly by modulating the expression of transcription factors.

Chromatin remodeling complexes catalyze the relocation of nucleosomes to modulate the accessibility of transcription factors and other regulatory proteins to DNA. There are multiple distinct families of chromatin- remodeling complexes [11]. For example, the SWItch/Sucrose Non-Fermentable (SWI/SNF) complex is an evolutionarily conserved multisubunit chromatin remodeling complex that are critical for differentiation and proliferation. SWI/SNF family exists in several subgroups and each subgroup contains one of two related ATPases, *Brahma-related gene-1 (*BRG1) or Brahma (BRM), and BRG1-associated factors (BAFs) per complex [12]. BAF and PBAF (Polybromo-associated BAF), SWI/SNF members, are structurally related but functionally distinct. As shown by multiple studies, when a cell differentiates from a pluripotent stem cell to a multi-potent progenitor leading ultimately into a terminally differentiated state the configuration of SWI/SNF complexes are changed. Brg1 null mouse embryos die shortly after E11.5 day post-conception when cardiomyocyte expansion and maturation begins [13]. This fact proves the importance of Brg1 during cardiac development. Thinning of the compact myocardium and failure to form an inter-ventricular septum are the cardiac defects seen in these embryos. These defects are harmonious with a loss of cardiac factor Bmp10 expression and up-regulation of cell cycle inhibitor p57 (Kip2) thus portraying a decrease in cardiac myocyte proliferation [14].

The BAF180 subunit of the PBAF chromatin remodeling complex facilitates the ligand dependent transcription of nuclear receptor genes such as Retinoid X receptor-alpha (RXRa), Vitamin D Receptor (VDR), and peroxisome proliferator-activated receptors (PPARs) in vitro [15]. BAF180 knockout mice are prone to be a mortality at around the 15th post-conceptional day. They display severe hypoplastic ventricular development. Ventricular wall thinning with ventricular septal defect is observed histologically in the hearts of BAF180 knockout mice [16]. BAF180 is indispensable for coronary vessel formation. During the embryogenesis of the heart BAF180 is expressed sufficiently especially in the pro-epicardium and epicardium [16].

There is a relationship between cardiac-specific transcription factors and histone modifiers that constitutes a mechanism for target specificity. During the early stages of cardiac development a key transcription factor called Nkx2.5 regulates the development of myocardial precursors [17]. Precise regulation of cardiac gene expression is possible with multiple layers of interactions between Nkx2.5 and epigenetic factors such as histone demethylase/methyltransferases. Defects in either Nkx2.5 or epigenetic factors that interact with Nkx2.5 have the potential to end up with congenital heart disease. Modulation of the expression of Jumonji, AT Rich Interactive Domain 2 (Jarid2/Jmj), and a component of a protein complex that contains histone demethylase and methyltransferases activities, in the second heart field by Nkx2.5 itself might contribute to the epigenetic regulation [18]. Nkx2.5 mutations have been linked with various septation and atrioventricular conduction defects in humans. Patients with congenital heart disease often showed decreased DNA binding and transcriptional activation by mutant Nkx2.5 when compared with wild type. With loss of Nkx2.5, the expression of *Atrial natriuretic factor* (ANF), brain natriuretic peptide, myosin light chain 2 ventricular (MLC2V), Myocyte Enhancer Factor 2C (MEF2C), *Heart And Neural Crest Derivatives Expressed 1 (*HAND1), Msh Homeobox 2 (Msx2), and iroquois homeobox 4 (Irx4) was significantly lost [19].

Congenital heart defects are among the most common of all medically significant birth defects. Several large population-based studies estimate the prevalence of congenital heart defects to range from 3 to 6 cases per 1000 live births [20]. Some syndromes, including Wolf–Hirschhorn syndrome, Holt–Oram syndrome, CHARGE syndrome, DiGeorge syndrome are related to congenital heart defects caused by the defects in epigenetic regulation. For example, CHARGE syndrome is defined by coloboma of the iris or retina, heart defects, atresia of the choanae, retardation of growth and/or development, genital, and ear abnormalities. Most of the children (71%) diagnosed with CHARGE have genetic mutations in the chromodomain helicase DNA-binding 7 (CHD7) gene. CHD7 gene is a member of the chromo domain helicase DNA binding family of ATP-dependent chromatin modifiers [21]. CHD7 is associated with all three forms of methylated histone H3 lysine 4 (H3K4), but most robustly with the mono-methylated (H3K4me1) and di-methylated (H3K4me2) sites as shown by analysis [19]. CHD7 binding sites were found to be located further away from the transcriptional start sites, overlapping with DNase hypersensitivity sites. This reveals that CHD7 may bind to and regulate putative enhancer regions or insulator elements for the deposition of epigenetic markers that define lineage specification [22].

2.1. DNA Base Modifications

Gametogenesis, hematopoiesis and stem cell differentiation, as well as various disease processes are several developmental processes in which DNA methylation patterns and dynamics have been described. The role of DNA methylation in regulation of heart development is under-studied. The role of DNA methylation in outflow tract septation through promoter regulation of the Van Gogh-like (VANGL2) gene has been suggested by recent scientific data [23]. VANGL2 is essential for the polarization of cells at the outflow tract. The homozygous mutation of the gene in mice results in a phenotype resembling Tetralogy of Fallot (overriding aorta, double-outlet right ventricle and ventricular septal defects). VANGL2 promoter methylation has been found to be elevated in patients with Tetralogy of Fallot when compared with healthy controls [24]. Genomic location within cells, cell types within the developing heart, as well as across various stages of development are factors which vary the role of DNA methylation in heart development.

Studies in embryonic stem cell (ESC) or animal models are the sources that provide our knowledge on DNA methylation. Inhibition of DNA methylation in ESCs by 5-azacytidine

(5-azaC) has been shown to induce cardiomyocyte differentiation [25]. However, recent studies on murine embryonic heart development have shown that heart development is associated with increases and decreases (greater increases than decreases) in DNA methylation, especially in cardiac-specific genes [26]. Studies with mice lacking key enzymes involved in the regulation of DNA methylation and demethylation were the first to suggest the gravity of DNA methylation in the development of the heart. The balance between DNA methylation and demethylation is imperative for proper mammalian development as portrayed by these results. The absence of DNMT1 led to global hypomethylation and early embryonic death as shown in these studies [27, 28]. DNMT3A-null mice survive until birth but die around 4 weeks of age, whereas DNMT3B-null mice die during gestation [29]. These findings are consistent with studies of postnatal cardiac development in rats, showing increases in DNA methylation in cardiomyocytes, as well as increased DNMT1, Methyl-CpG binding domain protein 1-3 (MBD1–3) and methyl-CpG–binding protein 2 (MeCP2), over the weeks and months following birth [30].

2.2. Histone Code

Epigenetic histone modifications have important roles in both normal and abnormal development as suggested by recent evidence. Factors that affect cardiac development are histone methylation and demethylation, acetylation and deacetylation status.

The status of histone methylation affects cardiac development. Being a histone methyltransferase, SET And MYND Domain Containing 1 (SMYD1) has been proposed to play a role in the development process of the heart. Mice lacking SMYD1 exhibit hypoplastic right ventricles and embryonic lethality [31]. In addition, Hand2 and Irx4 transcription factors, which are essential for normal cardiomyocyte development are found to be reduced in these mice [31, 32]. These changes suggest that SMYD1-mediated histone methylation is prerequisite for the expression of these essential cardiac transcription factors. Histone demethylase Jumonji family members have been proposed to have a role in septation. In the case of their deletion, ventricular septal defects and outflow tract defects are observed [33, 34]. Lysine (K)-specific demethylase 1A/Lysine-specific histone demethylase 1A (LSD-1/KDM1A) is another lysine demethylase that plays a role in the formation of ventricular septal defects [35].

The Creb-binding protein (Cbp) and p300 are two HATs and play critical roles in physiological and pathological growth of cardiac myocytes [36]. Cardiovascular defects and embryonic lethality are observed in p300 knockout mice, which die between days 9 and 11.5 of gestation and show reduced expression of muscle structural proteins such as β-MHC and α-actinin, as well as cardiac structural defects and reduced trabeculation [37, 38]. The potential connection between p300 regulatory complex and the actions of cardiac signals has led to the identification of transcription factors such as GATA binding protein 4 (Gata4), Mef2d, and myogenic differentiation (MyoD), as well as Mef2c involvement with transcriptional programs that result from signaling cues that control pathological gene expression in the heart [39].

Effects of histone deacetylation have been investigated in the development process of the heart. HDAC1 and 2 seem to have prominent roles in cardiac development. Double knockout mice show cardiac defects. Mice lacking HDAC1 and 2 are lost in the early periods of neonatal life. Their mortality is due to cardiac arrhythmias and dilated cardiomyopathy [40]. Being associated especially with cardiomyocyte metabolism and proliferation, HDAC3 has indeed been thought to have a role in heart development. HDAC3's loss leads to impairment of glucose metabolism, thus provides an inclination towards an increase in lipid metabolism [41]. Overexpression of HDAC3 promotes proliferation of cardiomyocytes by the reduction of cell cycle inhibitors. As a result of these data it can be concluded that histone modifications lead to major alterations in heart development.

2.3. MiRNAs and lncRNAs

Loss and gain-of function studies demonstrated that miR1 and miR133 families are an essential miRNAs in the cardiac development [42]. In mice, mostly expressed miRNAs in the heart are miR1 (miR1-1, miR1-2, and miR206) and miR133 (miR133a-1, miR133a-2, miR133b) [43]. MiR133 targets cyclin D2 and serum response factor (SRF) and negatively controls cardiomyocyte proliferation [44]. In addition, miR208 and miR499 are highly expressed in the heart [45]. Also, it is found decrease levels of miR133 in the infarcted areas of the heart [46].

MiR1 promotes cardiac specification [47]. It has been shown that increased miR1 expression was associated with decreased ventricular cardiomyocyte proliferation [42]. It is suggested that mir208a regulates mediator complex subunit 13 (MED13) expression and controls systemic energy homeostasis [48]. MiR499 enhances myocyte differentiation [49]. In addition, miRNAs could control cardiovascular differentiation [50].

Recent studies suggested the importance of IncRNAs as regulator molecules of transcriptional regulation in cardiovascular development. Braveheart (Bvht), one of the IncRNAs, is expressed in the early stages of embryonic heart development [51]. Bvht interacts with the epigenetic silencing complex, polycomb silencing complex 2 (PRC2), and affect the expression of critical genes during early cardiac development [52, 51]. PRC2 contains trimethylates histone H3 at Lys 27 that promotes tissue-specific differentiation [53, 54]. PRC2 has an important function during early and late cardiac development and homeostasis [55].

Another IncRNA, Foxf1 adjacent non-coding developmental regulatory RNA (Fendrr), is exclusively expressed in mesoderm and plays a role in the development of mouse heart [43]. Fendrr modulates the chromatin signatures of genes which forms and differentiates the lateral mesoderm [52]. Fendrr may also control the activating of HE3K4me3 mark on promoters. Several studies suggested that Fendrr inactivation causes to defects in vascularization [55].

Discovery of Bvht and Fendrr functions support the role of IncRNAs in cardiac development.

3. Epigenetic Mechanisms in CVDs

Our current scientific knowledge cannot fully describe the complex pathophysiology of CVD yet. Epigenetics provides a different path to approach human disease. The dynamic nature and response to environmental cues of epigenetic processes constitute a plan for apprehending the constraints to CVD research which focused solely on the static DNA code.

3.1. DNA Base Modifications

Atherosclerosis is characterized by chronic inflammation of the arterial wall. When atherosclerosis affects the arteries of the heart, it's called CHD. CHD is the most common form of CVD and also called coronary artery disease. Differentially methylated CpGs association with the onset of atherosclerosis onset together with endothelial and smooth muscle functionality was shown in a DNA methylation map of human atherosclerosis. In the atherosclerotic lesions, a gain of DNA methylation was observed. This process strikes the opportunity to develop DNA demethylating agents for therapeutic aims. The global DNA methylation profile of peripheral blood leukocytes may produce a suitable biomarker for increased CHD risk [56].

Homocysteine (Hcy) is one of the known risk factors for atherosclerosis and CVD. Its action may be caused by an epigenetic mechanism involving methylation. Hcy is biochemically linked to the principal epigenetic tag found in the DNA. High levels of Hcy in circulation are a risk factor for CVD. Nonetheless, in recent clinical trials where folate and/or other vitamin B therapies were used to lower Hcy; cardiovascular event rates did not decrease, making Hcy's direct causative role in vascular disease controversial [57]. In a lot of *in vivo* studies it was suggested that Hcy levels may modulate global DNA methylation. For example, in healthy human subjects, increased levels of plasma Hcy were linked with both increased S-adenosylhomocysteine concentrations and DNA hypomethylation in lymphocytes [58]. Vascular complications associated with high levels of circulating Hcy was partially caused by DNA hypomethylation in several studies [59]. Increased levels of plasma Hcy was observed with decreased DNA methylation in CVD patients. This situation supports the role for hyperhomocysteinemia (HHcy) in the modification of epigenetic mechanisms. In a study done by Ingrosso and colleagues increased levels of Hcy together and reduced global lymphocyte DNA methylation pattern was reported [60]. After the reduction of the plasma Hcy levels with folate administration both global DNA methylation patterns and allelic gene expression were normalized. Data suggests that HHcy is linked with reduction in global levels of CpG methylation and gene-specific methylation of promoters [61, 62]. Changes in the expression of specific genes are seen with these alterations. The increased expression of the nuclear factor-κB (NF-κB) transcription factor, interacts with the IRE1 and c-Jun NH2-terminal kinase family of proteins [63]. This increased expression may cause inflammatory-mediated vascular damage [64]. With the finding of recent data it was suggested that that alterations in lipid metabolism may play a role in vascular pathology associated with HHcy [64, 65]. Many studies propose that epigenetics may play a role in these processes. In mice with Hhcy, it was suggested that changes in DNA methylation have might be the potential mechanism for altered *apoA-I* and *apoA-IV* gene expression [66, 67]. Hcy is significantly and inversely correlated with HDL-bound cholesterol and apoA-I in both human and murine models of Hhcy [68].

In vascular smooth muscle cell (VSMC) phenotypic modulation in human atherosclerosis, certain atheroprotective genes like *ERα* and *ERβ* (*ESR1* and *ESR2*, respectively) which encode estrogen receptors (ERs) are hypermethylated [69]. ERs are found in the coronary arterial walls on both the smooth muscle cells (SMCs) and the endothelial cells (ECs). Their functions protect against atherosclerosis, especially in CHD. Deficiencies in ERα give way to accelerated atherosclerosis in humans [70]. Hypermethylation of ESR1 was also seen in in vitro senescing ECs and SMCs, together with ESR2 hypermethylation. With the light of these findings epigenetic changes in both ESR1 and 2 can influence vascular aging and atherosclerosis. Demethylation using 5-aza-2-deoxycytidine (DAC) could enhance the expression of silenced genes. These treatments increased the expression of ERα and ERβ in normal SMCs and ECs. DAC and/or trichostatin A treatments showed low toxicity in these cells [71]. Silencing of ERs in women, with epigenetic changes, might explain the failure of estrogen therapy to exert a cardioprotective effect. The combined use of epigenetic therapy and hormone replacement therapy could be suitable for the prevention of CVDs [71, 72].

Zhu and colleagues portrayed that methylation of the monocarboxylate transporter gene suppresses its transcription leading to the passage of smooth muscle cells and the progression of atherosclerosis in cultured tissues of aorta and coronary arteries with varying degrees of atherosclerosis [73]. The risk of vascular damage may increase with the loss of the atheroprotective factors expressed by these genes.

In patients with atherosclerosis, global DNA methylation decreases in the peripheral blood cells [74]. The era of DNA methylation in atherosclerosis was uncovered by two different groups [75, 76]. They portrayed coexistence of hypomethylation with atherosclerotic lesions in mice and rabbits. It was shown in other studies that hypomethylation could be detected at the early stages of atherosclerosis, before the appearance of the anatomical manifestations of the disease [77]. Baccarelli and colleagues revealed an association between hypomethylation of long interspersed nucleotide elements-1 (LINE-1) and higher incidence of ischemic heart disease [56]. Friso and colleagues showed that hypomethylation of the promoter of coagulation factor VII was linked with coronary artery disease in patients, who are angiographically proven to have coronary lesions [78].

Heart failure is a condition caused by the heart failing to pump enough blood around the body at the right pressure. The etiology of HF is quite complex depending on genetic predisposition and multiple environmental factors. Development of HF depends on several molecular and cellular mechanisms. These include reprogramming of the expression of certain critical cardiac genes. Downregulation of the alpha-myosin heavy chain (alpha-MHC) gene and sarcoplasmic reticulum Ca+ ATPase genes and reactivation of specific fetal cardiac genes such as atrial natriuretic factor and brain natriuretic peptide are examples of these cardiac gene alterations [79].

Changes in patterns of methylation and extent of regulatory regions of genes is known to modify their expression which correlates with status of HF. Without respect to the etiology of HF there are three angiogenesis-related genes that are found to have differential methylation. Hypermethylation of the 5'-regulatory region of platelet endothelial cell adhesion molecule-1 and hypomethylation of the angiomotin like protein-2 (AMOTL2) are noted in HF [80]. These are closely linked with reduced expression of these genes. Hypermethylation within the Rho-GTPase activating protein-24 gene (ARHGAP24) in failing hearts is associated with elevated expression of this gene [80]. Movassagh and colleagues showed that three angiogenesis-related genes underwent differential methylation, irrespective of the etiology in patients with HF [80]. *AMOTL2* gene was hypomethylated whereas the 5′ promoter region in the platelet/endothelial cell adhesion molecule gene and the gene body of *ARHGAP24* were hypermethylated [80]. The modified epigenomics of the three genes in endstage HF may reflect common epigenetic pathways in heart remodeling and vasculature. Dilated cardiomyopathy is seen in 33% of cases of HF. Nguyen and colleagues in a mouse model proved that cardiac specific knockout of telomeric silencing-1-like (DOT1L), the gene encoding for the HKMT Dot1L which catalyzes H3K79 methylation in mammals that gave way to the appearance of a phenotype similar to that seen in dilated cardiomyopathy [81].

Hypertension is a multifactorial disease whose pathogenesis is underlined by several mechanisms due to the complex interplay of genetic and environmental factors. Epigenetic mechanisms play a role in the pathogenesis of hypertension. Global 5mC levels of peripheral blood cells in patients with essential hypertension were found to be lower than healthy controls. The levels of 5mC correlates with the the stage of hypertension in these patients [82]. Valinluck et al. reported that 5hmC at CpG sites was discovered to suppress the binding of MeCP2 [83]. MeCP2 is a transcriptional repressor which probably has a potential role of 5hmC in gene expression regulation [83]. 5hmC may have a role in mediation of fetal heart and lung growth [84].

The role of gene-specific DNA methylation in the pathogenesis of hypertension has been recently proposed. 5hmC has a role in determining tissue-specific gene expression and regulating the expression of genes involved in cellular differentiation and pluripotency [84, 85]. In patients with essential hypertension, promoter methylation of 11βHSD2, the gene encoding 11 β-hydroxysteroid dehydrogenase 2 (11βHSD2), has been shown to be increased in the peripheral blood mononuclear cells [82]. 11βHSD2 breaks down cortisol to its biologically active form, cortisone. Reduction in the activity of 11βHSD2 promotes cortisol-mediated activation of mineralocorticoid receptors thus end up in increased blood pressure [82]. An increase in the risk of development of adult type hypertension and an increase in promoter methylation at hydroxysteroid (11-beta) dehydrogenase 2 (HSD11B2) site were observed in an intrauterine growth restriction (IUGR) rat model of hypertension [86]. Methylation levels at HSD11B2 promoter sites were significantly higher in IUGR newborns when compared with controls. This was related to lower HSD11B2 expression [87]. The pattern of methylation at different CpG sites were in conjunction with birth weight and ponderal index which are measures of fetal growth [87]. Methylation of the promoter at HSD11B2 site may associated with development of IUGR and the possible risk of future development of hypertension [87]. Another human study suggests that promoter methylation of HSD11B2 gene is closely linked to hypertension [88]. Methylation at HSD11B2 promoter in DNA of peripheral blood mononuclear cell (PBMCs) of hypertensive patients is inversely related to the function of the enzyme [88].

Somatic angiogensin-converting enzyme (sACE) is another major mediator of blood pressure regulation which is influenced by DNA methylation in human cell lines [89]. With the combination of knowledge from these it can be suggested that aberrant DNA methylation could lead to the formation of essential hypertension in a gene-specific manner. Overall reduction in global methylation is indicative of the progression of disease.

The role of membrane transporter genes for the known role of alterations in ion flux mechanisms in the pathogenesis of hypertension have been also evaluated. Promoter methylation of the $Na^+/K^+/2Cl^-$ cotransporter 1 (NKCC1) was studied by Lee et al. [90]. NKCC1 is a gene that encodes a solute carrier which is responsible for the transportation of sodium, potassium, and chloride through the cellular membrane. In spontaneously hypertensive rats, the methylation status of NKCC1 promoter were measured in the aorta and the heart [90]. In the study it was concluded that in a spontaneously hypertensive rodent model NKCC1 expression is upregulated by a mechanism induced by gene promoter hypomethylation [90].

Cardiomyopathies are defined as diseases of the myocardium associated with cardiac dysfunction. There are two types of cardiomyopathy: primary (idiopathic) or secondary to ischemia or myocardial infarction. They can also be defined on the type of remodeling. Ventricular dilation and hypertrophy take place as the patient develops HF.

DNA methylation patterns in the hearts of patients undergoing cardiac transplantation for primary or secondary cardiomyopathy were investigated in two current studies [91, 92]. Genomic profiling of DNA methylation in end-stage HF patients who have primary or secondary cardiomyopathy showed significant hypomethylation at the promoter regions and hypermethylation at the gene body regions when compared with healthy controls with normal hearts. There was no statistically significant difference in intergenic regions, including enhancer regions [92]. Haas et al. studied patients with primary cardiomyopathy and found that there was statistically significant difference in differential methylation in pathways related to cardiac disease, as well as differential methylation of several key genes when compared with healthy controls with normal hearts [91].

3.2. Histone Code

NOS3 is the well-characterized endothelial gene that has cardiovascular implications and it is controlled by histone modification. eNOS is a protein which is coded by the NOS3 gene. It catalyzes the formation of NO from L-arginine in blood vessels [93]. NO is a factor responsible for vasodilation. It has a major part in the regulation of vascular tone and protection against atherosclerosis [93]. The NOS3 gene is transcriptionally active in ECs and it is transcriptionally suppressed in VSMCs under normal conditions [94]. Acetyl H3K9, acetyl H4K12, and methyl H3K4 at the NOS3 proximal promoter site in ECs are activating histone modifications. When they are present they allow the recruitment of RNA polymerase II (RNAPII) [94]. In VSMCs, eNOS expression is suppressed by the absence of activating histone modifications at the NOS3 proximal promoter. It is further suppressed by the presence of cytosine methylation and MeCP2 recruitment [94]. The expression of eNOS is controlled by environmental stimuli. If short term hypoxia is experienced by the ECs, a reduction in the expression of eNOS is observed [95]. Significant reduction in NOS3 transcription observed in the ECs when faced with hypoxia is considered to happen via histone eviction, including processes that are in relation with transcriptional activation (acetyl H3K9, methyl H3K4, acetyl H4K12), at the NOS3 proximal promoter site [95]. H3 and H4 activating acetylation marks are activated by laminar shear stress. This is partially accomplished by the p300 HAT complex, at a shear stress reporter element in the human eNOS promoter [96]. The fusion of the activating acetylation marks via laminar shear stress provides for the chromatin at the eNOS shear stress reporter element site to stay in an open, transcriptionally-accessible state, permitting eNOS expression.

The activity of HDAC seems to have a major part in the prediction of the degree of myocardial ischemia and reperfusion damage especially after an MI [97]. There is proof that myogenesis and angiogenesis are stimulated in embryonic stem cell cultures with the inhibition of HDACs [98]. In vivo studies in mouse portrayed that by prevention of myocardial remodeling and a reduction in myocardial and serum tumour necrosis factor alpha, inhibition of HDAC improved functional myocardial recovery after MI [98]. Increases in the formation of myocytes and microvessels in the heart were seen after the inhibition of HDAC. After MI there is loss of myocardial performance due to stimulation of angiogenesis [98]. With the light of these findings it can be said that HDAC inhibition can make a reduction in the loss in myocardial performance by inhibiting angiogenesis [97]. The administration of the pharmacologic HDAC inhibitor TSA to atherosclerosis-prone Ldlr (-/-) mice exacerbated neointimal lesions [99]. With better understanding of epigenetic pathways with the help of animal models of atherosclerosis there is more to be discovered.

Curcumin (diferuloylmethane) is a polyphenol present in a curry spice. It is p300 HAT inhibitor which has a wide variety of molecular targets including transcription factors, growth factors and their receptors, cytokines, enzymes, and genes regulating cell proliferation and apoptosis. Curcumin has cardioprotective function [100]. In studies done in patients with atherosclerosis and healthy controls, its administration decreases LDL levels, increases high-density lipoprotein (HDL) levels [101, 102]. Moreover, both in a rat model of HF and primary cultured rat cardiac myocytes and fibroblasts, curcumin inhibits p300 HAT activity thus preventing ventricular hypertrophy, and preserving systolic function in rat model of HF and primary cultured rat cardiac myocytes, fibroblasts [103]. In the rat cardiomyocytes, curcumin is thought to act by inhibition of histone acetylation and hypertrophy-responsive transcription factors (e.g., GATA4), and by disruption of p300/GATA4 complex [104].

In the experimental animal CH models, CH was found to be closely associated with histone acetylation. HATs and HDACs are both playing major roles in this process. It has been shown that hypertrophic growth of cardiomyocytes could be induced by the depending on their HAT activity both overexpression of transcriptional co-activators CREB binding protein (CRB) or p300 itself may induce hypertrophy in cardiomyoctes [105]. Phenylephrine-induced hypertrophic cell growth may be suppressed by the suppression of these co-activators. Catalytic activity of p300 or overexpression of CBP may give rise to hypertrophy. On the contrary mutant forms of these without HAT activity do not end up with the same effects [105]. However, the activity of HDACs' was found to be involve both prohypertrophic and antihypertrophic pathways leading to controversial data. Class IIa type HDACs were discovered to inhibit CH [106]. However, in other animal in vivo studies HDAC inhibitors were found to have a protective role against hypertrophy. In the study by Liu and colleagues done on cardiac hypertrophy model in transgenic mice it was portrayed that injection of a specific HDAC inhibitor reverses atrial fibrosis and diminishes atrial fibrillation vulnerability following an electrical stimulation [107]. Deletion of both HDAC-1 and HDAC-2 from the myocardium lead to early rodent mortality from fetal arrhythmia in other experimental models [40]. Also, Chen and colleagues revealed that inhibition of HDAC suppresses myocardial remodeling [108].

New data uncovers the relationship between histone acetylation and vascular inflammation. Interleukin (IL)-1β, IL-6 and tumor necrosis factor (TNF)-α are some of the proinflammatory cytokine that are decreased by HDAC inhibitors as shown by multiple in vivo studies [109, 110]. The function of histone modifications in VSMC phenotypic modulation and proliferation in response to these cytokines have been studied in multiple other studies [111].

In a histone epigenetic modification study done by Wierda et al. it was shown that hypomethylation was linked with pathophysiological status [112]. Immunohistological examination of the samples taken from perirenal aortic tissues from clinical organ transplant cases, where morphological evidence of progression of atherosclerosis was more evident, showed that fewer nuclei exhibited trimethylation status at lysine 27 of histone H3 (H3K27Me3) [112].

Histone modifications play their role in HF. Kaneda and colleagues focused on genome-wide histone methylation of heart tissues [113]. They reported that tri-methylated histone H3H4 and H3K9 were altered in HF.

In animal models, especially H3K9 and H3K4 type, histone methylations are related with CH. Hypermethylation associated to hydroxymethylation of epidermal growth factor receptor (EGFR) gene was found to be linked with aortic valve calcification and subsequent ventricular hypertrophy in another study that utilized experimental mice model [114].

Many histone modifications are discovered to have a role in the pathophysiology of hypertension. Following pressure overload in the heart, in more than half of the differentially expressed genes methylation or acetylation alterations are seen [115]. Hypermethylation of H3 by disruptor of DOT1-Like Histone H3K79 Methyltransferase (DOT1L), a methyltransferase, enhances the expression of hypertension-promoting telomeric genes [116]. DOT1L also has a role as an antihypertensive. The antihypertensive role is played by the suppression of amiloride-sensitive renal epithelial sodium channel (ENaC) expression [117]. DOT1L is suppressed by aldosterone. As a result of this ENaC expression increases leading to an increase in blood pressure. In several animal models it was found that lack of LSD1 caused hypermethylation of H3 which in turn lead to increased inclination to hypertension [118]. When mice fed a high-salt diet, LSD1 deficiency was closely linked with arterial hypertension. LSD1 induces demethylation of H3K4 or H3K9 changing gene transcription [118]. In the high-salt diet group, LSD1 deficiency lead to hypertension, enhanced vascular contraction, and reduced relaxation via nitric oxide–cyclic guanosine monophosphate pathway [118]. This finding portrays that LSD1-mediated histone demethylation has a role in modifying NOS/guanylate cyclase gene expression and blood pressure.

Status of histone acetylation is involved in the control of blood pressure. Epigenetic modulation has a function in the beta2 adrenergic receptor–GR-WNK4 pathway. Data from murine studies suggest that this pathway is linked with the activation of HDAC8 [119]. HDAC8 reduces the binding of the acetylated histones 3 and 4 with the promoter of the WNK4 gene. This reduces the transcriptional activity of WNK4 through the recruitment of glucocorticoid receptors to a negative glucocorticoid response element [120]. WNK4 is a member of serine/threonine kinase family which down regulates expression of sodium chloride transporters and epithelial sodium channels [121].

Hyperglycemia with prolonged exposure induces specific chromatin alterations and transcriptional responses. These illustrate the role of epigenetic events controlling signaling pathways relevant to diabetic CVD [122]. Genome-wide analysis of human vascular cells demonstrates that histone modifications and DNA methylation take place at the same time under hyperglycemic conditions allowing for prediction of gene expression [123]. In the study done by Mutsktov and colleagues it was proved that insulin gene in human islet-derived precursor cells showed increased levels of histone modifications (H4 hyperacetylation and dimethylation of H3 lysine 4) which is indicative that a gene is active [124].

Alterations in epigenomics can happen in the inflammatory pathways with hyperglycemia. In the study performed by Villeneuve and colleagues, it was discovered that there was a rise in the expression of inflammatory genes in vascular cells cultured in high glucose. On the other hand, H3K9me3 expression was reduced at its promoter [125].

H3K9me3 is a histone H3 which prevents against the biochemical state of diabetic inflammation. El-Osta and colleagues portrayed that the slightest transient exposure to high levels of glucose induced sustained epigenomic alterations in the p65 subunit promoter of the NF-κB [126]. This change altered the epigenomics of cultured aortic endothelial cells. This could be in part the reason of the irreversibility of cardiovascular complications of long-standing uncontrolled diabetes which is usually the case in the routine clinical setting.

3.3. miRNAs and lncRNAs

Some of the miRNA and IncRNA molecules are under study to define the association between epigenetic control and the development of CVDs. We focus on the role of several miRNAs and lncRNAs in many CVDs.

Several miRNAs are known to be associated with atherosclerosis. It has been reported that atherosclerosis patients had reduced levels of miR-126, miR-17, miR92a (in endothelial cells), miR-145 (in smooth muscle cells), and miR-155 (in inflammatory response-associated cells) in plasma and serum [127]. Conversely, miR-133 and miR-92-a (in cardiac muscle) were significantly increased in plasma of atherosclerosis patients [127]. In addition to these findings, it is compared many miRNA levels in atherosclerosis patients with stable and unstable angina. And this research group showed that unstable angina patients had significantly increased levels of miR-134, miR-370, and miR-198 *vs* stable angina patients [128]. The other research group compared the platelet miRNA levels between control and patients with atherosclerosis. They revealed significantly increased miR-340 and miR-624 levels in atherosclerosis patients [129]. It has been recently shown that miR146a-miR146b are implicated in the pathogenesis of atherosclerosis [130].

Recently, it has been shown that increased expression of an antisense lncRNA (ANRIL, antisense non-coding RNA in the *INK4* locus) is associated with the severity of atherosclerosis [130]. Also, another study found that SNP rs10757278 in the *INK4* locus is related with loss of ANRIL expression and increased risk of atherosclerosis [131]. Cunnington and colleagues showed an association between SNP rs1333045 and susceptibility of atherosclerosis [132].

Several studies have been performed to define the association of miRNAs with MI. In a human study, Bostjancic and colleagues found an increased levels of miR-208 in MI patients compared with healthy controls [46]. MiR-1 is one of the most studied miRNA. It has been shown that plasma miR-1 levels were significantly increased in MI patients vs healthy controls [133, 134]. Furthermore, there was a significantly positive correlation between miR-1 expression and infarct size in rat tissues [135].

D'Alessandra and colleagues analyzed the pattern of miRNAs plasma in patients with acute MI and healthy controls. Onset of MI symptoms showed significantly increased miR-1, miR-133a/b, and miR-499-5p in plasma [136]. In consistent with these results, another research group revealed a markedly increased of plasma miR-499 in MI patients [137].

Little is known about the association between IncRNAs and MI. Myocardial infarction-associated IncRNA transcript (MIAT) is a one of the risk factor for MI. Vausort and collegues found higher MIAT levels in MI patients than healthy controls [138]. The same research group also showed an increased ANRIL expression in MI patients [138].

Many studies have determined a significant role of miRNAs in the pathogenesis of HF. The levels of several miRNAs showed alterations in animal models of HF and in patients with human cardiac [139]. While transgenic miR-195 mice had dilated cardiomyopathy, increased expression of miR-23a, miR-23b, miR-24, miR-195, and miR-214 induced hypertrophy in human cardiomyocytes [139]. In addition, overexpression of miR-133 decreased protein synthesis and inhibited hypertrophic growth of mouse cardiac myocytes [140]

Furthermore, Fukushima and colleagues has studied plasma miRNAs profiling in HF patients. They found that patients with HF had reduced miR-126 plasma concentration compared with healthy controls [141]. Also, the profiling of miRNAs has been analyzed by Corsten and colleagues, only miR-499 levels were significantly increased in HF patients compared to controls [142].

Recently, it has been defined markedly expression changes of IncRNAs in various mice and human models of HF. It was found that Pdk1 knockout mouse models had severe HF [143]. Also, one report revealed elevated levels of long intergenic noncoding RNA predicting cardiac remodeling (LIPCAR) in patients with HF [55]. In addition, many research studies suggested that it can be an association of two IncRNAs, Fendrr and Bvht, in HF [52, 51].

MiRNAs were characterized as important regulatory elements in differentiation and maintenance of nervous system [144]. In rat studies, miR-290, miR-292-5p, and miR-497 were upregulated in blood and brain, on the other hand miR-210, miR-215, miR-324-3P, miR-422b, miR-451, and miR-154 were found to be increased only in brain [145].

Another study demonstrated an increased expression of miR-10a, miR-182, miR-200b, and miR-298 in animal models with induced stroke [146]. In converse, the same research group found that miR-155, miR-362-3p, miR-223, and miR-210 were significantly reduced expression in ischemic brain [146]. MiRNA expression has been also investigated in a human study. It has been shown overexpression of five miRNAs in symptomatic, stroke-related, and atherosclerotic plaques [147].

It is considered that there is an association of ANRIL with stroke. ANRIL may increase the risk of stroke via regulation of caspase recruitment domain 8 (CARD8). ANRIL regulates the expression of its downstream gene CARD8 [148].

In conclusion, definition of the miRNA and IncRNA profiles could have an important impact on understanding the pathogenesis of CVDs. Identification of the biological functions of miRNAs and IncRNAs in the cardiovascular system will be essential for CVD prevention, diagnosis, and therapy. Also, these molecules may have the potential to be used as prognostic biomarkers.

4. Cardiovascular Risk Factors and Epigenetics

The majority of CVDs is caused by risk factors that can be controlled, treated or modified, such as diet and nutrition, tobacco use, obesity, lack of physical activity, high blood pressure, cholesterol and diabetes. However, there are also some major CVD risk factors (e.g., age, gender, family history) that cannot be controlled. When those risk factors are altered there could be a modification in epigenetics leading to a decline in the incidence of CVD.

Current evidence shows that epigenetic mechanisms can regulate nutritional effects. They can be the reason for the development of common complex or chronic diseases [149]. Nutritional factors are of major importance in CVD disease modification and prevention. Hypertension, atherosclerosis, many metabolic disorders, obesity, and weight-loss outcomes are affected by alterations in the epigenetic patterns. Histone modifications, DNA methylation, non-coding ribonucleic acid (ncRNA) expression, and chromatin remodeling resulting in a dynamic regulation of gene expression leading to control of the cellular phenotype are among the many different factors that show complex interaction with food components [150]. Changes in epigenetics seem to be more significantly linked with gene-nutrition and gene-environment interactions. The changes in lipid metabolism, inflammation, and other metabolic imbalances which resulted in CVD disease and obesity.

Obesity is often seen together with the metabolic syndrome. They both have major function in the establishment of CVD disease. Epigenetics have effects on both of them concerning the occurrence or prevention of CVD disease. In a study performed in pregnant mice, a high fat diet during pregnancy was found to be linked with epigenetic changes in the expression of adipocytokine genes [151]. Higher blood pressure, worse glucose tolerance, higher triglyceride levels, higher leptin levels and significantly lower adiponectin expression in white adipose tissue were seen in the offspring of the previously mentioned mice. The aforementioned relevance was with lower acetylation and higher methylation levels of histone H3 at lysine 9 of the adipose-tissue promoter of adiponectin [151].

Gene expression can be changed by epigenetic mechanisms imposed by nutritional factors. Recent data suggests that dietary lipids and their derivatives are dynamic modulators of pro- or anti-inflammatory gene expression pathways via nuclear receptors. These nuclear receptors take part in the regulation of various biological functions. Lipid metabolism, inflammatory mediator production, and vascular homeostasis are some examples of these biological functions. Selected dietary components may induce an atherosclerotic cellular phenotype. This could be partially done by producing epigenetic marks that shift differential gene activation and repression [74]. The PPAR receptor family is induced by small lipophilic ligands. These ligands are transcription factors involved in the regulation of genes in the cellular processes such as inflammation, beta-oxidation, and glucose homeostasis. They may be thought to be important targets for pharmacologic intervention in chronic diseases [74]. PPARs are major parts of a transcriptional network that imposes epigenetic marks. PPARs are also considered to be appropriate transcriptional response elements to specific ligands and molecules as factors that cross-talk with PPAR signaling. Fatty acids, intracellular inflammatory mediators, polyphenols, and multiple dietary factors that control the transcription of genes enrolled in the metabolic processes linked with inflammation and fatty acid metabolism are members of these factors.

Nutritional epigenetics may clear the way to understand the pathways of how nutrition can influence health with its effects on the genome. It is known that with administration of folate, homocysteine levels will be lowered. Nonetheless no reduction in the risk of CVD disease risk could be established. Some authors claim that a combination of vitamin B, folic acid, vitamin B6 and vitamin 12 could have harmful effects [152]. Especially during crucial times like fetal and postnatal development, nutritional effects on epigenetic programming could cause major effects to CVD disease [153]. An important therapeutic concept is that nutrients and medication could reciprocate epigenetic patterns. During special development milestones, epigenetic programming, or reprogramming could be effective mechanisms which may result in improvement in the health status of the patient. This could be achieved by dietary regulation [154].

Among the epigenetic modifications, DNA methylation is the most comprehensively investigated. Nutrition and other environmental stimuli can affect methylation. S-adenosylmethionine (SAM) is the prime methyl donor for DNA methylation. With the availability of specific dietary micronutrients such as folate, choline, betaine (trimethylglycine), and various B vitamins, this metabolism can be catalyzed by several enzymes. The relationship between nutrition and DNA methylation is endorsed by findings from animal studies. Epidemiological evidence from human studies is quite limited.

Folate affects the formation of SAM. SAM is a methyl donor for methylation of cytosines in DNA [155]. This kind of methylation is closely linked with gene silencing. Another part in gene silencing is played by the covalent attachment of biotin to histones. This reaction is also involved in any cellular response to DNA damage. Tryptophan and niacin are converted to nicotinamide adenine dinucleotide (NAD). NAD is a substrate for polyADP-ribosylation (PAR) of histones and other DNA-binding proteins. PAR is involved in DNA repair and apoptosis. Epigenetic changes seem to be a mechanism to mediate the effects of environmental exposures in the early periods of life and gene-environment interactions on the development of adult type disease [156]. Maternal dietary consumption of fat, folate, protein, and the amount of total energy intake may change the epigenetic regulation of specific genes in the offspring. This could end up in functional changes in the tissues [157].

Raising the amount of vitamin D consumption does not seem to have major disadvantages. Current evidence shows that vitamin D could have important effects on epigenetics during the intrauterine period leading to reduction in chronic disease later in life. Vitamin D also has many potential benefits to wellbeing including CHD [158].

Flavonoids are potent inhibitors of DNA methyltransferases in vitro. They have the power to reverse hypermethylation and reactivate DNA repair genes. Folate is found in high concentration in green leafy vegetables. It mediates DNA methylation by its generation of SAM. People who have diets with low levels of folate or who have low levels of folate in their blood, have a significantly increased risk of developing CHD disease [159]. Nonetheless there is controversy about the possible lack of value, or even harm, from direct supplementation with folate combined with vitamins B6 and B12 [152, 160]. Folate is important in preventing certain birth defects. Supplementation of folate with myo-inositol in the early periods of pregnancy to a potential mother could alter the epigenetics aiding to prevent environmentally induced birth defects which are closely linked with congenital heart disease [161].

Epigenetic modifications that are affected by the intrauterine environment are investigated with the current animal studies. In a study, offspring of low protein fed pregnant mothers was found to have hypomethylated angiotensin type 1b receptor (AT1bR) gene promoters in addition to increased adrenal expression of AT1bR [162]. From this finding it can be inferred that specific hypomethylation could have a role in the regulation of elevated blood pressure. Overexpression of the hepatic glucocorticoid (GR) and PPARα receptors in offspring were observed in the other studies that used a model of low protein diet receiving pregnant rats [163, 164]. A lot of different metabolic pathways could be affected by the key transcription factors GR and PPARα receptors. They have been considered to have a role in the pathogenesis of several disease conditions like obesity, diabetes, and atherosclerosis [165, 166].

Risk of adult CHD disease, osteoporosis, and type 2 DM are increased in people who were small for gestational age at birth [167]. They also have poor growth rates in infancy. The increased risk more apparent when the early growth is restricted and it is followed by increased childhood weight gain [167]. The previously mentioned aspects are linked with epigenetic processes altering the phenotype of the offspring, reflecting the developmental responses of the fetus and/or infant based on environment.

Barker and colleagues hypothesized that environmental factors in crucial periods of early life (during fetal development, for instance) can influence risks for cardiovascular and metabolic diseases later in life [168]. This concept is supported by a number of studies that have associated low birth weight in human populations with increased risk of cardiovascular disease. For example, individuals prenatally exposed to famine during the Dutch Hunger Winter (1944–45) experienced higher prevalence of obesity and coronary heart disease as adults, when compared to adults born before or conceived after that period [169]. In Barker's seminal work, it was found that low birth-weight babies who survived infancy had an increased risk of coronary heart disease later in life, and that increasing birth weight was associated with a graded decrease in risk [170].

In experimental animal models during the early stages of post-natal life exposure to different behavioral patterns was discovered to affect epigenetic modifications [171]. It could be hypothesized from this finding that the lifelong alterations are at the least partially affected by the epigenetic changes in the gene expression during the early periods of life [172]. When applied to the human population, social and environmental stresses during development that change epigenetic processes could be the reason of adult race-based US health differences seen in hypertension, diabetes, stroke, and coronary heart disease [173].

Wheatley et al. studied the hypothesis of modulation of adipose gene expression with the reversal of obese status by restriction of calories and exercise in 48 female mice [174]. 209 genes were found to be responsive to both calorie restriction and exercise in the gene expression microarray analysis of visceral white adipose tissue. 496 genes were affected by calorie restriction alone while just 20 genes were altered by exercise. 17 genes are related to carbohydrate metabolism and glucose transport from the genes which were responsive to calorie restriction. Glucose transporter 4 was among the genes affected by calorie restriction. Calorie restriction significantly increased acetylation of histone 4. This finding was in parallel with differential changes in adipose transcription with calorie restriction [174].

Long-term repetitive strenuous exercise is considered by many as having a positive effect on health, a reduction in aging, and leading to decreased incidence of disease. Modification of epigenetics can be achieved by physical exercise. Prevention from CVD disease by exercise can be explained by the exercise's power to modify epigenetics [175]. Possible benefits on health are caused by adaptations in skeletal muscle adaptations. These adaptations happen partly due to alterations in gene expression in the skeletal muscle. Chromatin remodeling by epigenetic histone modification seems to be the key regulatory mechanism modifying gene expression. Class IIa HDACs enzymes are closely linked with the adaptation in exercise. These enzymes inhibit histone acetylation. In their study McGee et al. portrayed that global histone 3 acetylation was increased at lysine 36, a site that was related with transcriptional elongation. They used cycling for 60 minutes as the method of exercise in their study [176]. They also discovered that that $HDAC_4$ and $HDAC_5$ were exported from the nucleus during exercise. The suppressive role of $HDAC_4$ and $HDAC_5$ was expelled.

In spite of the data that supports the positive influence of physical exercise on epigenetic mechanisms and an improvement in health, there is not a clear connection between exercise and epigenetics. Excessive and persistent physical exercise could in fact may have a negative influence on health. Gene expression may be affected by modified epigenetics that were changed by the body's physical adaptation to environmental conditions [175].

The functional genome in cardiac and peripheral vascular beds can be altered by exercise or physical activity. Reactive oxygen species (ROS) are formed during exercise but it also enhances anti-oxidative capacity. A deviation in the balance of cellular oxidative stress ensues. ROS have important regulatory function in muscle contraction, antioxidant protection, and repair of oxidative damage. DNA methylation and histone modification could have role in exercise-related ROS production thus leading to heritable conditions which are overseed by epigenetics. Many signal cascades controlling physiological and pathophysiological cardiac and peripheral vascular adaptations are modified [177]. The result is the modification in the molecular composition of the extracellular matrix. As a result several signaling cascades are changed.

There are many epigenetic signatures of environmental exposure pertaining to CVD as discovered from different studies (population cohorts for global and gene specific alterations). Oxidative stress, aging, and atherosclerosis are CVD disease outcome measures [178]. Exposure to traffic-related air pollution is closely linked with an increase in hospitalization rates and mortality from the perspective of CVD disease outcome measures. Metals affect health in many perspectives [179]. Under the light of new data, epigenetics could be a crucial pathway where metals produce their effects on health. Metallomics studies concerning environmental toxins are using epigenetics as a brand new major area to make their investigations. DNA methylation, modification of histone proteins, and RNA interference are three factors which are involved in CVD disease and other inherited conditions that are being studied by epigenetics [179]. Tobacco is a well-known CVD disease risk factor. It exerts its effects through epigenetics [180]. DNA methylation, histone modification, and microRNA alterations are environmental epigenetic mechanisms produced by tobacco. Tobacco consumption by the mother has been linked with changes in methylation of placental cytochrome P450A1 (CYP1A1) gene restriction [181]. This change leads to fetal growth restriction. Recently there has been a new finding between the prognosis for patients with stable CVD and smoking-related methylation patterns in the Coagulation factor II (thrombin) receptor-like 3 (F2RL3) gene [182].

Recent evidences have also shown the role of modulating or modifying miRNAs in established several risk factors for CVDs. Dyslipidemia is a common increased risk factor for developing of CVDs. Decreased levels of atherogenic lipoproteins are related with reduced CVD risk [183].

Numerous studies are performed for analyzing the biological functions of miRNAs on lipid metabolism. MiR122 that is an emerging miR in animal model hepatocytes targets many genes required in lipid metabolism [184]. Two studies demonstrated that miR122 inhibition reduced the levels of total serum cholesterol by increasing the fatty acid oxidation and hepatic lipid synthesis [185, 186]. Furthermore, it has been shown that miR122 reduced triglycerides and decreased the synthesis of the LDL receptor [187]. Rayner and colleagues found that another miR, miR33, regulates the cholesterol homeostasis by modulating either HDL biogenesis in the liver or cellular cholesterol efflux [188]. In African green monkeys, suppression of miR33 reduced serum triglycerides and increased HDL [188]. A genome-wide association study defined that presence of rare rs13702 C allele within the miR410 leads to lower triglycerides and higher HDL [189].

Diabetes and obesity which are another emerging risk factors for CVDs modified by miRNAs. It has been found an increased miR103 and miR107 in obese and diabetic mice liver [190, 191]. The treatment with Anti-miR103/miR107 extends glucose homeostasis and suppresses the size of adipocytes and total amount of visceral fat in these mice [191].

One research group demonstrated the association between antisense inhibition of miR208 and markedly repression of weight gain in mice which were bleeding with high-fat diet [192].

Tobacco smoke has lots of different chemical compounds that are CVD disease-causing compounds. Nicotine is one of the compound of cigarette smoke. It has been demonstrated that miR133 and miR590 progressed the remodeling of atrial in the present of nicotine treatment by modifying the TGF-β1 and TGF-β receptor II [193].

In addition, miR-197, miR-23a, and miR-509-5p were shown being correlated with body mass index, the risk factor for CVDs, in human study [194].

5. Epigenetics-Based Therapeutic Strategies in CVDs

Pharmacological therapy centered on the epigenome is a novel therapeutic concept. Pharmacoepigenomics uses epigenetic modification by medications as its basis of approach [195]. Pharmacoepigenomics might be able to more personally tailored treatment options. These kind of pharmaceutical agents are called epi-drugs. DNA methyltransferases, HDACs, HATs, histone mehtyltransferases, and histone demethylases are type of epi-drugs. Pharmacoepigenomics' major use was thought to be pharmacotherapy related to cancer [196]. However, currently new evidence shows that personally tailored oral antidiabetic treatment modalities could be more beneficial than standard treatments. In the future, pharmacoepigenomics could alter different individual response to oral antidiabetic treatment. Dysfunction of smooth muscle cells in diabetes could be the reason of increased incidence of macrovascular complications. Epigenetic modification of smooth muscle cell function may be the advantage of this treatment option as CHD risk in diabetic patients. Because, this treatment strategy does not require the tight glucose control in these patients [197].

CVD has a complex pathophysiological process and pathogenic pathways of CVDs remain unclear. Genetic approaches including epigenetics and personally tailored medical treatment established a novel approach in the treatment of CVDs. Chromatin structure has been chosen as a treatment target by the majority of epigenetic challenges. Most of the isoform-selective HDAC-inhibitors could have an importance in the treatment of cancer, Huntington's disease, sickle cell disease, or CVDs [198, 199].

Epigenetic tags can be affected by modifiers of methylation. DNA methylation status alter the function of target atheroprotective genes, such as ERα and ERβ genes in atherosclerosis [76]. Human coronary atherosclerotic tissues contain hypermethylated ERα and ERβ genes. This hypermethylation status could be demethylated by using epigenetic inhibitors such as DAC. DAC could upregulate the expression of silenced ERα and ERβ genes [75]. Therefore, therapeutic options could be used in the treatment of atherosclerosis by changing the epigenetic modifications.

DAC is a DNA methylation inhibitor. It has worked as a successful treatment option in hematological diseases and malignancies. As portrayed by cardiovascular research, DNA methylation inhibitor promotes the differentiation of cardiomyocytes from mesenchymal stem cells at low frequencies [200, 201]. DAC was given prime importance after this finding. Unluckily, myeloid suppression and induction of cardiomyocyte cell death are the toxic affects imposed by DAC on behalf of its potential benefits. Zebularine is another cytidine analog. It produces less toxicity than DAC [202]. This property of zebularine promotes its use in studies on cardiomyocyte differentiation from ESCs. In the recent study, Horrillo et al. demonstrated that zebularine promotes cardiomyocyte differentiation from ESCs with increased expression of cardiac-specific genes [203]. However, using of this knowledge in the development of stem-cell-based therapy for degenerative cardiac diseases is still elusive.

In addition, collagen type XV A1 (COL15A1) is hypomethylated in the process of aortic SMCs proliferation. DAC treatment induces aortic proliferation by inhibiting DNMTs and demethylating COL15A1 [204].

There are many documents which revealed that epigenetic mechanisms could be limited by HDAC and HAT inhibitors in CVDs. Trichostatin A and valproic acid, HDAC inhibitors, upregulated CYP7A1 expression and reduced plasma cholesterol levels in LDL-knockout mice [205]. Furthermore, it has been defined that Trichostatin A increased the ERα and ERβ levels in SMCs and ECs [76]. The other study demonstrated that reduction of HAT activity by inhibition of the co-activators nuclear cap-binding protein subunit 1 and p300 HAT reduced cardiac hypertrophy [105]. Curcumin is one of the inhibitor of HAT activity [206]. It has been found that Curcumin prevents ventricular hypertrophy and preserved systolic pressure by inhibiting p300 HAT activity in an animal model of HF [104].

MiRNA biological network in the cardiovascular system plays an important role for CVD prevention, diagnosis, and therapy [207]. MiRNAs seems have critical importance in the understanding of the diagnosis and therapy of CVDs, especially diagnostic rather than the therapeutic approach [208].

A number of miRNAs are supposed as biomarkers for the diagnosis of MI. Many reports showed that miR208a-b was upregulated in blood and heart tissue [209, 210]. Furthermore, miR1 and miR499 were demonstrated as biomarkers for MI [134, 210].

Fukushima and colleagues profiled plasma miRNA levels in patients with HF. They showed decreased plasma concentration of miR126 in patients vs healthy subjects, it is also negatively correlated with Brain Natriuretic Peptide (BNP) plasma levels. BNP is a biomarker of HF and so increased miR126 levels can be a biomarker for HF [141]. In addition, miR508-5p could be defined as novel diagnostic and therapeutic miRNAs for HF [211].

MiRNAs have been recently studying as likely therapeutic targets. It is supposed that miRNAs could be novel clinical therapeutic approaches for CVDs. Preclinical studies are being done on miRNAs and CVD treatment. Despite markedly advances, in vivo studies just remain scientific and therapeutic challenge [130].

Several animal models have been generated for therapeutic approaches in CVDs. It has been shown that treatment of antagomir-92a protects against endothelial activation and dysfunction and CAD in mice [212]. MiR155 induces the cardiac hypertrophy. So, miR155 inhibition might provide clinical therapeutic strategies for cardiac hypertrophy and HF [213]. Also, it has been shown negative correlation between silencing of miR329, miR487b, miR494, and miR495 and in arteriogenesis [214]. Recently, it is found that miR146a/b could be modulated by combined therapy [215]. Furthermore, pharmacological compounds can be used for modulation of miRNAs as an effective therapy [130]. Montgomery and colleagues silenced the miR208a by Anti-miR and improved heart function and surviving in HF progression in rats [216]. In addition, it has been shown that downregulation of miR1 could be therapeutic target against acute MI [217].

More human and animal studies are still few, so it is needed further investigations of miRNAs to improve therapeutic advance in CVDs.

At present IncRNAs have been studying to understand the function of these molecules in heart development and CVDs. So, future studies should reveal diagnostic and specific therapeutic benefits in CVDs.

Since 2012 when it was reported at the American Heart Association Scientific Sessions regenerative medicine is now a prominent actor of CVD research [218]. Genome wide association studies and epigenetics are major actors of both basic and clinical CVD research. There are ongoing trials and basic science research projects to establish the part of epigenetics in regenerative medicine. There is massive potential of induced pluripotent stem cells (IPSCs) in altering the way we approach CVD. Pluripotent stem cells could have a role in regenerative medicine and individualized CVD therapies [218].

In order to alter the programming of somatic cells in the production of IPSCs, Sun et al. proposed the use of methods using microRNA. Sustaining pluripotency, specific cell lineage and modification of chromatin by epigenetics requires microRNA [219].

The other advanced application of regenerative medicine is utilization of a distinct biologic framework as the foundation of an epigenetically-modified, surgically insertable graft of different tissues. Jungebluth et al. portrayed this in their case of human tracheobronchial transplantation. The utilization of this method for vascular or other CV graft may effortlessly be visualized [220].

Integration and solution of these kind of problems fits perfectly for the application of the principles of epigenetics in regenerative medicine.

Conclusion

CVDs have a complex pathophysiological basis. Multiple genetic and environmental risk factors are associated with CVDs. Modification of epigenetic marks are one of the risk factors of CVDs. Many epigenetic alterations may play key roles in initiation and progression of CVDs. Definition of the epigenetic mechanisms required to contribute to CVDs will identify strong significant contributions for novel and targeted therapeutic approaches in CVDs. Success of these clinical approaches could have an important impact on a large fraction of the human CVDs. Future studies will also clarify the detailed epigenetic modifications linked to CVDs and use this knowledge to prevent of CVDs.

References

[1] Bird AP. CpG-rich islands and the function of DNA methylation. *Nature* 1986;321:209-213.

[2] Portela A, Esteller M. Epigenetic modifications and human disease. *Nat Biotechnol* 2010;28:1057-1068.

[3] Tahiliani M, Koh KP, Shen Y, Pastor WA, Bandukwala H, Brudno Y, Agarwal S, Iyer LM, Liu DR, Aravind L, Rao A. Conversion of 5-Methylcytosine to 5-Hydroxymethylcytosine in Mammalian DNA by MLL Partner TET1. *Science* 2009;324:930-935.

[4] Lennartsson A, Ekwall K. Histone modification patterns and epigenetic codes. *Biochim Biophys Acta* 2009;1790:863-868.

[5] Chen HP, Zhao YT, Zhao TC. Histone Deacetylases and Mechanisms of Regulation of Gene Expression. *Crit Rev Oncog* 2015;20:35-47.

[6] Bannister AJ, Kouzarides T. Regulation of chromatin by histone modifications. *Cell Research* 2011;21:381-395.

[7] Lewis BP, Burge CB, Bartel DP. Conserved seed pairing, often flanked by adenosines, indicates that thousands of human genes are microRNA targets. *Cell* 2005;120:15-20.

[8] Jinek M, Doudna JA. A three-dimensional view of the molecular machinery of RNA interference. *Nature* 2009;457:405-412.

[9] Liang M, Liu Y, Mladinov D, Cowley AW, Trivedi H, Fang Y, Xu X, Ding X, Tian Z. MicroRNA: a new frontier in kidney and blood pressure research. *Am J Physiol Renal Physiol* 2009;297:553-558.

[10] Seema R. Lalani, John W. Belmont. Genetic basis of congenital cardiovascular malformations. *Eur J Med Genet* 2015;57:402-413.

[11] Paolo Piatti, Anette Zeilner, Alexandra Lusser. ATP-Dependent Chromatin Remodeling Factors and Their Roles in Affecting Nucleosome Fiber Composition. *Int J Mol Sci* 2011;12:6544-6565.

[12] Wang W, Cote J, Xue Y, Zhou S, Khavari PA, Biggar SR. Purification and biochemical heterogeneity of the mammalian SWI-SNF complex. *The EMBO journal* 1996;15:5370-5382.

[13] Hang CT, Yang J, Han P, Cheng HL, Shang C, Ashley E, Zhou B. Chromatin regulation by Brg1 underlies heart muscle development and disease. *Nature* 2010;466:62-67.

[14] Workman JL. Nucleosome displacement in transcription. *Genes Dev* 2006;20:2009-2017.

[15] Wang Z, Zhai W, Richardson JA, Olson EN, Meneses JJ, Firpo MT, Kang C, Skarnes WC, Tjian R. Polybromo protein BAF180 functions in mammalian cardiac chamber maturation. *Genes Dev* 2004;18:3106-3116.

[16] Huang X, Gao X, Diaz-Trelles R, Ruiz-Lozano P, Wang Z. Coronary development is regulated by ATP-dependent SWI/SNF chromatin remodeling component BAF180. *Dev Biol* 2008;319:258-266.

[17] Shirai M, Osugi T, Koga H, Kaji Y, Takimoto E, Komuro I, Hara J, Miwa T Yamauchi-Takihara K, Takihara Y. The Polycomb-group gene Rae28 sustains Nkx2.5/Csx expression and is essential for cardiac morphogenesis. *J Clin Invest* 2002;110:177-184.

[18] Barth JL, Clark CD, Fresco VM, Knoll EP, Lee B, Argraves WS, Lee KH. Jarid2 is among a set of genes differentially regulated by Nkx2.5 during outflow tract morphogenesis. *Dev Dyn* 2010;239:2024-2033.

[19] Bruneau BG, Bao ZZ, Tanaka M, Schott JJ, Izumo S, Cepko CL, Seidman JG, Seidma CE. Cardiac expression of the ventricle-specific homeobox gene Irx4 is modulated by Nkx2-5 and dHand. *Dev Biol* 2000;217:266-277.

[20] Hoffman JI, Kaplan S, Liberthson RR: Prevalence of congenital heart disease. *Am Heart J* 2004;147:425-439.

[21] Aramaki M, Udaka T, Kosaki R, Makita Y, Okamoto N, Yoshihashi H, Oki H. Phenotypic spectrum of CHARGE syndrome with CHD7 mutations. *J Pediatr* 2006;148:410-414.

[22] Schnetz MP, Bartels CF, Shastri K, Balasubramanian D, Zentner GE, Balaji R, Zhang X. Genomic distribution of CHD7 on chromatin tracks H3K4 methylation patterns. *Genome Res* 2009;19:590-601.

[23] Ramsbottom SA, Sharma V, Rhee HJ, Eley L, Phillips HM, Rigby HF, Dean C, Chaudhry B, Henderson DJ. Vangl2-regulated polarisation of second heart field-derived cells is required for outflow tract lengthening during cardiac development. *PLoS Genet* 2014;10:e1004871.

[24] Yuan Y, Gao Y, Wang H, Ma X, Ma D, Huang G. Promoter methylation and expression of the *VANGL2* gene in the myocardium of pediatric patients with Tetralogy of Fallot. *Birth Defects Res A Clin Mol Teratol* 2014;100:973-984.

[25] Abbey D, Seshagiri PB. Aza-induced cardiomyocyte differentiation of P19 EC-cells by epigenetic co-regulation and ERK signaling. *Gene* 2013;526:364-373.

[26] Chamberlain AA., Lin M, Lister RL, Maslov AA, Wang Y, Suzuki M, Wu B, Greally JM, Zheng D, Zhou B. DNA methylation is developmentally regulated for genes essential for cardiogenesis. *J Am Heart Assoc* 2014;3:e000976.

[27] Hirasawa R, Chiba H, Kaneda M, Tajima S, Li E, Jaenisch R, Sasaki H. Maternal and zygotic Dnmt1 are necessary and sufficient for the maintenance of DNA methylation imprints during preimplantation development. *Genes Dev* 2008;22:1607-1616.

[28] Kurihara Y, Kawamura Y, Uchijima Y, Amamo T, Kobayashi H, Asano T, Kurihara H. Maintenance of genomic methylation patterns during preimplantation development requires the somatic form of DNA methyltransferase 1. *Dev Biol* 2008;313:335-346.

[29] Okano M, Bell DW, Haber DA, Li E. DNA methyltransferases Dnmt3a and Dnmt3b are essential for de novo methylation and mammalian development. *Cell* 1999;99:247-257.

[30] Kou CY, Lau SL, Au KW, Leung PY, Chim SS, Fung KP, Waye MM, Tsui SK. Epigenetic regulation of neonatal cardiomyocytes differentiation. *Biochem Biophys Res Commun* 2010;400:278-83.

[31] Gottlieb PD, Pierce SA, Sims RJ, Yamagishi H, Weihe EK, Harriss JV, Maika SD, Kuziel WA, King HL, Olson EN, Nakagawa O, Srivastava D. Bop encodes a muscle-restricted protein containing MYND and SET domains and is essential for cardiac differentiation and morphogenesis. *Nat Genet* 2002;31:25-32.

[32] Srivastava D, Cserjesi P, Olson EN. A subclass of bHLH proteins required for cardiac morphogenesis. *Science* 1995;270:1995-1999.

[33] Lee Y, Song AJ, Baker R, Micales B, Conway SJ, Lyons GE. Jumonji, a nuclear protein that is necessary for normal heart development. *Circ Res* 2000;86:932-938.

[34] Jung J, Kim TG, Lyons GE, Kim HR, Lee Y. Jumonji regulates cardiomyocyte proliferation via interaction with retinoblastoma protein. *J Biol Chem* 2005;280:30916-30923.

[35] Wang J, Scully K, Zhu X, Cai L, Zhang J, Prefontaine GG, Krones A, Ohgi KA, Zhu P, Garcia-Bassets I, Liu F, Taylor H, Lozach J, Jayes FL, Korach KS, Glass CK, Fu XD, Rosenfeld MG. Wang J. Opposing LSD1 complexes function in developmental gene activation and repression programmes. *Nature* 2007;446:882-887.

[36] Mathiyalagan P, Chang L, Du XJ, El-Osta A. Cardiac ventricular chambers are epigenetically distinguishable. *Cell Cycle* 2010;9:612-617.

[37] Yao TP, Oh SP, Fuchs M, Zhou ND, Ch'ng LE, Newsome D, Bronson RT, Li E, Livingston DM, Eckner R. Gene dosage-dependent embryonic development and proliferation defects in mice lacking the transcriptional integrator p300. *Cell* 1998;93:361-372.

[38] Shikama N, Lutz W, Kretzschmar R, Sauter N, Roth JF, Marino S, Wittwer J, Scheidweiler A, Eckner R. Essential function of p300 acetyltransferase activity in heart, lung and small intestine formation. *EMBO J* 2003;22:5175-5185.

[39] Mathiyalagan P, Keating ST, Du XJ, El-Osta A. Chromatin modifications remodel cardiac gene expression. *Cardiovasc Res* 2014;103:7-16.

[40] Montgomery RL, Davis CA, Potthoff MJ, Haberland M, Fielitz J, Qi X, Hill JA, Richardson JA, Olson EN. Histone deacetylases 1 and 2 redundantly regulate cardiac morphogenesis, growth, and contractility. *Genes Dev* 2007;21:1790-1802.

[41] Montgomery RL, Potthoff MJ, Haberland M, Qi X, Matsuzaki S, Humphries KM, Richardson JA, Bassel-Duby R, Olson EN. Maintenance of cardiac energy metabolism by histone deacetylase 3 in mice. *J Clin Invest* 2008;118:3588-3597.

[42] Zhao Y, Samal E, Srivastava D. Serum response factor regulates a muscle-specific microRNA that targets Hand2 during cardiogenesis. *Nature* 2005;436:214-220.

[43] Philippen LE, Dirkx E, da Costa-Martins PA, De Windt LJ. Non-coding RNA in control of gene regulatory programs in cardiac development and disease. *J Mol Cell Cardiol* 2015 (in print).

[44] Cordes KR, Srivastava D, Ivey KN. MicroRNAs in cardiac development. *Pediatr Cardiol* 2010;31:349-356.

[45] Yang J, Xu WW, Hu SJ. Heart failure: advanced development in genetics and epigenetics. *Biomed Res Int* 2015 (in print).

[46] Bostjanjic E, Zidar N, Stajer D, GlavaC D. MicroRNAs miR-1, miR-133a, miR-133b and miR-208 are dysregulated in human myocardial infarction. *Cardiology* 2010;115: 163–169.

[47] Chen JF, Mandel EM, Thomson JM, Wu Q, Callis TE, Hammond SM, Conlon FL, Wang DZ. The role of microRNA-1 and microRNA-133 in skeletal muscle proliferation and differentiation. *Nat Genet* 2006;38:228-233.

[48] Grueter CE, van Rooij E, Johnson BA, DeLeon SM, Sutherland LB, Qi X, Gautron L, Elmquist JK, Bassel-Duby R, Olson EN. A cardiac MicroRNA governs systemic energy homeostasis by regulation of MED13. *Cell* 2012;149:671-683.

[49] Hosoda T, Zheng H, Cabral-da-Silva M, Sanada F, Ide-Iwata N, Ogórek B, Ferreira-Martins J, Arranto C, D'Amario D, del Monte F, Urbanek K, D'Alessandro DA, Michler RE, Anversa P, Rota M, Kajstura J, Leri A. Human cardiac stem cell differentiation is regulated by a mircrine mechanism. *Circulation* 2011;123:1287-1296.

[50] Ohtani K, Dimmeler S. Control of cardiovascular differentiation by microRNAs. *Basic Res Cardiol* 2011;106:5-11.

[51] Klattenhoff CA, Scheuermann JC, Surface LE, Bradley RK, Fields PA, Steinhauser ML, Ding H, Butty VL, Torrey L, Haas S, Abo R, Tabebordbar M, Lee RT, Burge CB, Boyer LA. Braveheart, a long noncoding RNA required for cardiovascular lineage commitment. *Cell* 2013;152:570-583.
[52] Grote P, Wittler L, Hendrix D, Koch F, Währisch S, Beisaw A, Macura K, Bläss G, Kellis M, Werber M, Herrmann BG. The tissue-specific lncRNA Fendrr is an essential regulator of heart and body wall development in the mouse. *Dev Cell* 2013;24:206-214.
[53] Margueron R, Reinberg D. The Polycomb complex PRC2 and its mark in life. *Nature* 2011;469:343-349.
[54] Simon JA, Kingston RE. Mechanisms of polycomb gene silencing: knowns and unknowns. *Nat Rev Mol Cell Biol* 2009;10:697-708.
[55] Rizki G, Boyer LA. Lncing Epigenetic Control of Transcription to Cardiovascular Development and Disease. *Circ Res* 2015;117:192-206.
[56] Baccarelli A, Wright R, Bollati V, Litonjua A, Zanobetti A, Tarantini L, Sparrow D, Vokonas P, Schwartz J. Ischemic heart disease and stroke in relation to blood DNA methylation. *Epidemiology* 2010;21:819-828.
[57] Antoniades C, Antonopoulos AS, Tousoulis D, Marinou K, Stefanadis C. Homocysteine and coronary atherosclerosis: from folate fortification to the recent clinical trials. *Eur Heart J* 2009;30:6-15.
[58] Yi P, Melnyk S, Pogribna M, Pogribny IP, Hine RJ, James SJ. Increase in plasma homocysteine associated with parallel increases in plasma S-adenosylhomocysteine and lymphocyte DNA hypomethylation. *J Biol Chem* 2000;275:29318-29323.
[59] Pogribny IP, Beland FA. DNA hypomethylation in the origin and pathogenesis of human diseases. *Cell Mol Life Sci* 2009;66:2249-2261.
[60] Ingrosso D, Cimmino A, Perna AF, Masella L, De Santo NG, De Bonis ML, Vacca M, D'Esposito M, D'Urso M, Galletti P, Zappia V. Folate treatment and unbalanced methylation and changes of allelic expression induced by hyperhomocysteinaemia in patients with uraemia. *Lancet* 2003;361:1693-1699.
[61] Jiang X, Yang F, Tan H, Liao D, Bryan RM Jr, Randhawa JK, Rumbaut RE, Durante W, Schafer AI, Yang X, Wang H. Hyperhomocystinemia impairs endothelial function and eNOS activity via PKC activation. *Arterioscler Thromb Vasc Biol* 2005;25:2515-2521.
[62] De Smet C, Lurquin C, Lethé B, Martelange V, Boon T. DNA methylation is the primary silencing mechanism for a set of germ line- and tumor-specific genes with a CpG-rich promoter. *Mol Cell Biol* 1999;19:7327-7335.
[63] Ron D. Hyperhomocysteinemia and function of the endoplasmic reticulum. *J Clin Invest* 2001;107:1221-1222.
[64] Jacobson TA, Miller M, Schaefer EJ. Hypertriglyceridemia and cardiovascular. risk reduction. *Clin Ther* 2007;29:763-777.
[65] Mikael LG, Wang XL, Wu Q, Jiang H, Maclean KN, Rozen R. Hyperhomocysteinemia is associated with hypertriglyceridemia in mice with methylenetetrahydrofolate reductase deficiency. *Mol Genet Metab* 2009;98:187–194.
[66] Devlin AM, Lentz SR. ApoA-I: a missing link between homocysteine and lipid metabolism? *Circ Res* 2006;98:431-433.

[67] Mikael LG, Genest J Jr, Rozen R. Elevated homocysteine reduces apolipoprotein A-I expression in hyperhomocysteinemic mice and in males with coronary artery disease. *Circ Res* 2006;98:564-571.

[68] Liao D, Tan H, Hui R, Li Z, Jiang X, Gaubatz J, Yang F, Durante W, Chan L, Schafer AI, Pownall HJ, Yang X, Wang H. Hyperhomocysteinemia decreases circulating high-density lipoprotein by inhibiting apolipoprotein A-I protein synthesis and enhancing HDL cholesterol clearance. *Circ Res* 2006;99:598-606.

[69] Kim GH, Ryan JJ, Archer SL. The role of redox signaling in epigenetics and cardiovascular disease. *Antioxid Redox Signal* 2013;18:1920-1936.

[70] Kim J, Kim JY, Song KS, Lee YH, Seo JS, Jelinek J, Goldschmidt-Clermont PJ, Issa JP. Epigenetic changes in estrogen receptor beta gene in atherosclerotic cardiovascular tissues and in vitro vascular senescence. *Biochim Biophys Acta* 2007;1772:72-80.

[71] Lund G, Zaina S. Atherosclerosis: an epigenetic balancing act that goes wrong. *Curr Atheroscler Rep* 2011;13:208-214.

[72] Castro R, Rivera I, Struys EA, Jansen EE, Ravasco P, Camilo ME, Blom HJ, Jakobs C, Tavares de Almeida I. Increased homocysteineand S-adenosyl homocysteine concentrations and DNA hypomethylation in vascular disease. *Clin Chem* 2003;49:1292-1296.

[73] Zhu S, Goldschmidt-Clermont PJ, Dong C. Inactivation of monocarboxylate transporter MCT3 by DNA methylation in atherosclerosis. *Circulation* 2005;112:1353-1361.

[74] Zaina S, Heyn H, Carmona FJ, Varol N, Sayols S, Condom E, Ramírez-Ruz J, Gomez A, Gonçalves I, Moran S, Esteller M. A DNA methylation map of human atherosclerosis. *Circ Cardiovasc Genet* 2014;7:692-700.

[75] Laukkanen M, Mannermaa S, Hiltunen M, Aittomäki S, Airenne K, Jänne. Local hypomethylation in atherosclerosis found in rabbit ec-sod gene. *Arterioscler Thromb Vasc Biol* 1999;19:2171-2178.

[76] Lund G, Andersson L, Lauria M, Lindholm M, Fraga M, Villar-Garea A. DNA methylation polymorphisms precede any histological sign of atherosclerosis in mice lacking apolipoprotein E. *J Biol Chem* 2004;279:29147-29154.

[77] Hiltunen M, Turunen M, Häkkinen T, Rutanen J, Hedman M, Mäkinen K. DNA hypomethylation and methyltransferase expression in atherosclerotic lesions. *Vasc Med* 2002;7:5-11.

[78] Friso S, Lotto V, Choi SW, Girelli D, Pinotti M, Guarini P, Udali S, Pattini P, Pizzolo F, Martinelli N, Corrocher R, Bernardi F, Olivieri O. Promoter methylation in coagulation F7 gene influences plasma FVII concentrations and relates to coronary artery disease. *J Med Genet* 2012;49:192-199.

[79] Backs J, Olson EN. Control of cardiac growth by histone acetylation/deacetylation. *Circ Res* 2006;98:15-24.

[80] Movassagh M, Choy MK, Goddard M, Bennett MR, Down TA, Foo RS. Differential DNA methylation correlates with differential expression of angiogenic factors in human heart failure. *PLoS One* 2010;5:e8564.

[81] Nguyen AT, Zhang Y. The diverse functions of Dot1 and H3K79 methylation. *Genes Dev* 2011;25:1345-1358.

[82] Smolarek I, Wyszko E, Barciszewska AM, Nowak S, Gawronska I, Jablecka A, Barciszewska MZ. Global DNA methylation changes in blood of patients with essential hypertension. *Med Sci Monit* 2010;16:149-155.

[83] Valinluck V, Tsai HH, Rogstad DK, Burdzy A, Bird A, Sowers LC. Oxidative damage to methyl-CpG sequences inhibits the binding of the methyl-CpG binding domain (MBD) of methyl-CpG binding protein 2 (MeCP2). *Nucleic Acids Res* 2004;32: 4100-4108.

[84] Kinney SM, Chin HG, Vaisvila R, BitinaiteJ, Zheng Y, Estève PO, Feng S, Stroud H, Jacobsen SE, Pradhan S. Tissue-specific distribution and dynamic changes of 5-hydroxymethylcytosine in mammalian genomes. *J Biol Chem* 2011;286:24685-24693.

[85] Wu H, Zhang Y. Tet1 and 5-hydroxymethylation: a genome-wide view in mouse embryonic stem cells. *Cell Cycle* 2011;10:2428-2436.

[86] Baserga M, Kaur R, Hale MA, Bares A, Yu X, Callaway CW, McKnight RA, Lane RH. Fetal growth restriction alters transcription factor binding and epigenetic mechanisms of renal 11beta-hydroxysteroid dehydrogenase type 2 in a sexspecific manner. *Am J Physiol Regul Integr Comp Physiol* 2010;299:334-342.

[87] Zhao Y, Gong X, Chen L, Li L, Liang Y, Chen S, Zhang Y. Site-specific methylation of placental HSD11B2 gene promoter is related to intrauterine growth restriction. *Eur J Hum Genet* 2013;22:734-740.

[88] Friso S, Carvajal CA, Fardella CE, Olivieri O. Epigenetics and arterial hypertension: the challenge of emerging evidence. *Transl Res* 2015;165:154-165.

[89] Riviere G, Lienhard D, Andrieu T, Vieau D, Frey BM, Frey FJ. Epigenetic regulation of somatic angiotensin-converting enzyme by DNA methylation and histone acetylation. *Epigenetics* 2011;6:478-489.

[90] Lee HA, Baek I, Seok YM, Yang E, Cho HM, Lee DY, Hong SH, Kim IK. Promoter hypomethylation upregulates Na+–K+–2Cl_ cotransporter 1 in spontaneously hypertensive rats. *Biochem Biophys Res Commun* 2010;396:252-257.

[91] Haas J, Frese KS, Park YJ, Keller A, Vogel B, Lindroth AM, Weichenhan D, Franke J. Alterations in cardiac DNA methylation in human dilated cardiomyopathy. *EMBO Mol Med* 2013;5:413-429.

[92] Movassagh M, Choy MK, Knowles DA, Cordeddu L, Haider S, Down T, Siggens L, Vujic A, Simeoni I, Penkett C, Goddard M, Lio P, Bennett MR, Foo RS. Distinct epigenomic features in end-stage failing human hearts. *Circulation* 2011;124:2411-2422.

[93] Sessa WC. eNOS at a glance. *J Cell Sci* 2004;117:2427-2429.

[94] Yan MS, Matouk CC, Marsden PA. Epigenetics of the vascular endothelium. *J Appl Physiol* 2010;109:916-912.

[95] Ho J, Man H, Marsden P. Nitric oxide signaling in hypoxia. *J Mol Med* (Berl) 2012;90:217-231.

[96] Chen W, Bacanamwo M, Harrison DG. Activation of p300 histone acetyltransferase activity is an early endothelial response to laminar shear stress and is essential for stimulation of endothelial nitric-oxide synthase mRNA transcription. *J Biol Chem* 2008;283:16293-16298.

[97] Granger A, Abdullah I, Huebner F, Stout A, Wang T, Huebner T, Epstein JA, Gruber PJ. Histone deacetylase inhibition reduces myocardial ischemia–reperfusion injury in mice. *FASEB J* 2008;22:3549-3560.

[98] Zhang L, Qin X, Zhao Y, Fast L, Zhuang S, Liu P, Cheng G, Zhao TC. Inhibition of histone deacetylases preserves myocardial performance and prevents cardiac

remodeling through stimulation of endogenous angiomyogenesis. *J Pharmacol Exp Ther* 2012;341:285-293.

[99] Choi JH, Nam KH, Kim J, Baek MW, Park JE, Park HY, Kwon HJ, Kwon OS, Kim DY, Oh GT. Trichostatin A exacerbates atherosclerosis in low density lipoprotein receptor-deficient mice. *Arterioscler Thromb Vasc Biol* 2005;25:2404-2409.

[100] Wongcharoen W, Phrommintikul A. The protective role of curcumin in cardiovascular diseases. *Int J Cardiol* 2009;133:145-151.

[101] Soni KB, Kuttan R. Effect of oral curcumin administration on serum peroxides and cholesterol levels in human volunteers. *Indian J Physiol Pharmacol* 1992;36:273-275.

[102] Ramírez-Boscá A, Soler A, Carrión MA, Díaz-Alperi J, Bernd A, Quintanilla C, Quintanilla Almagro E, Miquel J. An hydroalcoholic extract of Curcuma longa lowers the abnormally high values of human-plasma fibrinogen. *Mech Ageing Dev* 2000;114:207-210.

[103] Pan MH, Lai CS, Wu JC, Ho CT. Epigenetic and disease targets by polyphenols. *Curr Pharm Des* 2013;19:6156-6185.

[104] Morimoto T, Sunagawa Y, Kawamura T, Takaya T, Wada H, Nagasawa A, Komeda M, Fujita M, Shimatsu A, Kita T, Hasegawa K. The dietary compound curcumin inhibits p300 histone acetyltransferase activity and prevents heart failure in rats. *J Clin Invest* 2008;118:868–878.

[105] Gusterson RJ, Jazrawi E, Adcock IM, Latchman DS. The transcriptional co-activators CREB-binding protein (CBP) and p300 play a critical role in cardiac hypertrophy that is dependent on their histone acetyltransferase activity. *J Biol Chem* 2003;278:6838-6847.

[106] Cavasin MA, Demos-Davies K, Horn TR, Walker LA, Lemon DD, Birdsey N, Weiser-Evans MC, Harral J, Irwin DC, Anwar A, Yeager ME, Li M, Watson PA, Nemenoff RA. Selective class I histone deacetylase inhibition suppresses hypoxiainduced cardiopulmonary remodeling through an antiproliferative mechanism. *Circ Res* 2012;110:739-748.

[107] Liu F, Levin MD, Petrenko NB, Lu MM, Wang T, Yuan LJ, Stout AL, Epstein JA, Patel VV. Histone-deacetylase inhibition reverses atrial arrhythmia inducibility and fibrosis in cardiac hypertrophy independent of angiotensin. *J Mol Cell Cardiol* 2008;45:715-723.

[108] Chen Y, Du J, Zhao YT, Zhang L, Lv G, Zhuang S, Qin G, Zhao TC. Histone deacetylase (HDAC) inhibition improves myocardial function and prevents cardiac remodeling in diabetic mice. *Cardiovasc Diabetol* 2015;14:99.

[109] Fass DM, Shah R, Ghosh B, Hennig K, Norton S, Zhao WN, Reis SA, Klein PS, Mazitschek R, Maglathlin RL, Lewis TA, Haggarty SJ. Effect of Inhibiting Histone Deacetylase with Short-Chain Carboxylic Acids and Their Hydroxamic Acid Analogs on Vertebrate Development and Neuronal Chromatin. *ACS Med Chem Lett* 2010;2:39-42.

[110] Cardinale JP, Sriramula S, Pariaut R, Guggilam A, Mariappan N, Elks CM, Francis J. HDAC inhibition attenuates inflammatory, hypertrophic, and hypertensive responses in spontaneously hypertensive rats. *Hypertension* 2010;56:437-444.

[111] Findeisen HM, Kahles FK, Bruemmer D. Epigenetic Regulation of Vascular Smooth Muscle Cell Function in Atherosclerosis. *Curr Atheroscler Rep* 2013;15:319.

[112] Wierda RJ, Rietveld IM, van Eggermond MCJA, Belien JAM, van Zwet EW, Lindeman JHN. Global histone H3 lysine 27 triple methylation levels are reduced in vessels with advanced atherosclerotic plaques. *Life Sci* 2015;129:3-9.

[113] Kaneda R, Takada S, Yamashita Y, Choi YL, Nonaka-Sarukawa M, Soda M, Misawa Y, Isomura T, Shimada K, Mano H. Genome-wide histone methylation profile for heart failure. *Genes Cells* 2009;14:69-77.

[114] Barrick CJ, Roberts RB. Rojas M, Rajamannan NM, Suitt CB, O'Brien KD, Smyth SS, Threadgill DW. Reduced EGFR causes abnormal valvular differentiation leading to calcific aortic stenosis and left ventricular hypertrophy in C57BL/6J but not 129S1/SvImJ mice. *Am J Physiol Heart Circ Physiol* 2009;297:65-75.

[115] Papait R, Cattaneo P, Kunderfranco P, Greco C, Carullo P, Guffanti A, Viganò V, Stirparo GG, Latronico MV, Hasenfuss G, Chen J, Condorelli G. Genome-wide analysis of histone marks identifying an epigenetic signature of promoters and enhancers underlying cardiac hypertrophy. *Proc Natl Acad Sci U S A* 2013;110: 20164-20169.

[116] Rodriguez-Iturbe B. Arteriolar remodeling in essential hypertension: are connective tissue growth factor and transforming growth factor involved? *Kidney Int* 2006;69:1104-1105.

[117] Zhang D, Yu ZY, Cruz P, Kong Q, Li S, Kone BC. Epigenetics and the control of epithelial sodium channel expression in collecting duct. *Kidney Int* 2009;75:260-267.

[118] Pojoga LH, Williams JS, Yao TM, Kumar A, Raffetto JD, do Nascimento GR, Reslan OM, Adler GK, Williams GH, Shi Y, Khalil RA. Histone demethylase LSD1 deficiency during high-salt diet is associated with enhanced vascular contraction, altered NO-cGMP relaxation pathway, and hypertension. *Am J Physiol Heart Circ Physiol* 2011;301:1862-1871.

[119] Lee H, Rezai-Zadeh N, Seto E. Negative regulation of histone deacetylase 8 activity by cyclic AMP-dependent protein kinase A. *Mol Cell Biol* 2004;24:765-773.

[120] Li C, Li Y, Li Y, Liu H, Sun Z, Lu J, Zhao Y. Glucocorticoid repression of human with-no-lysine (K) kinase-4 gene expression is mediated by the negative response elements in the promoter. *J Mol Endocrinol* 2008;40:3-12.

[121] Mu S, Shimosawa T, Ogura S, Wang H, Uetake Y, Kawakami-Mori F, Marumo T, Yatomi Y, Geller DS, Tanaka H, Fujita T. Epigenetic modulation of the renal beta-adrenergic-WNK4 pathway in salt-sensitive hypertension. *Nat Med* 2011;17:573-580.

[122] El-Osta A. Glycemic memory. *Curr Opin Lipidol* 2012;23:24-29.

[123] Pirola L, Balcerczyk A, Tothill RW, Haviv I, Kaspi A, Lunke S, Ziemann M, Karagiannis T, Tonna S, Kowalczyk A, Beresford-Smith B, Macintyre G, Kelong M, Hongyu Z, Zhu J, El-Osta A. Genome-wide analysis distinguishes hyperglycemia regulated epigenetic signatures of primary vascular cells. *Genome Res* 2011;21:1601-1615.

[124] Mutskov V., Raaka B., Felsenfeld G., Gershengorn M. The human insulin gene displays transcriptionally active epigenetic marks in islet-derived mesenchymal precursor cells in the absence of insulin expression. *Stem Cells* 2007;25:3223-3233.

[125] Villeneuve L., Reddy M., Lanting L., Wang M., Meng L., Natarajan R. Epigenetic histone H3 lysine 9 methylation in metabolic memory and inflammatory phenotype of vascular smooth muscle cells in diabetes. *Proc Natl Acad Sci USA* 2008;105:9047-9052.

[126] El-Osta A, Brasacchio D, Yao D, Pocai A, Jones PL, Roeder RG, Cooper ME, Brownlee M. Transient high glucose causes persistent epigenetic changes and altered gene expression during subsequent normoglycemia. *J Exp Med* 2008;205:2409-2417.

[127] Fichtlscherer S, De Rosa S, Fox H, Schwietz T, Fischer A, Liebetrau C, Weber M, Hamm CW, Röxe T, Müller-Ardogan M, Bonauer A, Zeiher AM, Dimmeler S. Circulating microRNAs in patients with coronary artery disease. *Circ Res* 2010;107: 677-684.

[128] Hoekstra M, van der Lans CA, Halvorsen B, Gullestad L, Kuiper J, Aukrust P, van Berkel TJ, Biessen EA. The peripheral blood mononuclear cell microRNA signature of coronary artery disease. *Biochem Biophys Res Commun* 2010;394:792-797.

[129] Sondermeijer BM, Bakker A, Halliani A, de Ronde MW, Marquart AA, Tijsen AJ, Mulders TA, Kok MG, Battjes S, Maiwald S, Sivapalaratnam S, Trip MD, Moerland PD, Meijers JC, Creemers EE, Pinto-Sietsma SJ. Platelets in patients with premature coronary artery disease exhibit upregulation of miRNA340* and miRNA624*. *PLoS One* 2011;6:e25946.

[130] Schiano C, Vietri MT, Grimaldi V, Picascia A, Pascale MR, Napoli C. Epigenetic-related therapeutic challenges in cardiovascular disease. *Trends Pharmacol Sci* 2015;36:226-235.

[131] Liu Y, Sanoff HK, Cho H, Burd CE, Torrice C, Mohlke KL, Ibrahim JG, Thomas NE, Sharpless NE. INK4/ARF transcript expression is associated with chromosome 9p21 variants linked to atherosclerosis. *PLoS ONE* 2009;4:e5027.

[132] Cunnington MS, Santibanez Koref M, Mayosi BM, Burn J, Keavney B. Chromosome 9p21 SNPs associated with multiple disease phenotypes correlate with ANRIL expression. *PLoS Genet* 2010;6:e1000899.

[133] Long G, Wang F, Duan Q, Chen F, Yang S, Gong W, Wang Y, Chen C, Wang DW. Human Circulating MicroRNA-1 and MicroRNA-126 as Potential Novel Indicators for Acute Myocardial Infarction. *Int J Biol Sci* 2012;8:811-818.

[134] Ai J, Zhang R, Li Y, Pu J, Lu Y, Jiao J, Li K, Yu B, Li Z, Wang R, Wang L, Li Q, Wang N, Shan H, Li Z, Yang B. Circulating microRNA-1 as a potential novel biomarker for acute myocardial infarction. *Biochem Biophys Res Commun* 2010;391:73-77.

[135] Cheng Y, Tan N, Yang J, Liu X, Cao X, He P, Dong X, Qin S, Zhang C. A translational study of circulating cell-free microRNA-1 in acute myocardial infarction. *Clin Sci* (Lond) 2010;119:87-95.

[136] D'Alessandra Y, Devanna P, Limana F, Straino S, Di Carlo A, Brambilla PG, Rubino M, Carena MC, Spazzafumo L, De Simone M, Micheli B, Biglioli P, Achilli F, Martelli F, Maggiolini S, Marenzi G, Pompilio G, Capogrossi MC. Circulating microRNAs are new and sensitive biomarkers of myocardial infarction. *Eur Heart J* 2010;31:2765-2773.

[137] Adachi T, Nakanishi M, Otsuka Y, Nishimura K, Hirokawa G, Goto Y, Nonogi H, Iwai N. Plasma microRNA 499 as a biomarker of acute myocardial infarction. *Clin Chem* 2010;56:1183-1185.

[138] Vausort M, Wagner DR, Devaux Y. Long noncoding RNAs in patients with acute myocardial infarction. *Circ Res* 2014;115:668-677.

[139] Latronico MV, Condorelli G. MicroRNAs and cardiac pathology. *Nat Rev Cardiol* 2009;6:419-229.

[140] Soifer HS, Rossi JJ, Saetrom P. MicroRNAs in disease and potential therapeutic applications. *Mol Ther* 2007;15:2070-2079.

[141] Fukushima Y, Nakanishi M, Nonogi H, Goto Y, Iwai N. Assessment of plasma miRNAs in congestive heart failure. *Circ J* 2011;75:336-340.

[142] Corsten MF, Dennert R, Jochems S, Kuznetsova T, Devaux Y, Hofstra L, Wagner DR, Staessen JA, Heymans S, Schroen B. Circulating MicroRNA-208b and MicroRNA-499 reflect myocardial damage in cardiovascular disease. *Circ Cardiovasc Genet* 2010;3:499-506.

[143] Mora A, Davies AM, Bertrand L, Sharif I, Budas GR, Jovanović S, Mouton V, Kahn CR, Lucocq JM, Gray GA, Jovanović A, Alessi DR. Deficiency of PDK1 in cardiac muscle results in heart failure and increased sensitivity to hypoxia. *EMBO J* 2003;22:4666-4676.

[144] Qureshi IA, Mehler MF. The emerging role of epigenetics in stroke: II. RNA regulatory circuitry. *Arch Neurol* 2010;67:1435-1441.

[145] Jeyaseelan K, Lim KY, Armugam A. MicroRNA expression in the blood and brain of rats subjected to transient focal ischemia by middle cerebral artery occlusion. *Stroke* 2008;39:959-966.

[146] Liu DZ, Tian Y, Ander BP, Xu H, Stamova BS, Zhan X, Turner RJ, Jickling G, Sharp FR. Brain and blood microRNA expression profiling of ischemic stroke, intracerebral hemorrhage, and kainate seizures. *J Cereb Blood Flow Metab* 2010;30: 92-101.

[147] Cipollone F, Felicioni L, Sarzani R, Ucchino S, Spigonardo F, Mandolini C, Malatesta S, Bucci M, Mammarella C, Santovito D, de Lutiis F, Marchetti A, Mezzetti A, Buttitta F. A unique microRNA signature associated with plaque instability in humans. *Stroke* 2011;42:2556-2563.

[148] Bai Y, Nie S, Jiang G, Zhou Y, Zhou M, Zhao Y, Li S, Wang F, Lv Q, Huang Y, Yang Q, Li Q, Li Y, Xia Y, Liu Y, Liu J, Qian J, Li B, Wu G, Wu Y, Wang B, Cheng X, Yang Y, Ke T, Li H, Ren X, Ma X, Liao Y, Xu C, Tu X, Wang QK. Regulation of CARD8 expression by ANRIL and association of CARD8 single nucleotide polymorphism rs2043211 (p.C10X) with ischemic stroke. *Stroke* 2014;45: 383-388.

[149] McKay JA, Mathers JC. Diet induced epigenetic changes and their implications for health. *Acta Physiol* (Oxf) 2011;202:103-118.

[150] Milagro FI, Mansego ML, De Miguel C, Martinez JA. Dietary factors, epigenetic modifications and obesity outcomes: progresses and perspectives. *Mol Aspects Med* 2012;34:782-812.

[151] Masuyama H, Hiramatsu Y. Effects of a high-fat diet exposure in utero on the metabolic syndrome-like phenomenon in mouse offspring through epigenetic changes in adipocytokine gene expression. *Endocrinology* 2012;153:2823–2830.

[152] Bonaa KH, Njolstad I, Ueland PM, Schirmer H, Tverdal A, Steigen T, Wang H, Nordrehaug JE, Arnesen E, Rasmussen K. Homocysteine lowering and cardiovascular events after acute myocardial infarction. *N Engl J Med* 2006;354:1578–1588.

[153] Junien C, Gallou-Kabani C, Vige A, Gross MS. [Nutritionnal epigenomics: consequences of unbalanced diets on epigenetics processes of programming during lifespan and between generations]. *Ann Endocrinol* (Paris) 2005;66:2S19-28.

[154] Hanley B, Dijane J, Fewtrell M, Grynberg A, Hummel S, Junien C, Koletzko B, Lewis S, Renz H, Symonds M, Gros M, Harthoorn L, Mace K, Samuels F, van Der Beek EM.

Metabolic imprinting, programming and epigenetics–a review of present priorities and future opportunities. *Br J Nutr* 2010;104:1-25.

[155] Oommen AM, Griffin JB, Sarath G, Zempleni J. Roles for nutrients in epigenetic events. *J Nutr Biochem* 2005;16:74-77.

[156] Hong X, Wang X. Early life precursors, epigenetics, and the development of food allergy. *Semin Immunopathol* 2012;34:655-669.

[157] Burdge GC, Hoile SP, Lillycrop KA. Epigenetics: are there implications for personalised nutrition? *Curr Opin Clin Nutr Metab Care* 2012;15:442-447.

[158] Hossein-Nezhad A, Holick MF. Optimize dietary intake of vitamin D: an epigenetic perspective. *Curr Opin Clin Nutr Metab Care* 2012;15:567-579.

[159] Duthie SJ. Epigenetic modifications and human pathologies: cancer and CVD. *Proc Nutr Soc* 2011;70:47-56.

[160] Whayne TH. Epigenetics in the development, modification, and prevention of cardiovascular disease. Mol Biol Rep 2015;42:765-776.

[161] Linask KK, Huhta J. Folate protection from congenital heart defects linked with canonical Wnt signaling and epigenetics. Curr Opin Pediatr 2010;22:561-566.

[162] Bogdarina I, Welham S, King PJ, Burns SP, Clark AJ. Epigenetic modification of the renin-angiotensin system in the fetal programming of hypertension. *Circ Res* 2007;100:520-526.

[163] Lillycrop KA, Phillips ES, Jackson AA, Hanson MA, Burdge GC. Dietary protein restriction of pregnant rats induces and folic acid supplementation prevents epigenetic modification of hepatic gene expression in the offspring. *J Nutr* 2005;135: 1382-1386.

[164] Lillycrop KA, Slater-Jefferies JL, Hanson MA, Godfrey KM, Jackson AA, Burdge GC. Induction of altered epigenetic regulation of the hepatic glucocorticoid receptor in the offspring of rats fed a protein-restricted diet during pregnancy suggests that reduced DNA methyltransferase-1 expression is involved in impaired DNA methylation and changes in histone modifications. *Br J Nutr* 2007;97:1064-1073.

[165] Das SK, Chakrabarti R. Role of PPAR in cardiovascular diseases. *Recent Pat Cardiovasc Drug Discov* 2006;1:193–209.

[166] Walker BR. Glucocorticoids and cardiovascular disease. *Eur J Endocrinol* 2007;157:545-559.

[167] *Godfrey KM, Inskip HM, Hanson MA. The long-term effects of prenatal development on growth and metabolism. Semin Reprod Med 2011;29:257-265.*

[168] Barker DJ. The developmental origins of well-being. *Philos Trans R Soc Lond B Biol Sci* 2004;359:1359–1366.

[169] Painter RC, Roseboom TJ, Bleker OP. Prenatal exposure to the Dutch famine and disease in later life: an overview. *Reprod Toxicol* 2005;20:345-352.

[170] Godfrey KM, Barker DJ. Fetal programming and adult health. *Public Health Nutr* 2001;4:611-624.

[171] Weaver IC, Cervoni N, Champagne FA, D'Alessio AC, Sharma S, Seckl JR, Dymov S, Szyf M, Meaney MJ. Epigenetic programming by maternal behavior. *Nat Neurosci* 2004;7:847-854.

[172] Gluckman PD, Hanson MA, Buklijas T, Low FM, Beedle AS. Epigenetic mechanisms that underpin metabolic and cardiovascular diseases. *Nat Rev Endocrinol* 2009;5:401-408.

[173] Kuzawa CW, Sweet E. Epigenetics and the embodiment of race: developmental origins of US racial disparities in cardiovascular health. *Am J Hum Biol* 2009;21:2-15.

[174] Wheatley KE, Nogueira LM, Perkins SN, Hursting SD. Differential effects of calorie restriction and exercise on the adipose transcriptome in diet-induced obese mice. *J Obes* 2011;265417.

[175] Sanchis-Gomar F, Garcia-Gimenez JL, Perez-Quilis C, Gomez-Cabrera MC, Pallardo FV, Lippi G. Physical exercise as an epigenetic modulator. Eustress, the "positive stress" as an effector of gene expression. *J Strength Cond Res* 2012;26:3469-3472.

[176] McGee SL, Fairlie E, Garnham AP, Hargreaves M. Exercise-induced histone modifications in human skeletal muscle. *J Physiol* 2009;587:5951-5958.

[177] Bloch W, Suhr F, Zimmer P. Molecular mechanisms of exercise-induced cardiovascular adaptations. Influence of epigenetics, mechanotransduction and free radicals. *Herz* 2012;37:508-515.

[178] Baccarelli A, Ghosh S. Environmental exposures, epigenetics and cardiovascular disease. *Curr Opin Clin Nutr Metab Care* 2012;15:323-329.

[179] Wrobel K, Caruso JA. Epigenetics: an important challenge for ICP-MS in metallomics studies. *Anal Bioanal Chem* 2009;393:481-486.

[180] Breitling LP. Current genetics and epigenetics of smoking/tobacco-related cardiovascular disease. *Arterioscler Thromb Vasc Biol* 2011;33:1468–1472.

[181] Ordovas JM, Smith CE. Epigenetics and cardiovascular disease. *Nat Rev Cardiol* 2010;7:510-519.

[182] Suter M, Ma J, Harris A, Patterson L, Brown KA, Shope C, Showalter L, Abramovici A, Aagaard-Tillery KM. Maternal tobacco use modestly alters correlated epigenome-wideplacental DNA methylation and gene expression. *Epigenetics* 2011;6:1284-1294.

[183] NCEP. Third Report of the National Cholesterol Education Program (NCEP) Expert Panel on Detection, Evaluation, and Treatment of High Blood Cholesterol in Adults (Adult Treatment Panel III) final report. *Circulation* 2002;106:3143-3421.

[184] Flowers E, Aouizerat BE. MicroRNA associated with dyslipidemia and coronary disease in humans. *Physiol Genomics* 2013;45:1199-1205.

[185] Esau C, Davis S, Murray SF, Yu XX, Pandey SK, Pear M, Watts L, Booten SL, Graham M, McKay R, Subramaniam A, Propp S, Lollo BA, Freier S, Bennett CF, Bhanot S, Monia BP. miR-122 regulation of lipid metabolism revealed by in vivo antisense targeting. *Cell Metab* 2006;3:87-98.

[186] Krützfeldt J, Rajewsky N, Braich R, Rajeev KG, Tuschl T, Manoharan M, Stoffel M. Silencing of microRNAs in vivo with 'antagomirs.' *Nature 2005;438:685-689.*

[187] Flowers E, Singh K, Molina C, Mathur A, Aouizerat BE. MicroRNA associated with atherogenic dyslipidemia in South Asian men. *Int J Cardiol* 2013;168:4884-4885.

[188] Rayner KJ, Suárez Y, Dávalos A, Parathath S, Fitzgerald ML, Tamehiro N, Fisher EA, Moore KJ, Fernández-Hernando C. MiR-33 contributes to the regulation of cholesterol homeostasis. *Science* 2010;328:1570-1573.

[189] Richardson K, Nettleton JA, Rotllan N, Tanaka T, Smith CE, Lai CQ, Parnell LD, Lee YC, Lahti J, Lemaitre RN, Manichaikul A, Keller M, Mikkila V, Ngwa J, van Rooij FJ, Ballentyne CM, Borecki IB, Cupples LA, Garcia M, Hofman A, Ferrucci L, Mozaffarian D, Perala MM, Raitakari O, Tracy RP, Arnett DK, Bandinelli S, Boerwinkle E, Eriksson JG, Franco OH, Kahonen M, Nalls M, Siscovick DS, Houston DK, Psaty BM, Viikari J, Witteman JC, Goodarzi MO, Lehtimaki T, Liu Y, Zillikens

MC, Chen YD, Uitterlinden AG, Rotter JI, Fernandez-Hernando C, Ordovas JM. Gain-of-function lipoprotein lipase variant rs13702 modulates lipid traits through disruption of a microRNA-410 seed site. *Am J Hum Genet* 2013;92:5-14.

[190] Herrera BM, Lockstone HE, Taylor JM, Ria M, Barrett A, Collins S, Kaisaki P, Argoud K, Fernandez C, Travers ME, Grew JP, Randall JC, Gloyn AL, Gauguier D, McCarthy MI, Lindgren CM. Global microRNA expression profiles in insulin target tissues in a spontaneous rat model of type 2 diabetes. *Diabetologia* 2010;53:1099-1109.

[191] Trajkovski M, Hausser J, Soutschek J, Bhat B, Akin A, Zavolan M, Heim MH, Stoffel M. MicroRNAs 103 and 107 regulate insulin sensitivity. *Nature* 2011;474:649-653.

[192] van Rooij E, Quiat D, Johnson BA, Sutherland LB, Qi X, Richardson JA, Kelm RJ Jr, Olson EN. A family of microRNAs encoded by myosin genes governs myosin expression and muscle performance. *Dev Cell* 2009;17:662-673.

[193] Shan H, Zhang Y, Lu Y, *Zhang Y, Pan Z, Cai B, Wang N, Li X, Feng T, Hong Y, Yang B*. Downregulation of miR-133 and miR-590 contributes to nicotine-induced atrial remodelling in canines. *Cardiovasc Res* 2009;83:465-472.

[194] Karolina DS, Tavintharan S, Armugam A, Sepramaniam S, Pek SL, Wong MT, Lim SC, Sum CF, Jeyaseelan K. Circulating miRNA profiles in patients with metabolic syndrome. *J Clin Endocrinol Metab* 2012;97:2271-2276.

[195] Manolopoulos VG, Ragia G, Tavridou A. Pharmacogenomics of oral antidiabetic medications: current data and pharmacoepigenomic perspective. *Pharmacogenomics* 2011;12:1161-1191.

[196] Mai A, Altucci L. Epi-drugs to fight cancer: from chemistry to cancer treatment, the road ahead. *Int J Biochem Cell Biol 2009*;41:199-213.

[197] Porter KE, Riches K. The vascular smooth muscle cell: a therapeutic target in Type 2 diabetes? *Clin Sci* (Lond) 2013;125:167-182.

[198] Bieliauskas AV, Pflum MK. Isoform-selective histone deacetylase inhibitors. *Chem Soc Rev* 2008;37:1402-1413.

[199] Pons D, de Vries FR, van den Elsen PJ, Heijmans BT, Quax PH, Jukema JW. Epigenetic histone acetylation modifiers in vascular remodelling: new targets for therapy in cardiovascular disease. *Eur Heart J* 2009;30:266-277.

[200] Naeem N, Haneef K, Kabir N, Iqbal H, Jamall S, Salim A. DNA methylation inhibitors, 5-azacytidine and zebularine potentiate the transdifferentiation of rat bone marrow mesenchymal stem cells into cardiomyocytes. *Cardiovasc Ther* 2013;31:201-209.

[201] Antonitsis P, Ioannidou-Papagiannaki E, Kaidoglou A, Papakonstantinou C. In vitro cardiomyogenic differentiation of adult human bone marrow mesenchymal stem cells. The role of 5-azacytidine. *Interact Cardiovasc Thorac Surg* 2007;6:593–597.

[202] Zhou L, Cheng X, Connolly BA, DickmanMJ, Hurd PJ, Hornby DP. Zebularine: a novel DNA methylation inhibitor that forms a covalent complex with DNA methyltransferases. *J Mol Biol* 2002;321:591-599.

[203] Horrillo A, Pezzolla D, Fraga MF, Aguilera Y, Salguero-Aranda C, Tejedo JR, Martin F, Bedoya FJ, Soria B, Hmadcha A. Zebularine regulates early stages of mESC differentiation: effect on cardiac commitment. *Cell Death Dis* 2013;4:e570.

[204] Connelly JJ, Cherepanova OA, Doss JF, Karaoli T, Lillard TS, Markunas CA, Nelson S, Wang T, Ellis PD, Langford CF, Haynes C, Seo DM, Goldschmidt-Clermont PJ, Shah SH, Kraus WE, Hauser ER, Gregory SG. Epigenetic regulation of COL15A1 in

smooth muscle cell replicative aging and atherosclerosis. *Hum Mol Genet* 2013;22:5107-5120.

[205] Mitro N, Godio C, De Fabiani E, Scotti E, Galmozzi A, Gilardi F, Caruso D, Vigil Chacon AB, Crestani M. Insights in the regulation of cholesterol 7alpha-hydroxylase gene reveal a target for modulating bile acid synthesis. *Hepatology* 2007;46:885-897.

[206] Balasubramanyam K, Varier RA, Altaf M, Swaminathan V, Siddappa NB, Ranga U, Kundu TK. Curcumin, a novel p300/CREB-binding protein-specific inhibitor of acetyltransferase, represses the acetylation of histone/nonhistone proteins and histone acetyltransferase-dependent chromatin transcription. *J Biol Chem* 2004;279:51163-51171.

[207] Bauersachs J, Thum T. Biogenesis and regulation of cardiovascular microRNAs. *Circ Res* 2011;109:334-347.

[208] Udali S, Guarini P, Moruzzi S, Choi SW, Friso S. Cardiovascular epigenetics: from DNA methylation to microRNAs. *Mol Aspects Med* 2013;34:883-901.

[209] Widera C, Gupta SK, Lorenzen JM, Bang C, Bauersachs J, Bethmann K, Kempf T, Wollert KC, Thum T. Diagnostic and prognostic impact of six circulating microRNAs in acute coronary syndrome. *J Mol Cell Cardiol* 2011;51:872-875.

[210] Zile MR, Mehurg SM, Arroyo JE, Stroud RE, DeSantis SM, Spinale FG. Relationship between the temporal profile of plasma microRNA and left ventricular remodeling in patients after myocardial infarction. *Circ Cardiovasc Genet* 2011;4:614-619.

[211] Qiang L, Hong L, Ningfu W, Huaihong C, Jing W. Expression of miR-126 and miR-508-5p in endothelial progenitor cells is associated with the prognosis of chronic heart failure patients. *International Journal of Cardiology* 2013;168:2082–2088.

[212] Loyer X, Potteaux S, Vion AC, Guérin CL, Boulkroun S, Rautou PE, Ramkhelawon B, Esposito B, Dalloz M, Paul JL, Julia P, Maccario J, Boulanger CM, Mallat Z, Tedgui A. Inhibition of microRNA-92a prevents endothelial dysfunction and atherosclerosis in mice. *Circ Res* 2014;114:434-443.

[213] Seok HY, Chen J, Kataoka M, Huang ZP, Ding J, Yan J, Hu X, Wang DZ. Loss of MicroRNA-155 protects the heart from pathological cardiac hypertrophy. *Circ Res* 2014;114:1585-1595.

[214] Welten SM, Bastiaansen AJ, de Jong RC, de Vries MR, Peters EA, Boonstra MC, Sheikh SP, La Monica N, Kandimalla ER, Quax PH, Nossent AY. Inhibition of 14q32 MicroRNAs miR-329, miR-487b, miR-494, and miR-495 increases neovascularization and blood flow recovery after ischemia. *Circ Res* 2014;115:696-708.

[215] Takahashi Y, Satoh M, Minami Y, Tabuchi T, Itoh T, Nakamura M. Expression of miR-146a/b is associated with the Toll-like receptor 4 signal in coronary artery disease: effect of renin-angiotensin system blockade and statins on miRNA-146a/b and Toll-like receptor 4 levels. *Clin Sci* (Lond) 2010;119:395-405.

[216] Montgomery RL, Hullinger TG, Semus HM, Dickinson BA, Seto AG, Lynch JM, Stack C, Latimer PA, Olson EN, van Rooij E. Therapeutic inhibition of miR-208a improves cardiac function and survival during heart failure. *Circulation* 2011;124:1537-1547.

[217] Yang B, Lin H, Xiao J, Lu Y, Luo X, Li B, Zhang Y, Xu C, Bai Y, Wang H, Chen G, Wang Z. The muscle-specific microRNA miR-1 regulates cardiac arrhythmogenic potential by targeting GJA1 and KCNJ2. *Nat Med* 2007;13:486-491.

[218] Fujita J. Report of the American Heart Association (AHA). Scientific sessions 2012, Los Angeles. *Circ J* 2013;77:35-40.

[219] Sun X, Fu X, Han W, Zhao Y, Liu H. Can controlled cellular reprogramming be achieved using microRNAs? *Ageing Res Rev* 2010;9:475-483.

[220] Jungebluth P, Alici E, Baiguera S, Le Blanc K, Blomberg P, Bozoky B, Crowley C, Einarsson O, Grinnemo KH, Gudbjartsson T, Le Guyader S, Henriksson G, Hermanson O, Juto JE, Leidner B, Lilja T, Liska J, Luedde T, Lundin V, Moll G, Nilsson B, Roderburg C, Stromblad S, Sutlu T, Teixeira AI, Watz E, Seifalian A, Macchiarini P. Tracheobronchial transplantation with a stem-cell-seeded bioartificial nanocomposite: a proof-of-concept study. *Lancet* 2011;378:1997-2004.

In: Epigenetics: Mechanisms and Clinical Perspectives
Editor: Yildiz Dincer
ISBN: 978-1-63484-490-1

Chapter 6

INSIGHTS FROM EPIGENOMICS ANALYSIS OF SPERM: SPERM DEVELOPMENT AND MALE INFERTILITY

Dheepa Balasubramanian*[1], *PhD,
Eisa Tahmasbpour*[2], *PhD,
Ashok Agarwal*[3,*] *PhD, HCLD, ABB, EMB, ACE
[1]Case Western Reserve University, Cleveland, OH, US
[2]Chemical Injury Research Center, Baqiatallah Medical Science University, Tehran, Iran
[3]Center for Reproductive Medicine, Glickman Urological and Kidney Institute, Cleveland Clinic and Obstetrics and Gynecology and Women's Health Institute, Cleveland, Ohio, US

ABSTRACT

Extensive characterization has been performed on the genes and genetic mutations that are involved in spermatogenesis and male infertility, but the vast role of the sperm's transcriptome and epigenome in male reproduction has yet to be completely explored. Recent research has established that epigenetic remodeling of the sperm is necessary for development and for its function following fertilization. The histone- retained regions of the sperm have been recently shown to carry the bivalent marks of activating histone H3 lysine K4 trimethylation and the repressive H3 lysine K27 trimethylation. These bivalent histone marks facilitate the dynamic changes in stage-specific gene expression during sperm development. The paternal epigenome bears unique and important epigenetic modifications determined to be potentially important to the developing embryo, but the complete scope of its contribution is still emerging. Additionally, advances in assisted reproductive techniques have also suggested that alterations in the epigenetic profile in infertile men are transmitted to the developing embryo. This review will highlight the latest advances in epigenome profiling of the chromatin modifications during the development of immature to a mature sperm as well as provide a glimpse into the future

* Correspondence author: agarwaa@ccf.org.

role of epigenetic mechanisms in the generation of new germ cells/gametes from induced pluripotent stem cells, to treat male infertility.

Keywords: epigenetics, male infertility, epigenome, histone modification

Abbreviations

AR	Androgen receptor
ART	Assisted reproductive technology
AZF	Azoospermia a/b/c
BRDT	Bromodomain testis-specific protein
CFTR	Cystic fibrosis transmembrane conductance regulator
DAZ	Deleted in Azoospermia
DMR	differentially methylated regions
DNMT1/3a/3b	DNA methyltransferase
ESC	embryonic stem cells
HAT	histone acetyltransferases
HDAC	histone deacetylases
ICSI	Intra-cytoplasmic sperm injection
IGF2	Insulin-like growth factor
iPS/iPSCs	Induced pluripotent stem cells
IVF	In vitro fertilization
KLF4	Krueppel-like factor 4
MTHFR	methylene tetrahydrofolate reductase
MIWI2	PIWI protein
OCT4	octamer-binding transcription factor 4
PRM1/2	Protamine 1/2
PGC	Primordial germ cells
PGCLC	primordial germ cell-like cells
SAM	S-adenosyl methionine
SCOS	sertoli cell-only syndrome
SSC	spermatogonial stem cells
SNRPN	small nuclear ribonucleoprotcin polypeptide
TP	transition proteins

Introduction

1. Sperm Cell Development

In the past, it was believed that the sperm is but a vehicle to transfer the male genome into the oocyte, but increasing evidence now suggests that sperm cells play a substantial role in initiating embryo development, and environmental influences on the paternal genome can adversely affect male fertility as well as babies conceived via assisted reproductive technology (ART) [1].

Spermatogenesis involves the precise and well-controlled differentiation of germinal cells from spermatogonia to spermatozoa [2, 7]. This highly orchestrated process is a continuous process, characterized by a pre-meiotic, meiotic and a post-meiotic stage. Spermatogonia divide by mitosis, followed by meiosis through the formation of primary spermatocytes. As spermatocytes, they replicate DNA and undergo meiosis. During the meiosis stages, pairing of homologous chromosome, recombination, replacement of somatic histones with variants occur, and finally, at the post-meiotic stage, spermatids form into compact spermatozoa [3, 4].

Active transcription occurs during the stages of spermatocytes and round spermatids, but the later stages of spermatogenesis are relatively transcriptionally inert due to the compaction of the chromatin. Transcriptional, post-transcriptional, epigenetic regulatory networks direct the multiple stages of germ cell specification, differentiation and development [5]. Sperm chromatin is essential for sperm function as well as for embryonic development; defective sperm chromatin leads to abortion, decreased fertility, loss of genomic integrity or assisted reproductive failure [6].

2. Epigenetic Reprogramming of Male Germ Cells

Epigenetics is dominantly involved in the preservation of cellular identity and lineage fidelity, through the different mechanisms of alterations in chromatin structure, modifications of DNA and histones, remodeling of nucleosomes etc. Global reprogramming of epigenetic marks, in addition to meiotic chromatin organization and recombination as well as compaction of sperm genome, in the primordial germ cells (PGC) occurs during gametogenesis and in zygote following fertilization (Figure 1) [2]. The first reprogramming event occurs in the developing gonad. The reversible nature of epigenetic processes allows for the resetting of epigenetic information in the germ cells, in order to equip the sperm and oocyte to direct embryonic and post-natal development [7].

Dynamic epigenetic alteration is central to differentiation of mammalian sperm; however, the nature of these changes largely remains unknown. Spermatogenesis is a continuous and precisely controlled process that is accompanied by a dramatic reorganization of chromatin from a nucleosomal histone-based structure to a structure largely based on protamines [8]. During the spermatogenesis process, upon histone H4 hyperacetylation, nucleosomes are disassembled, DNA breaks occur and incorporation of non-canonical histone variants is made possible [9]. Histones are widely replaced by highly basic proteins, first by transition proteins and subsequently by the two protamines PRM1 and PRM2 [10].

Upon fertilization, the products of germ-cell development, the oocyte and sperm cell, fuse to form a zygote, a totipotent structure that can develop into a whole new organism [11]. Following fertilization, histones seem to remain associated with the paternal genome, in order, to contribute to zygotic chromatin despite the extensive reorganization of the sperm chromatin. Also, the sperm DNA is decondensed from its highly compacted and transcriptionally quiescent state, and expands to the inducible state found in the paternal pronucleus [12]. The paternal histones could therefore serve as a template for the incorporation of newly synthesized histones during replication in the zygote. Accordingly, the programmatic chromatin packaging in sperm could potentially deliver epigenetic information to the oocyte and the zygote post-fertilization [13].

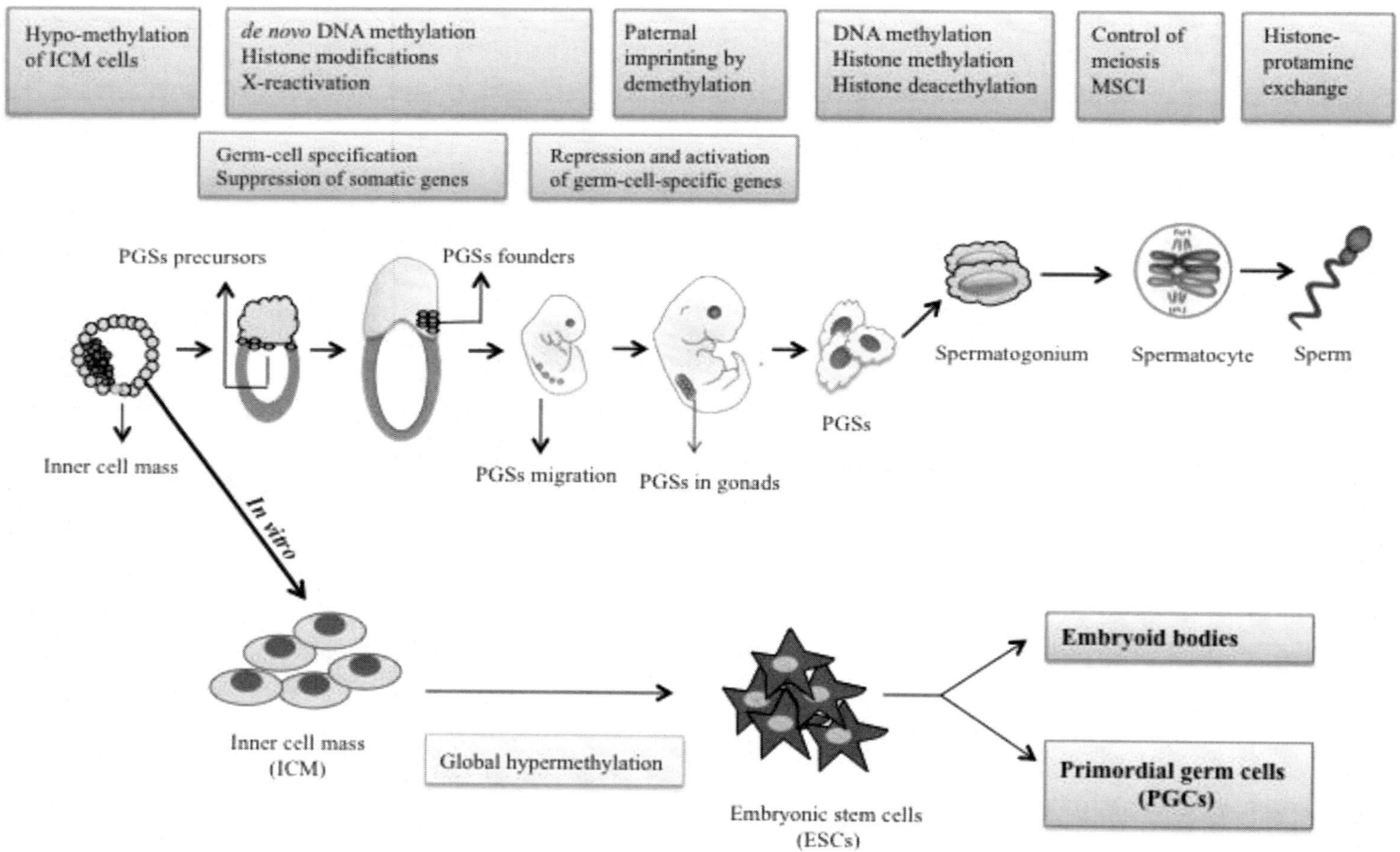

Figure 1. Epigenetics govern the regulation of processes from the initial stages of spermatogenesis to embryo development.

2.1. DNA Methylation

DNA methylation occurs largely on CpG islands located upstream of genes [14]. It is a process in which a methyl group from S-adenosyl methionine (SAM) is transferred to a cytosine residue to create 5-methyl cytosine (5mC). This process is catalyzed by a family of closely related DNA methyl transferases (DNMT1, DNMT3a and DNMT3b) [15]. The passive process takes place during replication of newly synthesized DNA strands by DNMT1.

The purpose of DNA demethylation in PGCs is to prevent transmission of inappropriate epigenetic information to the next generation. It is interesting to note that genome-wide methylation patterns in sperm differ markedly from that of somatic cells, highlighting the sperm-specific role of DNA methylation [8]. DNA methylation patterns are first acquired during gametogenesis. Evidence for the significance of DNA methylation for male germ cell development has been demonstrated in the mouse, where gene targeting of enzymes involved in catalyzing DNA methylation, the DNA (cytosine-5)-methyltransferases (DNMTs), results in loss of germ cell methylation, and failure of meiosis and infertility [16, 17, 18]. Recent studies have shown that mono methyl, dimethyl and trimethyl modifications of H3K4, H3K9 or H3K27 display tightly controlled temporal expression and ensure proper progression through spermatogenesis [19]. The level of H3K4 methylation peaks in the spermatogonial stem cell stage, and a targeted loss of H3K4 methylation, caused by reduction of Mll2 (an H3K4 methyl transferase) activity results in a dramatic reduction in the number of spermatocytes, suggesting that H3K4 methylation is essential for the exit from the stem cell stage and commitment to become a spermatocyte [20, 21, 22]. In contrast, H3K9 methylation and H3K27 methylation are low in the stem cell and increase during meiosis, persisting long after meiosis is complete, presumably to ensure gene-silencing [23]. Methylation on lysine 9 of histone H3 (H3K9me) is also associated with the sex chromosomes, euchromatin and heterochromatin in the late pachytene stage; however, the levels of H3K9 methylation drop upon completion of meiosis [24, 25]. This reduction in H3K9me is concurrently associated with an increase in H3K4me levels [7].

2.2. Histone Modifications

Histone modifications include the various post-translational alterations on the lysine-rich tail region of histones, especially of H3 and H4 histones [26]. However, histone acetylation, methylation and phosphorylation are more relevant during sperm development, spermatogenesis, fertilization and embryo development (Figure 2) [27].

The compactness of DNA is determined by histone proteins that are bound to DNA [28]. During the post-meiotic maturation of spermatids into spermatozoa or spermiogenesis, the genome of the male germ cell is entirely restructured. Indeed, while the core of the spermatid extends and condenses, most somatic histones are removed and replaced progressively by proteins called "transition" or TP, which are then replaced with specific nuclear proteins of sperm nucleus known as protamines [29, 30]. Protamines aid in the tight packaging of chromatin and are protective against the effects of harmful physical or chemical stressors [29]. In humans, histone to protamine transformation is considered to be incomplete, and the residual histones are highly acetylated and located in the annular region [29]. However, a small amount of histone proteins is retained in spermatids, and chemical modification of these histones play an important role in sperm formation.

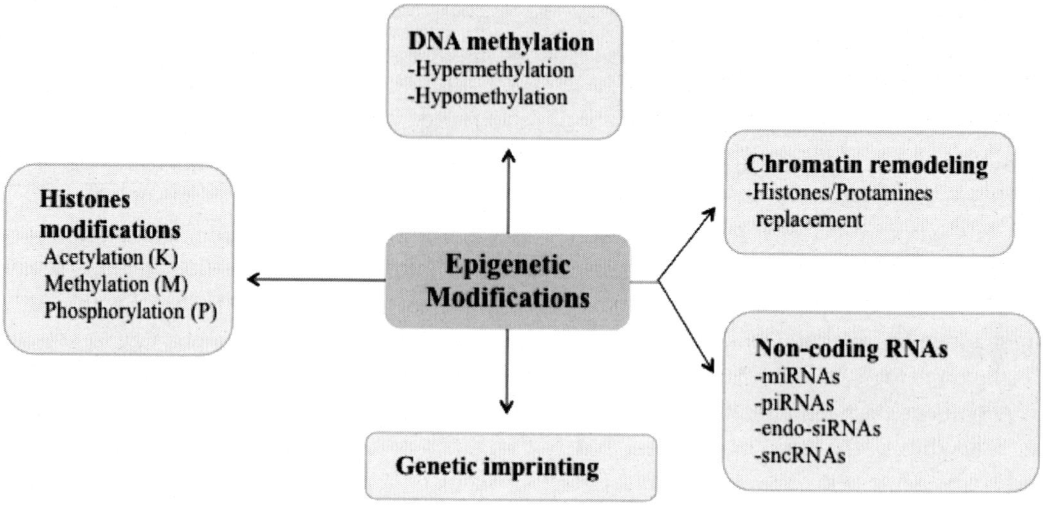

Figure 2. Epigenetic modifications involved in Sperm development.

During spermiogenesis, only 4% of the genome is retained in nucleosomes, while the rest of the histones are replaced with protamines. But the retained histones are overwhelmingly found at CpG islands near key developmental genes. The fetal germ cells contain several genes in a poised conformation, with both the H3K4me3 and H3K27me3 modifications [23, 25, 31]. Based on these findings, it can be presumed that the poised state of chromatin in germ cells are necessary for germ cell identity, to prepare for totipotency following fertilization and possibly to prevent DNA methylation at key developmental promoters [22].

The acetylated histone marks H3K4ac and H3K39ac are associated with transcriptional activation. There is growing interest in histone acetyl transferase (HAT) and histone deacetylase (HDAC) proteins involved in histone acetylation and spermatogenesis because they modulate gene expression and serve as a trigger for chromatin reorganization and remodeling [19, 32]. Waves of histone acetylation occur throughout human spermatogenesis. For example, hyperacetylation of histone-H4 was observed in spermatogonia, pre-leptotene spermatocyte and spermatids [19]. It facilitates the interaction of non-histone chromosomal proteins, such as transcription factors, with the chromatin. In spermatids, hyperacetylation of histone-H4 recruits bromodomain protein (BRDT) that triggers nuclear reorganization. Recent studies have demonstrated a statistically significant reduction in the percentage of spermatogonia with Hypac-H4 or Lys12ac-H4 [33]. Furthermore, hypac-H4 is believed to trigger nuclear reorganization and probably facilitate exchange of histones by protamines that allow condensation of the spermatid nucleus [33]. Another study demonstrated increased acetylation of histone H4 may be associated with a histone-to-protamine substitution, suggesting specific relationship between histone H4 modification and gene expression during spermatogenesis [19].

Every stage of spermatocyte development shows specific patterns of H4 modification. In pre-leptotene spermatocytes, the expression of H4me, H4K20me2, H4K5ac, H4K8ac and H4K12ac was high, and that of H4me3 was relatively low. Whether acetylation itself is only associated with gene activation is still not certain because it has been recently shown that protamines also contain acetyl groups. Studies have also alluded to other post-translational modifications, such as S42 and K49 acetylation of protamine-1 and K64 acetylation on protamine-2 [34]. The fact that protamines carry marks associated with transcriptional activation is intriguing, as these proteins are thought to ensure tight packaging of the DNA in sperm cells and contribute to transcriptional silencing. Although histone acetylation is a characteristic feature of transcriptional active genes, it is known that spermatozoa are transcriptionally inactive. This obvious inconsistency resulted in the hypothesis that histone acetylation represents an epigenetic mark that is transmitted from sperm to oocyte and involved in the regulation of gene expression in the early embryo [34].

2.3. Chromatin Remodeling

Chromatin remodeling is a process in which ATP-dependent chromatin remodeling complexes (SWI/SNF, ISW1 and MI-2) use energy to alter the location and structure of nucleosomes [35]. The remodeling process activates or represses gene expression, as required for proper meiotic development and maturation of gametes [28]. It is a critical step in gamete development, and the compacted structure of the chromatin transmits vital information to the embryo to guide it through development. The lack of proper DNA packaging in sperm has been associated with infertility in mice [36]. During the chromatin repackaging process in spermatogenesis, about 85% of the histones are replaced by protamines [12]. In the initial stages of spermiogenesis, histones are hyperacetylated and undergo other modifications [36]. Subsequently, at the final stage of spermatogenesis, the nucleosomal structure is progressively disassembled, then replaced by transition proteins (TPs) and finally by protamines [37].

The incorporation of protamines into sperm chromatin induces DNA compaction that is important for the formation of spermatozoa [30]. However, it is known that protamines are phosphorylated before binding to DNA and that substantial dephosphorylation takes place concomitant with nucleo-protamine maturation. Indeed, mutation of the calmodulin-dependent protein kinase Camk4 that phosphorylates protamine 2, results in defective spermiogenesis and male infertility [38]. At fertilization, the mature sperm cells are packaged densely with protamines, whereas maternal genome is packaged with histones. Subsequently, upon fertilization, the highly compact nucleoprotamine structure must be unpacked and reorganized into a nucleosomal structure [39]. Recent studies have shown that epigenetic errors in these processes are a possible cause of male infertility. Both protamines 1 and 2 are essential for sperm function, and the haploinsufficiency of either P1 or P2 results in a reduced amount of the respective protein, abnormalities in the structure of the chromatin, DNA damage and infertility [28]. Furthermore, it is known that the optimal protamine-1 to protamine-2 ratio (P1/P2 ratio) is critical and well regulated [40]. Indeed, the P1/P2 ratio in fertile men ranges from 0.8–1.2 and deviation from this ratio has been shown to lead to infertility [41]. A change in either direction of this ratio adversely affects semen quality, DNA integrity and fertility in men. Patients with abnormally depressed or elevated P1/P2 ratios are characterized by poor sperm concentration, motility and morphology as well as decreased fertilization capabilities. Impaired spermatogenesis has been reported to be associated with aberrant H4 acetylation [42]. A study by Sonnack et al., showed that men exhibiting qualitative or quantitative infertility have significantly decreased levels of histone H4 acetylation associated with impaired spermatogenesis [43]. Hyperacetylation of histone H4 is required in the transition from histones to protamines. This step decreases the affinity of the interaction between the sperm histones and DNA to allow for the exchange of transition proteins to occur [28]. H4 hyperacetylation was also observed in infertile men exhibiting sertoli-cell-only syndrome (SCOS) [44]. A significant difference in transcript expression of chromatin remodeling factors between normal spermatogenesis and the stage of round spermatid maturation arrest has been found in a previous study, suggesting impaired epigenetic information and aberrant transcription during sperm development. This could represent one possible reason for the developmental arrest of round spermatids [45].

2.4. Small RNAs

Spermatozoa contain numerous populations of RNAs. Small RNAs that have been enriched in sperm include microRNA (miRNA), endogenous small interfering RNAs (endo-siRNA), small non-coding RNA (sncRNA) and PIWI-interacting RNAs (piRNAs). Alterations in any one of the populations of RNAs in sperm indicate sperm abnormalities [3]. The importance of piRNAs in germline is very evident through their role of silencing of transposable elements during DNA demethylation in PGCs [46]. Mice knockout studies, where Dicer 1 was specifically deleted in spermatogonia, revealed prominent defects during haploid differentiation, specifically, in nuclei elongation and chromatin organization. Speculated roles for endo-siRNAs in germ cells include silencing and regulation of heterochromatin formation and maintenance. Among the piRNAs, MILI and MIWI2 together with pre-pachytene piRNAs, participate in silencing of transposable elements in fetal and neonatal germ cells. In the absence of these piRNAs in mice, it was shown that transposons were uncontrollably expressed, leading to meiotic arrest and sterility eventually [47].

3. Epigenetic Dysregulation: Male Infertility

3.1. Male Infertility

Human infertility, caused largely by a deficiency of sperm cells, is a major health problem that affects ~15% of couples globally [48]. It is a multifactorial syndrome encompassing a wide variety of disorders (Figure 3). Anatomic defects underlying male infertility include varicocele, vesicular damage due to torsion, obstruction of testicular sperm passage and ejaculatory failures, genital tract infections, gametogenesis dysfunction, molecular genetics disorders, endocrine disturbances and immunologic problems [48-50]. Additionally, factors such as life style, environment and smoking have also been reported to affect gamete and embryo development [51, 52]. In order to understand potential regulatory mechanisms involved in disease pathogenesis, it is essential to identify the molecular genetic factors involved in the etiology and physiology of male infertility.

Deletions in the Azoospermia Factor a, b and c (AZFa, AZFb and AZFc) regions of the long arm of the Y chromosome have been identified as the most common cause of Y-linked male factor infertility, particularly spermatogenic failure [53, 54]. In addition to Y-linked genes, many autosomal and X-linked genes such as CFTR, androgen receptor (AR) genes, estrogen receptors (ESR) genes, ubiquitin-specific protease-26 (USP26) gene, are also necessary for normal sexual development, testis determination, testis descent, spermatogenesis and eventually fertility in men [48, 55].

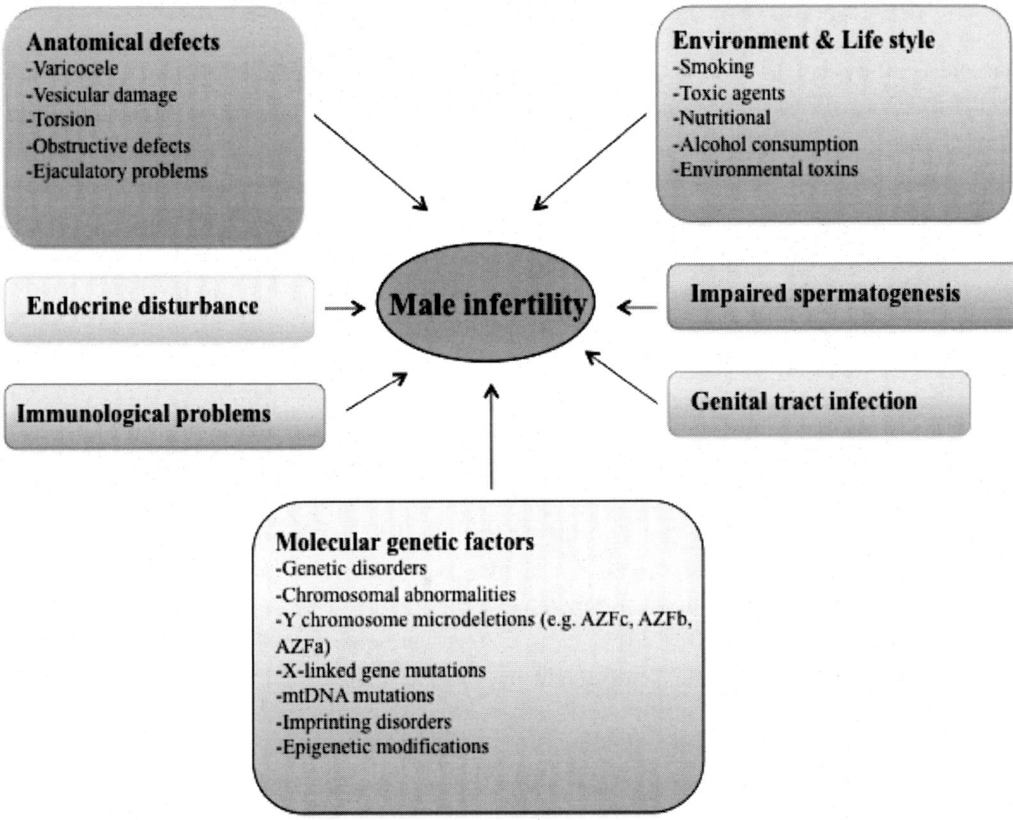

Figure 3. Most common causes of male infertility. Male Infertility is a multifactorial disease, with varied causative factors including genetic, molecular, epigenetic, immunological, environmental and life styles.

The advent of new technologies has spurred research on the role of epigenetics in spermatogenesis and male infertility. Epigenetics refer to process of gene regulation devoid of changes in DNA sequence [56]. Epigenetics is dominantly involved in the preservation of cellular identity and lineage fidelity, through the different mechanisms of alterations in chromatin structure, modifications of DNA and histones, remodeling of nucleosomes etc. Studies have demonstrated that numerous genes can be regulated through epigenetic mechanisms in the testes, indicating a direct influence of epigenetic mechanisms on the process of spermatogenesis, sperm development or maturation, fertilization process as well as male fertility [27]. It has become increasingly clear that epigenetic regulation of gene expression is critical during spermatogenesis and fertilization process [57].

Studies have made it abundantly clear that, in fathers exposed to environmental toxins or pesticides, sperm quality and chromatin integrity can be negatively affected through the primordial germ cells (PGCs), where there are some protected portions of the genome containing repetitive elements and single copy sequences that have the potential to transmit DNA methylation across generations or during differentiation, from PGCs to spermatogonia, de novo methylation occurs at imprinted loci or during development from spermatogonium to spermatocytes and during reprogramming that happens at fertilization, through the 4-10% of retained histones [58-60].

3.2. Genomic Imprinting

Genomic imprinting, namely, the expression of alleles in a sex-specific manner, is characterized by epigenetic modifications including differential methylation in the CpG islands in either the maternal or paternal alleles [28]. It determines which genes from the parental and maternal genomes are expressed in the embryo and is critical for normal development [61]. Parental imprints are erased in the PGCs of the developing embryo, and re-established during gametogenesis in a sex-dependent manner, maintained through fertilization, pre- and post-implantation embryonic development [62]. Imprint re-establishment occurs at late fetal stages in male germ ells and after birth in growing oocytes [64].

In humans, fetal spermatogonia seem to be mostly unmethylated at *H19* differentially methylated regions, although spermatogonia in adult testis demonstrate significant methylation in this region [28]. Imprinted genes, including the paternally imprinted *GTL2* and *H19* loci, have been previously examined by Kobayashi et al., in 97 infertile men. The importance of genomic imprinting during spermatogenesis has been reinforced by the association of decreased methylation of the paternal *IGF2/H19* imprinting control region 1 (ICR1) and *GTL2* imprints in spermatozoa of men with disturbed spermatogenesis [60]. In another study by Poplinski et al., fertile men had a high degree of *IGF2/H19* ICR1 and a low degree of *MEST* methylation [64]. Low sperm counts were clearly associated with *IGF2/H19* ICR1 hypomethylation and, even stronger, with MEST hypermethylation. These results suggest that idiopathic male infertility is strongly associated with imprinting defects at *IGF2/H19* ICR1 and MEST, with aberrant MEST methylation being a strong indicator for sperm quality. Previous studies have also suggested that Prader-Willi Syndrome and Angelman syndrome are caused by the specific loss of paternally expressed genes [2].

ART techniques such as ICSI and round spermatid injection (ROSI) may increase the incidence of imprinting disorders and adversely affect embryonic development by using immature spermatozoa that may not have established proper imprints or global methylation yet. Andrologists have voiced concern about concealing reproductive defects through ART that might have negative consequences at the epigenetic level [2].

3.3. DNA Methylation

DNA methylation occurs largely on CpG islands found upstream of genes [14]. It is a process in which a methyl group from S-adenosyl methionine (SAM) is transferred to a cytosine residue to create 5-methyl cytosine (5mC). This process is catalyzed by a family of closely related DNA methyl transferases (DNMT1, DNMT3a and DNMT3b) [65].

The establishment and maintenance of correct DNA methylation patterns is essential for fertility, embryo development and viability of the offspring. Abnormal methylation patterns in human spermatozoa have been reported in men with infertility and low sperm counts [66]. Recent studies have demonstrated that CpG methylation occurs in selected promoters of the fertile group, while methylation levels in subfertile patients vary considerably [64]. Sperm DNA methylation levels are related to conventional sperm parameters, as well as, sperm chromatin and DNA integrity [67]. Recently, hypermethylation of promoters of several genes including *MTHFR*, *PAX8*, *NTF3*, *SFN*, *HRAS*, *JHM2DA*, *IGF2*, *H19*, *RASGRF1*, *GTL2*, *PLAG1*, *D1RAS3*, *MEST*, *KCNQ1*, *LIT1* and *SNRPN* have been reported to be associated with poor semen parameters or male infertility [28]. Improper DNA methylation of various genes has been implicated in abnormal semen parameters, as well as several instances of male infertility. Houshdaran et al., demonstrated for the first time, that poor sperm concentration, motility and morphology were associated with broad DNA hypermethylation across a number of loci including *PLAG1*, *DIRAS3*, *MEST*, *PAX8*, *NTF3*, *SFN* and *HRAS*. They suggested that the underlying mechanism for these epigenetic changes may be improper erasure of DNA methylation during epigenetic reprogramming of the male germ line [68]. In another study by Khazamipour et al., 53% of men with nonobstructive azoospermia revealed hypermethylation of the *MTHFR* promoter, while none of the men with obstructive azoospermia exhibited hypermethylation of this gene promoter, suggesting that *MTHFR* hypermethylation is a specific epigenetic aberration that may specifically contribute to certain types of male infertility [69]. In addition, Wu et al., in a study reported that hypermethylation of *MTHFR* gene promoter in sperm was associated with idiopathic male infertility [70]. Abnormal DNA methylation of *H19* and *MEST* imprinted genes has also been shown to be associated with oligozoospermia, suggesting that spermatogenesis may be particularly vulnerable to changes in the methyl pool brought about by deficiency in MTHFR enzyme [71]. Hammoud et al., examined CpG methylation patterns in infertile (oligozoospermic and those with abnormal protamine ratios) and fertile donors at seven imprinted loci including *LIT1*, *MEST*, *SNRPN*, *PLAGL1*, *PEG3*, *H19*, and *IGF2* [72]. At six of the seven imprinted genes, the overall DNA methylation patterns were significantly altered in both infertile patient populations. They demonstrated a link between abnormal spermatogenesis and abnormal methylation of genes. Contradictory reports demonstrated high methylation at the *H19/IGF2* ICR1 locus and low methylation at the *MEST* locus in normozoospermia in one study, while another study revealed low methylation of the *H19/IGF2* ICR1 locus and high methylation of the *MEST* locus in association with low sperm concentrations [64]. Similarly, Boissonnas et al., found that many patients with teratozoospermia and oligoasthenoteratozoospermia exhibited hypomethylation at variable CpG islands at the *H19* locus [73]. However, recent studies have shown that hypermethylation at *MEST* was more strongly linked with poor sperm quality than hypomethylation at *H19/IGF2* ICR1 [28]. They suggested that sperm from infertile patients, especially those with oligospermia, may carry a higher risk of transmitting incorrect primary imprints to their offspring, highlighting the need for more research into epigenetic changes, when considering ART. The X-linked RHOX cluster encodes a set of homeobox genes that are selectively expressed in the male and female reproductive tract. Some members of the *Rhox* cluster are crucial for normal spermatogenesis, germ cell survival and male fertility. A recent study has demonstrated that the human RHOX gene cluster serves as an excellent marker for idiopathic male infertility. This study revealed a strong association between human RHOX cluster hypermethylation and the severity of poor sperm quality [74]. It appears that, DNA methylation directly represses the transcription of human RHOX gene family members. Incorrect DNA methylation of the DAZL promoter CpG island has been shown to be associated with defective human sperm [66]. DAZ gene family is crucial for normal spermatogenesis, as deletion of DAZ is estimated for 10% of cases of men with spermatogenic defect [75]. Targeted disruption of DNMTs genes in mice resulted in embryonic lethality and infertility [76]. Hammoud et al., showed a significant association between male factor infertility and alterations in sperm DNA methylation at imprinted loci [72]. They reported that patients with male factor infertility had significantly increased methylation alteration at LIT1, MEST, SNRPN, PLAGL1, PEG3, H19, and IGF2 genes [72]. In another study, DNA methylation at MEST gene was significantly associated with oligozoospermia, decreased bi-testicular volume and increased FSH levels [79]. Increased abnormal DNA methylation in the ART sample (41%) has been reported previously [80]. The reduced DNA methylation of the H19 ICR has been reported negatively correlated with sperm concentration [81]. A significant decrease in DNA methylation at the H19 DMR (differentially methylated regions) was found in testicular sperm of azoospermic men compared with proven fertile men [16]. In another study by Vieweg et al., sperm DNA methylation of normozoospermic men displayed low standard deviation, while a higher variability in the percentage of DNA methylation was detected in the promoters of subfertile men [68]. These data suggest that abnormalities in DNA methylation level in human sperm not only impact on sperm parameters quality and male infertility, but it could also represent a new approach to investigate the fertilizing ability of sperm in an assisted reproduction procedure, especially when sperm samples with normal characteristics are used [82].

4. Epigenetics in ART (Assisted Reproductive Techniques)

Assisted Reproductive technology (ART) includes all artificial methods used to achieve pregnancy [83]. Although a boon to infertile couples, ART-related manipulations to oocytes and embryos, have to be reviewed and designed with caution, as they coincide with the development of sex-specific genomic imprints [84]. The invasive as well as the non-invasive procedures performed during ART, including hormonal stimulation, egg retrieval, intra-cytoplasmic sperm injection (ICSI), micro-manipulation of gametes, *in vitro* fertilization (IVF), *in vitro* oocyte maturation, exposure to culture medium and centrifugation could potentially harm the early embryos and gametes. The process of ART coincides with crucial developmental phases of the pre-implantation embryo, and is therefore, implicated in inducing epigenetic changes that could affect fertility in subsequent generations and result in the birth of children with a higher risk of infertility, congenital abnormalities and morbidity [49]. It is thus imperative for the clinicians involved in the treatment of these couples to initiate genetic evaluation and counseling prior to any therapeutic procedures.

Until recently, the main evidence that ART like IVF/ICSI causes abnormalities in offspring, was corroborated by low birth weight babies, or premature births [83, 84]. Zhang et al., utilized microarray studies to show that poor perinatal outcomes in ART children arise from changed expression of genes involved in immune response, cell differentiation, compared to control patients [85]. But more recent investigations have shed light into differential expression of DNA methylation-associated genes that have been detected in offspring conceived by ART procedures. The commonly investigated list of imprinted genes for methylation-dependent disruptions include KCNQ1OT1, IGF2, SNRPN, MEST/PEG1, PEG10, PEG3, L3MBTL [87]. Children conceived from infertile couples, through ART have a higher propensity to develop imprinting disorders, since it has been shown that imprinted genes are more susceptible to environmental stresses [87]. A small but statistically significant subset of genes have altered DNA methylation profiles in children derived through ART, and the authors [87] propose that these significant differences in gene expression could be accounted for, by either the abnormal epigenetic marks borne by infertile couples or from the ART process itself. But, overall, conflicting studies on the effects of ART on offspring have made it impossible to conclusively point to major differences between ART and normal children [88, 89].

In addition to short-term outcomes, long-term outcomes, with regard to ART-conceived children is also important to scrutinize. It is interesting to note that several studies have observed taller ART children, at the ages of 6-10 years, probably due to the pre- or early implantation factors. But, with respect to gonadal development or fertility status in ART-conceived children, there was basically no change between ART offspring and spontaneously conceived children [90, 91, 92].

5. Epigenetic Make-Up of Induced Pluripotent Stem Cells/PGC from Adult Cells

In order to enhance our understanding of male reproduction and its associated diseased condition, male infertility, it is crucial to recapitulate gametogenesis and embryogenesis, *in vitro* [93, 94]. The last decade in reproductive research has seen important strides in this field using induced germ cells. The *in vitro* induction of germ cells from autologous iPS (induced pluripotent cells) and embryonic stem cells will also function as an additional route to conceive, for infertile couples, who cannot produce any gametes of their own [94]. Needless to say, the differentiation of human iPS cells will also enable in the molecular elucidation of diseases in gynecology, pediatrics and obstetrics.

Human IPS cells can be generated from somatic cells by the ectopic expression of OCT4, SOX2, KLF4 and c-MYC [95]. Recently, additional reports have demonstrated that human iPS cells can be induced to differentiate into the PGCs, gonocytes, spermatocytes, and into round spermatid –like cells [94]. It is however, important to note that PGCs derived from iPS cells have shown inefficient imprinting, as well as a propensity to acquire trisomies in chromosomes 12 and 17, possibly hinting to epigenetic instabilities. Additionally, human iPS cells seem to acquire higher number of CNVs, and revealed copy number variants at early passages compared to human ES cells. High-resolution methylation studies have also suggested that some iPS cell lines possess somatic memory, raising concerns about aberrant methylation and imprinting [10, 11, 18].

As a landmark discovery in this field, Saitou et al., accomplished the difficult task of generating PGC-like cells (PGCLC) from mouse ESC/iPSC cells. In order to prove that these PGCLCs could generate germ cells, they placed the PGCLCs into neonatal mice lacking their own germ cells. Very interestingly, mature sperms were produced from the testes, and could even be fertilized to produce their own offspring [95]. Though the PGCLCs could form functional sperm, there were minor differences in PGCLC-derived sperms compared to PGC-derived sperms [97]. But it is necessary to proceed with caution, since it is very likely that, they will be subject to genetic/epigenetic instabilities during iPS cell generation process, since the generation of such germ cells seem to occur against the Weismann barrier (hereditary information moves only from germ cells to somatic cells) [94].

An impairment of epigenetic reprogramming in germ cells can lead to loss of germ cells, faulty embryos or defectively developed offspring from these faulty embryos. In mice, during reprogramming of PGCs, they initiate a transcriptional network for pluripotency by predominantly acquiring epigenetic marks such as histone H3 lysine 9 dimethylation (H3K9me2), histone H3 lysine 27 trimethylation (H3K27me3) and demethylation of 5-methylcytosine, in addition to, silencing somatic programs [97, 98]. As a result, the PGCs acquire low levels of genome-wide 5-methylcytosine.

Recent progress in germ cell induction could lead to a new form of ART, where patients with no viable gametes will have the opportunity to produce fertile spermatozoa from autologous iPS cells, which can subsequently be used for IVF/ICSI. Utilizing spermatogonial stem cells (SSCs) from autologous iPS cells could also initiate in vivo spermatogenesis in otherwise infertile males [99].

CONCLUSION

The latest and most promising development is the creation of the first human artificial primordial germ cells (PGC) from reprogrammed skin cells by stem cell biologist Mitinori Saitou's group [100]. Initial studies performed on these artificial PGC's showed that their epigenetic and protein profiles were very similar to those of primordial germ cells obtained from aborted fetuses. In the next step in these studies, sperm or eggs were generated from these artificial PGC's in mice. It is interesting to envisage the future where these iPSc will be utilized to treat male infertility [101].

A conceivable future for the fertility treatment field will be to obtain sperm cells or eggs cells derived from skin cells of sterile men or women, which could then be a viable option to enable infertile couples to conceive, albeit, with the associated ethical concerns [102]. This technology will greatly benefit patients suffering from idiopathic male infertility, as well as the male cancer survivors. But, in order, to maximally benefit from these exciting techniques, it is imperative to gain a better understanding of the epigenetic modifications that occur during the generation of these iPSc cultures, as well as better investigate the imprints of the sperm chromatin, on a long-term basis in offspring [82].

REFERENCES

[1] Schagdarsurengin U., Paradowska A. and Steger K. Analysing the sperm epigenome: roles in early embryogenesis and assisted reproduction. *Nature Reviews Urology.* 2012;9:609-19.

[2] Kota SK and Feil R. Epigenetic transitions in germ cell development and meiosis. *Developmental Cell.* 2010;19:675-86.

[3] Rousseauz S., Caron C., Govin J., Lestrat C., Faure AK. and Khocbin S. Establishment of male-specific epigenetic information. 2005; 345:39-153.

[4] Hess RA and De Franca LR. Spermatogenesis and cycle of the seminiferous epithelium. *Adv Exp Med Biol.* 2008;636:1-15.

[5] Lopez JG., Alonso L., Cardenas DB., Alvarez HA., Hourcade JD., Martinez S., Enriquez MAB and Del Mazo J. Diversity and functional convergence of small non coding RNAs in male germ cell differentiation and fertilization. *RNA*. 2015;21:946-62.

[6] Kumar M., Kumar K., Jain S., Hassan T and Dada R. Novel insights into the genetic and epigenetic paternal contribution to the human embryo. *Clinics.* 2013;68:5-14.

[7] Shaoqin G, Zhenghui Z, Xueqian Z and Yuan H. Epigenetic modifications in human spermatozoon and its potential role in embryonic development. *Hereditas/Zhongguo yi chuan xue hui bian ji.* 2014;36:439-46.

[8] Meikar O., Ros MD and Kotaja N. Epigenetic regulation of male germ cell differentiation. *Sub-cellular biochemistry*. 2013; 61:119-138.

[9] Yuen BT, Bush KM, Barrilleaux BL, Cotterman R and Knoepfler PS. Histone H3.3 regulates dynamic chromatin states during spermatogenesis. *Development*. 2014;141:3483-94.

[10] Zamudio NM, Chong S and O'Bryan MK. Epigenetic regulation in male germ cells. *Reproduction*. 2008;136:131-46.

[11] Nazor KL., Altun G., Lynch C., Tran H., Harness JV., Slavin I., Garitaonandia I., Muller FJ., Wang YC. and Boscolo FS. Recurrent variations in DNA methylation in human pluripotent stem cells and their differentiated derivatives. *Cell Stem Cell*. 2012;10:620-34.

[12] Lister R., Pelizzola M., Kida YS., Hawkins RD., Nery JR., Hon G., Antosiewicz-Bourget J., O' Malley R., Castanon R. and Klugman S. Hotspots of aberrant epigenomic reprogramming in human induced pluripotent stem cells. *Nature*. 2011;471:68-73.

[13] Oliva R. Protamines and male infertility. *Human reproduction update*. 2006;12:417-3.

[14] Benchaib M., Ajina M., Braun V., Niveleau A and Guerin JF. DNA sperm methylation in assisted reproductive techniques. *Gynecologie, obstetrique & fertilite*. 2006;34:836-9.

[15] Cisneros FJ. DNA methylation and male infertility. Frontiers in bioscience: *a journal and virtual library*. 2004;9:1189-200.

[16] Minor A., Chow V. and Ma S. Aberrant DNA methylation at imprinted genes in testicular sperm retrieved from men with obstructive azoospermia and undergoing vasectomy reversal. *Reproduction*. 2011;141:749-57.

[17] Miller D., Brinkworth M, and Iles D. Paternal DNA packaging in spermatozoa: more than the sum of its parts? DNA, histones, protamines and epigenetics. *Reproduction*. 2010;139:287-301.

[18] Ohi Y., Qin H., Hong C., Blounin L., Polo JM., Guo T., Qi Z., Downey SL., Manos PD. and Rossi DJ. Incomplete DNA methylation underlies a transcriptional memory of somatic cells in human iPS cells. *Nat. Cell Biol*. 2011;13: 541-49.

[19] Shirakata Y., Hiradate Y., Inoue H., Sato E. and Tanemura K. Histone H4 modification during mouse spermatogenesis. *The Journal of reproduction and development*. 2014;60:383-7.

[20] Boissonnas CC., Jouannet P. and Jammes H. Epigenetic disorders and male subfertility. *Fertility and sterility*. 2013;99:624-31.

[21] Castillo J., Estanyol JM., Ballesca JL. and Oliva R. Human sperm chromatin epigenetic potential: genomics, proteomics, and male infertility. *Asian journal of andrology*. 2015;17:601-9.

[22] Lesch BJ and Page DC. Poised chromatin in the mammalian germ line. *Development*. 2014; 141:3619-26.

[23] Vavouri T. and Lehner B. Chromatin organization in sperm may be the major functional consequence of base composition variation in the human genome. *PLoS Genet*. 2011;7:e1002036.

[24] Drewell RA., Bush EC., Remnant EJ., Wong GT., Beeler SM., Stringham JL., Lim J. and Oldroyd BP. The dynamic DNA methylation cycle from egg to sperm in the honey bee Apis mellifera. *Development*. 2014;141:2702-11.

[25] Erkek S., Hisano M, Liang CY., Gill M., Murr R., Dierker J., Schubeler D., van der Vlag J., Stadler MB, and Peters AHFM. Molecular determinants of nucleosome retention at CpG-rich sequences in mouse spermatozoa. *Nat. Struct. Mol. Biol.* 2013; 20: 868-75.

[26] Yuen BT., Bush KM., Barrilleaux BL., Cotterman R. and Knoepfler PS. Histone H3.3 regulates dynamic chromatin states during spermatogenesis. *Development.* 2014;141:3483-94.

[27] Rousseaux S., Reynoird N., Escoffier E., Thevenon J., Caron C and Khochbin S. Epigenetic reprogramming of the male genome during gametogenesis and in the zygote. *Reproductive biomedicine online.* 2008;16:492-503.

[28] Rajender S., Avery K. and Agarwal A. Epigenetics, spermatogenesis and male infertility. *Mutation research.* 2011;727:62-71.

[29] Oliva R and Ballesca JL. Altered histone retention and epigenetic modifications in the sperm of infertile men. *Asian journal of andrology.* 2012;14:239-40.

[30] Gaucher J., Reynoird N., Montellier E., Boussouar, Rousseaux S and Khochbin S. From meiosis to postmeiotic events: The secrets of histone disappearance. *FEBS J.* 2010;277:599-604.

[31] Vastenhouw NL and Scheir AF. Bivalent histone modifications in early embryogenesis. *Curr Opin Cell Biol.* 2012;24:374-386.

[32] Pan XY., Li ZX., Wang ZC., Wang XN., Huang BY., Dou ZH and Sun YM. Histone modifications during spermatogenesis and male infertility. *Zhongguo yi xue ke xue yuan xue bao Acta Academiae Medicinae Sinicae.* 2014;36:108-13.

[33] Kleiman SE., Bar-Shira Maymon B., Hauser R., Botchan A., Paz G., Yavetz H. and Yogev L. Histone H4 acetylation and AZFc involvement in germ cells of specimens of impaired spermatogenesis. *Fertility and sterility.* 2008;89:1728-36.

[34] Vieweg M., Dvorakova-Hortova K., Dudkova B., Waliszewski P., Otte M., Oels B., Hajimohammad A., Turley H., Schorsch M., Schuppe HC., Weidner W., Steger K. and Paradowska-Dogan A. Methylation analysis of histone H4K12ac-associated promoters in sperm of healthy donors and subfertile patients. *Clinical epigenetics.* 2015;7:31.

[35] Zamudio NM., Chong S and O'Bryan MK. Epigenetic regulation in male germ cells. *Reproduction.* 2008;136:131-46.

[36] Oliva R., Bazett-Jones D., Mezquita C and Dixon GH. Factors affecting nucleosome disassembly by protamines in vitro. Histone hyperacetylation and chromatin structure, time dependence, and the size of the sperm nuclear proteins. *The Journal of biological chemistry.* 1987;262:17016-25.

[37] Oliva R and Dixon GH. Vertebrate protamine genes and the histone-to-protamine replacement reaction. *Progress in nucleic acid research and molecular biology.* 1991;40:25-94.

[38] Wu JY and Means AR. Ca(2+)/calmodulin-dependent protein kinase IV is expressed in spermatids and targeted to chromatin and the nuclear matrix. *The Journal of biological chemistry.* 2000;275:7994-9.

[39] Oliva R and Mezquita C. Histone H4 hyperacetylation and rapid turnover of its acetyl groups in transcriptionally inactive rooster testis spermatids. *Nucleic acids research.* 1982;10:8049-59.

[40] Torregrosa N., Dominguez-Fandos D., Camejo MI., Shirley CR., Meistrich ML., Ballesca JL. and Olivia R. Protamine 2 precursors, protamine 1/protamine 2 ratio, DNA

integrity and other sperm parameters in infertile patients. *Human reproduction.* 2006;21:2084-9.

[41] Aoki VW., Liu L and Carrell DT. Identification and evaluation of a novel sperm protamine abnormality in a population of infertile males. *Human reproduction.* 2005;20:1298-306.

[42] Faure AK., Pivot-Pajot C., Kerjean A., Hazzouri M., Pelletier R., Peoc'h M., Sele B., Khochbin S. and Rousseaux S. Misregulation of histone acetylation in Sertoli cell-only syndrome and testicular cancer. *Molecular human reproduction.* 2003;9:757-63.

[43] Sonnack V., Failing K., Bergmann M and Steger K. Expression of hyperacetylated histone H4 during normal and impaired human spermatogenesis. *Andrologia.* 2002;34:384-90.

[44] Dada R., Kumar M., Jesudasan R., Fernandez JL., Gosalvez J and Agarwal A. Epigenetics and its role in male infertility. *Journal of assisted reproduction and genetics.* 2012;29:213-23.

[45] Steilmann C., Cavalcanti MC., Bartkuhn M., Pons-Kuhnemann J., Schuppe HC., Weidner W., Steger K and Paradowska A. The interaction of modified histones with the bromodomain testis-specific (BRDT) gene and its mRNA level in sperm of fertile donors and subfertile men. *Reproduction.* 2010;140:435-43.

[46] Siomi MC., Sato K., Pezic D and Aravin AA. PIWI-interacting small RNAs: the vanguard of genome defence. *Nature Reviews Molecular Cell Biology.* 2011;12: 246-58.

[47] Carmell MA., Girard A., van de Kant HJ., Bourc'his D., Bestor TH., de Rooji DG. and Hannon GJ. MIWI2 is essential for spermatogenesis and repression of transposons in the mouse male germ line. *Dev. Cell.* 2007;4: 503-14.

[48] Tahmasbpour E., Balasubramanian D and Agarwal A. A multi-faceted approach to understanding male infertility: gene mutations, molecular defects and assisted reproductive techniques (ART). *Journal of assisted reproduction and genetics.* 2014;31:1115-37.

[49] Agarwal A., Mulgund A., Hamada A and Chyatte MR. A unique view on male infertility around the globe. Reproductive biology and endocrinology. *RB&E.* 2015;13:37.

[50] Ko EY., Sabanegh ES, Jr. and Agarwal A. Male infertility testing: reactive oxygen species and antioxidant capacity. *Fertility and sterility.* 2014;102:1518-27.

[51] Benoff S., Jacob A and Hurley IR. Male infertility and environmental exposure to lead and cadmium. *Human reproduction update.* 2000;6:107-21.

[52] Colagar AH., Jorsaraee GA and Marzony ET. Cigarette smoking and the risk of male infertility. Pakistan journal of biological sciences. *PJBS.* 2007;10:3870-4.

[53] Zhang YS., Dai RL., Wang RX., Zhang ZH., Fadlalla E and Liu RZ. Azoospermia factor microdeletions: occurrence in infertile men with azoospermia and severe oligozoospermia from China. Andrologia. 2014;46(5):535-40.

[54] Ambulkar PS., Sigh R., Reddy M., Varma PS., Gupta DO., Shende MR. And Asoke KP. Genetic Risk of Azoospermia Factor (AZF) Microdeletions in Idiopathic Cases of Azoospermia and Oligozoospermia in Central Indian Population. Journal of clinical and diagnostic research : JCDR. 2014;8:88-91.

[55] Lo Giacco D., Chianese C., Ars E., Ruiz-Castane E., Forti G and Krausz C. Recurrent X chromosome-linked deletions: discovery of new genetic factors in male infertility. Journal of medical genetics. 2014;51:340-4.

[56] Shaoqin G., Zhenghui Z., Xueqian Z and Yuan H. Epigenetic modifications in human spermatozoon and its potential role in embryonic development. Hereditas/Zhongguo yi chuan xue hui bian ji. 2014;36:439-46.

[57] Xu YN., Cui XS., Tae JC., Jin YX. and Kim NH. DNA synthesis and epigenetic modification during mouse oocyte fertilization by human or hamster sperm injection. *Journal of assisted reproduction and genetics.* 2011;28:325-33.

[58] Miranda CL., Gomez PR., Rojas G., Cruz I., Berrueta L., Salmen S., Colmenares M., Barreto S., Baiza A., Zavala L., Morales Y., Molina Y., Valeri L., Contreras CA and Osuna JA. Occupational exposure to organophosphate and carbamate pesticides affects sperm chromatin integrity and reproductive hormone levels among Venezuelan farm workers. *J Occup. Health.* 2013; 55: 195-203.

[59] Martenies SE and Perry MJ. Environmental and occupational pesticide exposure and human sperm parameters: a systematic review. *Toxicology.* 2013; 307:66-73.

[60] Soubry A., Hoyo C., Jirtle RL. and Murphy SK. A paternal environmental legacy: Evidence for epigenetic inheritance through the male germ line. *Bioessays.* 2014;36:359-371.

[61] Kobayashi H., Sato A., Otsu E., Hiura H., Tomatsu C., Utsunomiya T., Sasaki H., Yaegashi N. and Arima T. Aberrant DNA methylation of imprinted loci in sperm from oligospermic patients. *Human molecular genetics.* 2007;16:2542-51.

[62] Reik W and Walter J. Genomic imprinting :parental influence on the genome. *Nat Rev Gnet.* 2001;2:21-32.

[63] Delaval K and Feil R. Epigenetic regulation of mammalian genomic imprinting. *Curr Opin Genet Dev.* 2004;14:188-95.

[64] Poplinski A., Tuttelmann F., Kanber D., Horsthemke B. and Gromoll J. Idiopathic male infertility is strongly associated with aberrant methylation of MEST and IGF2/H19 ICR1. *International journal of andrology.* 2010;33:642-9.

[65] Cisneros FJ. DNA methylation and male infertility. Frontiers in bioscience: *a journal and virtual library.* 2004;9:1189-200.

[66] Navarro-Costa P., Nogueira P., Carvalho M., Leal F., Cordeiro I., Calhaz-Jorge C. goncalves J. and Plancha CE. Incorrect DNA methylation of the DAZL promoter CpG island associates with defective human sperm. *Human reproduction.* 2010;25:2647-54.

[67] Montjean D., Zini A., Ravel C., Belloc S., Dalleac A., Copin H., Boyer P., McElreavey K and Benkhalifa M. Sperm global DNA methylation level: association with semen parameters and genome integrity. *Andrology.* 2015;3:235-40.

[68] Houshdaran S., Cortessis VK., Siegmund K., Yang A., Laird PW. and Sokol RZ. Widespread epigenetic abnormalities suggest a broad DNA methylation erasure defect in abnormal human sperm. *PloS One.* 2007;2:e1289.

[69] Khazamipour N., Noruzinia M., Fatehmanesh P., Keyhanee M. and Pujol P. MTHFR promoter hypermethylation in testicular biopsies of patients with non-obstructive azoospermia: the role of epigenetics in male infertility. *Human reproduction.* 2009;24:2361-4.

[70] Wu W., Shen O., Qin Y., Niu X., Lu C., Xia Y., Song L., Wang S and Wang X. Idiopathic male infertility is strongly associated with aberrant promoter methylation of methylenetetrahydrofolate reductase (MTHFR). *PloS one.* 2010;5:e13884.

[71] Kumar R., Venkatesh S., Kumar M., Tanwar M., Shasmsi MB., Kumar R., Gupta NP. Sharma RK., Talwar P and Dada R. Oxidative stress and sperm mitochondrial DNA mutation in idiopathic oligoasthenozoospermic men. *Indian journal of biochemistry & biophysics.* 2009;46:172-7.

[72] Hammoud SS., Purwar J., Pflueger C., Cairns BR and Carrell DT. Alterations in sperm DNA methylation patterns at imprinted loci in two classes of infertility. *Fertility and sterility.* 2010;94:1728-33.

[73] Boissonnas CC., Abdalaoui HE., Haelewyn V., Fauque P., Dupont JM., Gut I., Vaiman D., Jouannet P., Tost J and Jammes H. Specific epigenetic alterations of IGF2-H19 locus in spermatozoa from infertile men. European journal of human genetics: *EJHG.* 2010;18:73-80.

[74] Richardson ME., Bleiziffer A., Tuttelmann F., Gromoll J. and Wilkinson MF. Epigenetic regulation of the RHOX homeobox gene cluster and its association with human male infertility. *Human molecular genetics.* 2014;23:12-23.

[75] Kuo PL., Wang ST., Lin YM., Lin YH., Teng YN and Hsu CC. Expression profiles of the DAZ gene family in human testis with and without spermatogenic failure. *Fertility and sterility.* 2004;81:1034-40.

[76] Webster KE., O'Bryan MK., Fletcher S., Crewther PE., Aapola U., Craig J., Harrison DK., Aung H., Phutikanit N., Lyle R., Meachem SJ., Antonarakis SE., de Krestser DM., Hedger MP., Peterson P., Carroll BJ and Scott HS. Meiotic and epigenetic defects in Dnmt3L-knockout mouse spermatogenesis. Proceedings of the National Academy of Sciences of the United States of America. 2005;102:4068-73.

[77] Klaver R., Tuttelmann F., Bleiziffer A., Haaf T., Kliesch S. and Gromoll J. DNA methylation in spermatozoa as a prospective marker in andrology. *Andrology.* 2013;1:731-40.

[78] Kobayashi H., Hiura H., John RM., Sato A., Otsu E., Kobayashi N. Suzuki R., Suzuki F., Hayashi C., Utsunomiya T., Yaegashi N., Arima T. DNA methylation errors at imprinted loci after assisted conception originate in the parental sperm. *European journal of human genetics : EJHG. 2009*;17:1582-91.

[79] Li JB., Li B., Liang XX., Wang J., Ma YF., Zhang YQ., Liu Z., Min Bh., Ma XH and Wang XH. Correlation of DNA methylation status of imprinted gene H19 ICR with oligozoospermia and asthenozoospermia *National journal of andrology.* 2013;19:511-7.

[80] Benchaib M., Braun V., Ressnikof D., Lornage J., Durand P., Niveleau A. and Guerin JF. Influence of global sperm DNA methylation on IVF results. *Human reproduction.* 2005;20:768-73.

[81] Nagano M., Patrizio P. and Brinster RL. Long-term survival of human spermatogonial stem cells in mouse testes. *Fertil. Steril.* 2002; 78:1225-33.

[82] Easley CA., Phillips BT., McGuire MM., Barringer JM., Valli H., Hermann BP., Simerly CR., Rajkovic A., Miki T., Orwig KE. and Schatten GP. Direct differentiation of human pluripotent stem cells into haploid spermatogenic cells. *Cell.* 2012;2: 440-6.

[83] Henningsen AK., Pinborg A., Lidegaard O., Vestergaard C., Forman JL. and Nyboe AA. Perinatal outcome of singleton siblings born after assisted reproductive technology

and spontaneous conception: Danish national sibling-cohort study. *Fertil Steril.* 2011; 95: 959-63.

[84] Lu YH., Wang N and Jin F. Long-term follow-up of children conceived through assisted reproductive technology. *J. Zhejiang Univ Sci. B.* 2013;14:359-71.

[85] Zhang Y., Cui Y., Zhou Z., Sha J., Li Y. and Liu J. Altered global expressions of human placentae subjected to assisted reproductive technology treatments. *Placenta.* 2010; 31:251-258.

[86] Uyar A and Seli E. The impact of assisted reproductive technologies on genomic imprinting and imprinting disorders. *Curr. Opin. Obstet. Gynecol.* 2014; 26:210-21.

[87] Katari S., Turan N., Bibikova M., Erinie O., Chalian R., Foster M., Gaughan JP, Coutifaris C. and Sapienza C. DNA methylation and gene expression differences in children conceived in vitro or in vivo. *Hum Mol Genet.* 2009;18: 3769-78.

[88] Tierling S., Souren NY., Gries J., Lo Porto C., Groth M., Lutsik P., Neitzel H., Utz-Billing I., Gillessen-Kaesbach G., Kentenich H., Griesinger G., Sperling K., Schwinger E. and Walter J. Assisted reproductive technologies do not enhance the variability of DNA methylation imprints in human. *J Med Genet* 2010;47:371-76.

[89] Rancourt RC., Harris HR. and Michels KB. Methylation levels at imprinting control regions are not altered with ovulation inductions or in vitro fertilization in a birth cohort. *Hum Reprod.* 2012; 27:2208-16.

[90] Belva F., Henriet S., Liebaers I., van Steirteghem A., Celestin-Weistreich S. and Bonduelle M. Medical outcome of 8-year old singleton ICSI children (born ≥32 weeks gestation) and a spontaneously conceived comparison group. *Hum Reprod.* 2007;22:506-15.

[91] Belva F., Roelants M., Painter R., Bonduelle M., Devroey P and de Schepper J. Pubertal development in ICSI children. *Human Reprod.* 2012;27:1156-61.

[92] de Schepper J., Belva F., Schiettecatte J., Anckaert E., Tournaye H. and Bonduelle M. Testicular growth and tubular function in prepubertal boys conceived by intracytoplasmic sperm injection. *Horm Res.* 2009;71:359-63.

[93] Ishii T. Human iPS cell-derived germ cells: Current status and clinical potential. *J. CLin. Med.* 2014; 3: 1064-83.

[94] Hayashi K., Ohta H., Kurimoto K., Aramaki S. and Saitou M. Reconstitution of the mouse germ cell specification pathway in culture by pluripotent stem cells. *Cell.* 2011; 146:519-32.

[95] Takahashi K., Tanabe K., Ohnuki M., Narita M., Ichisaka T., Tomoda K. and Yamanaka S. Induction of pluripotent stem cells from adult human fibroblasts by defined factors. *Cell.* 2007; 131:861-72.

[96] Reik W and Walter J. Genomic imprinting: Parental influence on the genome. *Nat Rev Genet.* 2001;2:21-32.

[97] Seisenberger S., Andrews S., Krueger F., Arand J., Walter J., Santos F., Popp C., Thienpont B., Dean W. and Reik W. The dynamics of genome-wide DNA methylation reprogramming in mouse primordial germ cells. *Mol. Cell.* 2012. 48:849-62.

[98] Seki Y., Hayashi K., Itoh K., Mizugaki M., Saitou M and Matsui Y. Extensive and orderly reprogramming of genome-wide chromatin modifications associated with specification and early development of germ cells in mice. *Develop. Biol.* 2005. 278:440-58.

[99] Nagano M., Patrizio P. and Brinster RL. Long-term survival of human spematogonial stem cells in mouse testes. *Fertil. Steril.* 2002; 78:1225-33.

[100] Saitou M., Kagiwada S. and Kurimoto K. Epigenetic reprogramming in mouse pre-implantation development and primordial germ cells. *Development.* 2012. 139:15-31.

[101] Gafni O., Weinberger L., Mansour AA., Manor YS., Chomsky E., Ben-Yosef D., Kalma Y., Viukov S., Maza I., Zviran A., Rais Y., Shipony Z., Mukamel Z., Krupalnik V., Zerbib M., Geula S., Caspi I., Schneir D., Schwartz T., Gilad S., Amann-Zalcenstein D., Benjamin S., Amit I., Tanay A., Massarwa R., Novershtem N. and Hanna JH. Derivation of novel human ground state naïve pluripotent stem cells. *Nature.* 2013; 504:282-6.

In: Epigenetics: Mechanisms and Clinical Perspectives
Editor: Yildiz Dincer
ISBN: 978-1-63484-490-1

Chapter 7

EPIGENETICS, INFLAMMATION AND INFLAMMATION-ASSOCIATED DISEASES: A GENERAL LOOK

Müge Sayitoğlu* and Yücel Erbilgin
Istanbul University, Institute of Experimental Medicine,
Department of Genetics, Istanbul, Turkey

ABSTRACT

Epigenetic is defined as the study of mitotically and/or meiotically heritable changes in the gene function without changing DNA sequence and playing crucial roles both in normal development and human diseases. The molecular basis of epigenetic process consists of histone modifications, DNA methylation, positioning of histone variants, and non-coding RNAs. Genome-wide patterns of DNA and chromatin modifications together called as ‘epigenomes’. Epigenomes undergo precise, coordinated, reversible changes through the developmental stages and so contributes to the lineage and tissue specific expression of genes. In addition to the tissue specific impact, environmental factors such as nutrients, toxins, infections and hypoxia can also influence epigenomes. Distinct or global changes in the epigenetic landscape are hallmarks of chronic inflammation associated diseases. Alteration in methylation status of CpG sites, monoallelic silencing, and other epigenetic regulatory mechanisms have been observed in key inflammatory response genes. Epigenetic changes including DNA methylation, histone modification and noncoding RNA expression, were found associated with acute and chronic inflammatory disorders. Recently, therapies targeting epigenetic mechanisms are trending options for the treatment of chronic and degenerative disorders. Epigenetic drugs that are applied on animal models and some clinical trials are displaying positive therapeutic effects. Not only mono therapies but also combined usage of HDAC or DNMT inhibitors could be the next step for the epigenetic therapeutic modulations. Scope of this chapter is to provide an overview of the epigenetic modifications in inflammation and inflammation driven diseases.

* Corresponding author: mugeay@istanbul.edu.tr.

Abbreviations

5-azaC	5-azacytidine cytosine
ACPA	against citrullinated peptide antigens
CD	Crohn's disease
CpG	cytosine-guanine nucleotides
CVD	cardiovascular diseases
DMV	DNA methylation valleys
DNA	deoxyribonucleic acid
DNMT	DNAmethyltransferases
EBV	Epstein Barr virus
HATs	histone acetyltransferases
HDACs	histone deacetylases
HDMs	demethylases
HIV	human immunodeficiency virus
HMTs	histone methyltransferases
IBD	inflammatory bowel disease
IFN	type 1 interferon
JAK	Janus-activated kinase
LncRNAs	Long noncoding RNAs
MAP	mitogen-activated protein
MBP	myelin basic protein
miRNA	microRNA
MS	multiple sclerosis
PADI2	peptidyl arginine deiminase type 2
PBMCs	peripheral blood mononuclear cells
RA	rheumatoid arthritis
RASFs	RA synovial fibroblasts
RNA	ribonucleic acid
RNS	reactive nitrogen species
ROS	reactive oxygen species
SIRS	systemic inflammatory response syndrome
SLE	systemic lupus erythematous
SSI	severe systemic inflammation
T1D	type 1 diabetes
UC	ulcerative colitis
UVB	ultraviolet B

1. Introduction

Inflammation (Latin, *inflammatio*) is a part of the complex biological protective response of the body that involves immune cells, blood vessels, and molecular mediators. Inflammation develops acute or chronic response of immune system and it is modulated by genetic, epigenetic, macrobiotic and environmental events. Although many variations exist among inflammatory diseases, a consistency occurs in local and systemic acute inflammation (like abscess and sepsis) and chronic inflammation like metabolic syndrome of obesity, cancer, atherosclerosis, autoimmunity, and aging.

One of the earliest reports which is suggested by a Roman physician named Cornelius Celsus, therein described patient symptoms against to tissue injury were *rubor* (redness), *tumor* (swelling, caused by increased permeability of the microvasculature), *calor* (heat), and *dolor* (pain). After centuries, scientists have begun to understand the physiological and cellular basis of inflammation. The inflammatory cells are not only important for host defense against pathogens; but also they play beneficial roles in tissue maintenance and homeostasis through the process of phagocytosis. Acute inflammation occurs over seconds, minutes, hours, and days, and chronic inflammation occurs over longer times. Inflammation is mainly described by redness, heat, pain and swelling, against pathogens, damaged cells, irritant and/or tissue repair/healing. Inflammatory response can be vascular or cellular. During vascular response histamine mediated hyperemia occurs, blood vessels dilate, fluid moves into tissue platelets forms a clot to trap the injurious agent. In the cellular response, white blood cells move to the site of the injury (called chemotaxis), neutrophils, monocytes/macrophages, eosinophils, basophils and lymphocytes involve in the inflammation process [1, 2]. Inflammation has a slow onset, persists for a long period of time, and becomes chronic. The symptoms in chronic inflammation are not as severe as in acute inflammation, but the condition is persistent. Chronic inflammation is associated with a group of multifactorial complex disorders such as cancer, diabetes, neurological, cardiovascular and pulmonary disorders that will be discussed further.

Stages of inflammation can change in hours so the process needs a strict gene expression regulation. During the basal state, genes are repressed until a diverse receptor system, like toll-like receptors (TLRs), senses a threat. Whenever a threat is noticeable, cells rapidly transcribe pro-inflammatory genes to induce acute inflammation. If the threat is transient, genes return to the basal state within hours. If it is severe, the incitement phase is replaced within 4-6 hours by a gene-specific epigenetic reprogram that may endure days to weeks. Duration may extent up to 21 days in human. This sustained epigenetic regulation period produces variable inflammatory clinical phenotypes and predicts outcome [2].

Nuclear factor kappa B (NF-κB) pathway is the first outstanding molecular mechanism, when we focus on to the molecular bases of inflammation. Canonical NF-κB pathway has complex roles in innate immunity, inflammation and oncogenesis, which is conserved in all multicellular animals [3]. As a part of the immune defense, NF-κB activation leads to target and eliminate transformed cells [4]. Microbial infections, TLR activation and proinflammatory cytokines such as interleukine (IL)-1 and Tumor Necrosis Factor-alpha (TNF-α) activate NF-κB pathway. This activation induces the expression of inflammatory cytokines, adhesion molecules, key enzymes of prostaglandin synthase pathway (COX-2), nitric oxide (NO) synthase, angiogenic factors and antiapoptotic genes [5]. Termination of the transcriptional activity of NF-κB is mainly achieved by NF-κB, up regulating its own inhibitors such as IκBα, A20 and CYLD [6, 7]. IκBα enters the nucleus and removes NF-κB from the DNA and sends it to the cytosol. In acute inflammation, this self-regulation usually ends with complete deactivation of NF-κB. However, in chronic inflammatory states, the persistent presence of NF-κB activating stimuli seems to outperform the inhibitory feedback circuits, rooting to an increased constitutive activity of NF-κB. Depending on the accessibility of the genome regulated by epigenetic and genomic alterations, different transcription factors can be expressed after eNF-κB activation. The crosstalk between different pathways leads development of cancers and other diseases [1, 3].

The importance of the interconnection between germ line DNA and epigenetic modifications in inflammation is still emerging. The inflammatory regulatory response requires complex regulatory networks at genetic and epigenetic levels. Thousands of inflammation and innate immunity genes contribute directly or indirectly in acute or chronic inflammatory processes. Transcription factors of the NF-κB, FOXP3, IRF, STAT families including epigenetic phenomena, play critical roles in the regulation of inflammatory genes. The proposed epigenetic mechanisms regulate acute and chronic inflammation in different ways. The epigenetic silencing state does not exist during chronic inflammation and the reason for this paradox is unknown, but it has important therapeutic implications. Therefore, we investigate the effect of epigenetic modifications in inflammation and related diseases as below.

1.1. DNA Methylation and Inflammation

DNA methylation, a dynamic process involving methylation and demethylation events, occurs in different regions of the genome and is extremely important for embryogenesis, cellular proliferation and differentiation [8]. In eukaryotes, DNA methylation only occurs at cytosine residues and catalyzed by the DNA methyltransferase (DNMT1, DNMT3a, and DNMT3b) enzymes [9]. In DNA methylation, cytosine is converted to methyl-cytosine at CpG site and about 60% of human genes have CpG islands at their promoters. The changes in DNA methylation, such as hypomethylation mostly associated with chromosome instability and activation of transposable elements, beside that hypermethylation of promotor associated CpGs often linked with suppression of gene expression and cancers [10]. DNA demethylation is necessary for the epigenetic reprogramming of somatic nuclei and active demethylation is associated with the activation of immune cells [11, 12].

DNA methylation is the most widely studied mechanism in autoimmune diseases. A range of disease associated with chronic inflammation, including systemic lupus erythematous (SLE) and rheumatoid arthritis (RA), display global hypomethylation associated with a concurrent decreased expression of methylation-related genes. Moreover hypomethylation patterns were also observed at interferon-regulated genes, such as *IRF5, IFIT2, STAT1* and *USP18*, in $CD4^+$ T cells in the context of SLE. However, not all epigenetic alterations in lymphocytes are subject to hypomethylation. It has been reported that regulatory regions of *FOXP3* in peripheral blood $CD4^+$ T cells from patients with SSc, RA and T1DM is hypermethylated, affecting the expression of key transcription factor which is required for the generation of regulatory T cells [13, 14].

Alterations in DNA methylation are known to cooperate with genetic events and to be involved in various diseases [15]. Therefore, understanding the roles of DNA methylation is important for understanding disease progression. Applications of next generation techniques have led to accumulate more data on DNA methylation in both normal and disease cells. Methylation is a reversible epigenetic mark and highly informative to described gene regulation in normal and diseased cells, and it can potentially function as a biomarker [16].

1.2. Histone Modifications and Inflammation

Histones are highly alkaline proteins that package and order the DNA into structural unites named nucleosomes. Nucleosomes consist of about 146 bp of DNA wrapped 1.65 turns around an octameric histone core, comprised of 2 copies each of histones H2A, H2B, H3 and H4. Nucleosomes are stabilized by linker histones H1 and H5 [17]. The core histones have a globular domain and protruding N-terminal tails, both of which are subjected to post-translational modifications [18, 19].

There are three well studied mechanism that modulate nucleosome structure and chromatin dependent cellular processes; core histone post-translational modifiactions, chromation remodeling and composition of nucleosomes. Core histone post-translational modifications, including acethylation, ubiquitination, phosphorylation, methylation, sumoylation and ribosylation affect histone interactions, change dynamics of nucleosome and subsequently regulate gene expression. In general, acetylation of lysine in the promoter regions of histone triggers the gene expression on the other hand; histone deacetylation, biotinylation, and sumoylation inhibit the gene expression. Methylation and ubiquitination mechanisms can be activator or repressor of gene expression depending on target histone residues [20, 21]. DNA methylation and histone methylation are tightly controlled entwined events. For instance, DNA methylation requires the histone H3 N-terminal tail with an unmethylated lysine 4 (H3K4). Tri-methylation of histones H3 on lysine 9 (H3K9me3) participates in DNA methylation mediated by DNMT1 [22].

Epigenetic modifications in chromatin's structure affect gene expression such loosely packed form of DNA which is called euchromatin, shows higher transcriptional activity in contrast, tightly packed form of DNA, heterochromatin presents lower transcriptional activity [23]. Various chromatin remodeling complexes can disrupt and remodel the nucleosome structure by increasing DNA accessibility [24]. Two groups of chromatin-modifying complexes have been described; ATP-dependent complex, covalent histone modifications. The balance of histone modifications is established and maintained by a range of enzymes including, but not limited to histone methyltransferases (HMTs) and demethylases (HDMs), histone acetyltransferases (HATs) and deacetylases (HDACs). Histone modifications are important regulators of gene transcription and modulate dynamic processes that affect nucleosomes. Long noncoding RNAs have also been shown to be necessary for targeting histone-modifying activities [25].

Aberrant histone modifications contribute to human diseases such as cancer, neurodegenerative diseases and infections. Including inflammation (IL-1β, TNF-α, LPS ect.) many stimulants can promote histone acetylation. Histone acetyl transferases (HATs) activate inflammatory genes, whereas histone deacetylase (HDACs) activity represses the inflammatory gene expression. Promoters of pro-inflammatory cytokines (such as IL-1, IL-2, IL-8, and IL-12) are rapidly acetylated and become transcriptionally active. HDACs regulate transcription of both pro-inflammatory and anti-inflammatory cytokines via their co-repressor complexes and transcription factors such as *FOXP3, STATs, GATAs, ZEB1*, and *NF-κB* [22].

In addition to nuclear function, histones also act like an endogenous signal when they locate at extra nuclear space. As a response to stress, immune cells, cerebellar neurons, Schwann cells and microglia present H2A, H2B, H3, H4, and H1 on their cell surface or cytoplasm. Levels of circulating histones as well as nucleosomes are increased in cancer, inflammation, and infection, which suggest that histones are therapeutic targets for infectious and inflammatory disorders [26, 27].

1.3. Non-Coding RNAs and Inflammation

1.3.1. Long Non-Coding RNAs

Long noncoding RNAs (LncRNAs) vary widely in size, genomic localization, and biogenesis and can be classified according to their molecular mechanism of action. LncRNAs function through a variety of different molecular mechanisms, and most commonly play a scaffolding role to promote proper recruitment and positioning of protein regulators both in the nucleus and in the cytoplasm. Loss-of-function approaches have found that lncRNAs regulate the biology of both innate and adaptive immune cells during inflammatory responses. Some nuclear lncRNAs that contribute to the regulation of DNA methylation in the context of inflammation have been identified. Some lncRNAs implicated in histone modification, particularly methylation, that play role in processes such as cell cycle, inflammation, and senescence [28].

Some inflammation related lncRNAs and their functions are summarized below;

NeST is involved in inflammation during microbial infection, binds to the H3K4 methyltransferase component WDR5 (WD repeat domain 5) and recruits it to the *IFNγ* locus. *IFNγ* methylation increases with advancing age, *NeST* could contribute to the inflammatory response and infection in elderly.

17A is up regulated in cerebral tissues derived from Alzheimer disease patients as well as in response to inflammatory stimuli such as *IL-1α*. *17A* is encoded within the *G-protein-coupled receptor 51* (*GPR51*) gene. *17A* provides an interesting link between GPRs and age-associated neurodegeneration.

Lethe, in mouse embryonic fibroblasts (MEFs), lncRNAs are found to be regulated following treatment with TNF-α, a pro-inflammatory cytokine, which is associated with inflammation, aging, age-related diseases, and cellular senescence. Among these, *Lethe* was found to be particularly important for the pro-inflammatory state of aging tissues by providing a key negative feedback loop: binding of *Lethe* to the NF-κB subunit RelA reduce the production of inflammatory proteins.

THRIL, TNF-α was also found induced by lncRNA *THRIL* (TNF-α- and hnRNPL-related immunoregulatory lincRNA). This lncRNA interacts with the TNF-α promoter and regulates its expression in *THP1* macrophages. Together, lncRNAs *lethe* and *THRIL* are involved in inflammation in a TNF-α-dependent manner.

Lnc-IL7R, was identified as a regulator of the lipopolysaccharide (LPS)-induced inflammatory response. Depletion of *Lnc-IL7R* reduces trimethylation of histone H3 at lysine 27 (H3K27me3) causing a reduction in the levels of inflammatory mediators including E-selectin, VCAM-1, IL-6 and IL-8.

The TGF-β/Smad3 pathway is involved in the inflammatory response. It has been identified that Smad3 associated lncRNAs are related to renal inflammation. The analysis identified several lncRNAs altered in Smad3 knockout mice. As this pathway is impaired with age and Smad3 regulates senescence phenotype, this network of proteins and lncRNAs differentially expressed may play a role in aging related inflammation (‘inflammaging’).

1.3.2. Micro RNAs

MicroRNAs are small, single-stranded ncRNAs and function to repress gene expression and influence virtually all organ systems in vertebrates. As an epigenetic regulatory mechanism, miRNAs, enable increased human complexity despite a genome size that is similar to less complex organisms. This appears to include critical roles in establishing proper inflammatory set points and facilitating optimal responses and resolution by our immune system.

Recent studies demonstrated that aberrantly expressed miRNAs exert a significant role in inflammatory genes. RNA interference mechanism culminates with the up or down regulation of hundreds of immune response genes, an essential first step in coordinating inflammatory responses. miRNAs act as a modulator of inflammation signals or as mediators between inflammation and cancer. The experimental data indicated that miRNAs have been shown to be expressed in immune cells, to target proteins involved in inflammation regulation, and consequently to affect the severity of the response. There are two major ways for the regulations: To impacting the development of inflammatory cell subsets (e.g., Th2 versus Th17) or to establish the level of immune cell function (e.g., controlling how much cytokine is made by DCs).

Certain fundamental attributes of miRNAs make them ideally suited to regulate chronic inflammatory conditions. MicroRNAs modulate immune cell differentiation and responses. Inflammation related miRNAs play important roles in various inflammatory processes. The opposing effects of miRNAs on inflammatory responses indicate that, the immune system utilizes multiple miRNAs to properly balance its functional capacity, creating a tension between activation and repression that can be precisely tuned [29]. It is still unknown how multiple miRNAs are in cooperation to balance inflammatory responses, and how miRNA networks can work together to optimize responses.

miRNAs have been implicated in regulating macrophage lineage skewing during inflammatory responses. Macrophages can be skewed toward either pro-inflammatory subtypes (M1), or less inflammatory subtypes (M2). *MiR-125* has been shown to repress M1 skewing while promoting the M2 fate. *MiR-223* has also been implicated in control of granulocyte activation, and *miR-223-/-* mice display overactive immune responses and develop inflammatory lung pathology. Age dependent inflammatory phenotypes are being explored among miRNAs. It has been discovered that mice lacking *miR-146a* develop an age-dependent, chronic inflammatory disease that is spontaneous, life shortening, associated with inflammatory cytokines and autoantibodies, and that involves a variety of hematopoietic abnormalities and malignancies associated with the aging [30].

Details of epigenetic mechanisms, which play driver roles in inflammation and related diseases, will be discussed as below.

2. Epigenetics and Inflammatory Disorders

Epigenetic modifications in inflammatory autoimmune diseases have been recognized and attracted recently. Although there are many components (such as genomic variations and environmental factors), which lead predisposition of autoimmunity and related phenotypes, there are several reported epigenetic alterations as well.

The inflammatory response is a double-edged sword; immune responses are necessary for immune defense, wound repair and tissue homeostasis, if these responses become deregulated, they initiate a chronic reaction leading to chronic status. This condition, referred as *chronic low-grade inflammation*, and adversely contributes to many diseases such as aging, obesity, type 1 diabetes (T1D), rheumatoid arthritis (RA), systemic lupus erythematous (SLE), neurodegeneration, and cardiovascular diseases (CVD) [13].

Epigenetic alterations contribute immune response and those alterations related autoimmune diseases are summarized in Table 1. In the last decade, epigenetic studies provide new insights into autoimmune diseases. The identification of specific epigenetic markers may lead to lighten uncharacterized mechanisms. Identification of clinical significance of the epigenetic markers may be utilized for diagnostic biomarkers and therapeutic benefits for the autoimmune disorders. In this section we will give some examples to the epigenetic regulations, malfunctions and markers of known inflammatory disorders;

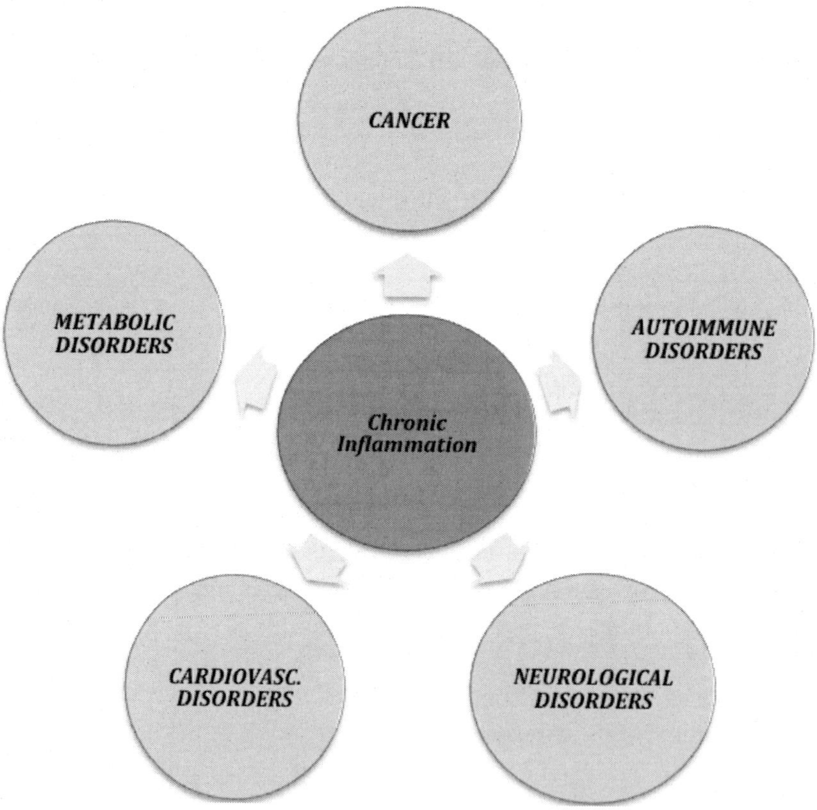

Figure 1. Inflammation mediated diseases.

2.1. Inflammatory Bowel Disease (IBD)

Inflammatory bowel disease (IBD) is a group of inflammatory conditions of the colon and small intestine. IBD is classified as an autoimmune disease, and Crohn's disease (CD) and ulcerative colitis (UC) is the main types of the IBD. Deregulated immunologic response causes IBD and subsequently long lasting active IBD evokes to development of colorectal carcinoma in many cases. So far the data supports the notion that epigenetic factors can act as fine-tuners of gene expression in diseases such as in IBD. Particular combinations of epigenetic changes could result in a pathogenic phenotype through activating genes that promote chronic inflammation or inhibiting anti-inflammatory gene expression in IBD. Regarding the interaction between environmental, genetic and epigenetic mechanisms seem to have great importance in the IBD. Intestinal micro biota are likely the most important environmental factor in IBD as targets of the inflammatory response [31]. It is a challenge to measure the impact of the environmental factors that potentially can contribute to IBD; thus far epigenetic factors can mediate gene-environment interactions involved in pathogenesis.

Intestinal bacteria can regulate the epithelial gene expression and the immune response by epigenetic mechanisms. Bacterial and host epigenetics can also affect host genetics to trigger inflammatory processes. A number of studies have similarly reported the bacteria induce histone modifications (see in part 2.6), thereby inflammatory response modulation and other cellular processes of the epithelium in the gastrointestinal tract [32].

The first evidence of DNA methylation, as an epigenetic mechanism in UC was reported in 1996 and found that incorporation of the 3^{H}-methyl groups into DNA was 10-fold higher in UC patients than controls and significantly higher in histologically active than inactive disease. A progressive increase in methylation status was found associated with neoplastic progression in UC [33]. Another reported promoter methylation on the *E-CADHERIN* gene, which is coding a specific calcium ion-dependent cell adhesion protein, was detected in 93% of dysplastic samples from UC. The result was confirmed by *E-cadherin* synthesis levels, found reduced in samples with dysplasia and normal in samples without dysplasia, suggesting that long standing inflammation is related to hypermethylation of the gene promoter and that DNA methylation may be used as a biomarker for detecting high risk patients for developing colorectal cancer [34].

Both targeted gene methylation and methylome array based studies have detected numerous colonic mucosal associates of inflammation in IBD [35, 36]. Several studies confirmed that *ER, MYOD, GSPG2* and *p16* genes were found hyper methylated in dysplastic than normal epithelium. Severe disease phenotypes of UC was found related to *MDR1* hypermethylation [37]. A tumor suppressor gene death-associated protein kinase (*DAPK)* hypermethylation was also found to be associated to the inflammation of mucosa in UC patients and a correlation is observed between higher methylation and severe inflammation [38]. DNA methylation has been found to be related to many different clinical aspects, such as disease severity, disease duration, disease phenotype, disease extent, steroid use, steroid dependence or refractoriness, age of onset, number of hospitalizations and finally active inflammation and dysplasia [39, 40].

There is limited data on concerning the contribution of DNA methylation status in CD pathogenesis. Lin et al. compared normal to inflamed tissue from CD and UC patients and found significant differences in DNA methylation of several CpG loci such as S*ERPINA5, TNFRSF1A, AATK, GABRA5, MAPK10* and *STAT5A [35]*. One of the first genome wide methylation studies analyzed the methylation profiles of peripheral blood samples from women and children with CD. Several genes identified differentially methylated in patients with IBD and controls. Ontology analysis highlighted several pathways associated with IBD, including immune system processes, immune response, and host response to bacteria, whereas canonical pathway analysis indicated the involvement of Th17 cell pathways [41].

Histone modifications are complex processes and less extensively studied in IBD than DNA methylation. The patterns of histone acetylation in the colon of rats with colitis and humans with CD have been described and increased acetylation of H4 (at lysine residues 8 and 12) was determined in the inflamed mucosa and in Peyer's patches [42]. A gene-specific pattern of histone modification is observed for collagen type I, the most abundant component of the fibrotic extracellular matrix, can be induced by fibrosis relevant cytokines (IL1b, transforming growth factor beta, and tumor necrosis factor alpha) suggesting that epigenetic factors regulate fibrotic gene transcription relevant to IBD with a fibrostenosing phenotype [32] Several anti-inflammatory mechanisms of HDACi have been proposed, one of these is T-regulatory cells, which mediate immune tolerance and repress the excessive inflammation, is linked to FOXP3 gene expression. Administration of HDACi to increased T-regulatory cell differentiation and suppression of bowel inflammation, as a result of acetylation of lysines in the Foxp3 [43]. Another mechanism could involve their effects on acetylation of proteins in the NF-kB pathway [44].

Studies in animal models have shown that intestinal miRNAs regulate gut homeostasis [45]. Studies reported miRNA expression profiles in the peripheral blood and gut biopsy specimens of healthy controls versus adult/pediatric patients with active and inactive IBD. The miRNA expression profiles found deregulated at both the tissue level and in peripheral blood of UC and CD patients. Moreover IBD-related glucocorticoid therapy can modify the expression profile of different miRNAs [46, 47]. Specific miRNA profiles and identification of their targets might provide additional insight into IBD pathogenesis and help to predict IBD susceptibility, progression, relapse, and response to therapy. The miRNAs associated with IBD could also serve also as fine tuners of gene expression. The levels of specific mRNAs shown to be tightly controlled by miRNAs, alterations of the threshold mRNA levels results in marked changes in protein synthesis. Deregulation of the intestinal inflammatory response that results from the disruption in the balance between miRNA activity and its specific target mRNAs, e.g., genes with important functions in intestinal homeostasis [48].

Changes in miRNAs in human IBD were first described in in sigmoid biopsy specimens from patients with active UC. *MiR-192*, which is expressed in colonic epithelial cells, was found significantly reduced in tissues of patients with active UC. MiR-*192* levels were found increased in colon tissues of patients with UC leading to the expression of *macrophage inhibitory peptide 2a*, a CXC chemokine expressed in the epithelial cells [49]. *MiR-15*0 is up regulated in mice with dextran sulfate sodium induced colitis in colon tissues from patients with UC; its levels correlate inversely with those of its target c-Myb, which has a role in apoptosis [50]. Another miRNA, *miR-21*, which promotes inflammation, has been reported in several studies of patients with active UC and CD colitis, along with *miR- 155* [51]. *MiR-196* is overexpressed in the inflamed epithelium of patients with CD [52]. Several micro RNAs have been found to be significantly up-regulated or down-regulated in 2 or more studies, including *miRs-16, miR-28, miR-149, miR-151, miR-199*, and *miR-534* in UC and CD [53] There are several other miRNAs found related to intestinal fibrosis; *miR-29, miR-200b, miR-21, miR-192* or associated with epithelial barrier and immune function in IBD pathogenesis; *miR-192, miR-21, miR-126, miR-155, miR-106a* [48].

The importance of epigenetic factors in IBD derives from the mechanisms of action of some therapeutics used for the treatment of IBD. There is increasing evidence that glucocorticoids change the chromatin structure of genes that code mediators of inflammation.

Glucocorticoids can increase transcription of anti-inflammatory genes which results from the uncoiling of DNA wound around histone by acetylation of the histone residues. Glucocorticoids may also lead to deacetylation of histones on the genes of pro-inflammatory cytokines, resulting in tighter coiling of DNA and reduced access of transcription factors to their binding sites, and suppressing the target gene expression [54].

2.2. Rheumatoid Arthritis (RA)

Another chronic auto inflammatory disease, rheumatoid arthritis (RA), is characterized by synovial hyperplasia, chronic inflammation and progressive destruction of the synovial joints. RA has an unknown etiology and affecting about 1% of the population [55]. There are two subsets of RA, according to the existence of the antibody against citrullinated peptide antigens (ACPA). The presence or absence of ACPA is one of the best clinical predictors for disease outcome. The most important genetic risk factors, accounting for up to 50% of the overall risk for RA, are mainly confined to the human leukocyte antigen locus [56]. All kinds of epigenetic alterations occur in RA including DNA methylation, histone acetylation, and miRNAs has contribution [57] and continuation of the disease [58, 59].

The RA synovial fibroblasts (RASFs) play a major role in the initiation and continuation of the disease [58, 59]. These cells display an activated, aggressive and invasive phenotype [60, 61]. A global hypomethylation of the genes on these cells could be responsible for the overexpression of inflammatory cytokines in synovial fluid leading to disease phenotype [62] and these epigenetic biomarkers might provide the missing link between RA, risk factors and a lack of therapy response. Global DNA hypomethylation has also been observed in T cells and peripheral blood mononuclear cells (PBMCs) from RA patients [63].

L1 is one of the major classes of repetitive elements in the genome. Since they are methylated in the normal synovial tissue they used as biomarkers. In synovial tissue from patients with RA, L1 is hypomethylated as a consequence of reduced expression of *DNMTs*. This reduction of methylation in inflammatory response promoter genes causes an overexpression of growth factors and receptors, adhesion molecules, and cytokines, and cause irreversible phenotypic changes in synovial fibroblasts. The other hypomethylated gene, *IL-6*, is a proinflammatory cytokine that participates in B cell response. When it is hypomethylated, an overexpression of *IL-6* will occur and trigger the overexpression of pro-inflammatory cytokines at the same time and this is associated with a local hyper activation of the inflammation circuit [64, 65]. Another hypomethylated gene, *IL-1* is associated with elevated *IL-10* expression in RA and by regulating *IL-10* transcription it can be responsible for the pathogenesis of RA. Not only global but also single gene hypomethylations, such as *chemokin ligand 12 (CXCL-12)* in RASFs and *(Interleukin 1 receptor, type II) IL1R2*, has also been reported in RA [66, 67]. Genome-wide evaluation revealed many hypomethylated loci in RASFs including *CHI3L1, CASP1, STAT3, MAP3K5, MEFV*, and *WISP3* [12].

Besides hypomethylation, CpG hypermethylations in the regulatory regions of some genes lead to aberrant transcription in RA. Hypermethylation of *EBF3* and *IRX1* genes interacts with transforming growth factor beta (TGF-β) pathway components and reduce mRNA expression in RASFs [68]. The promoter region of the *death receptor 3 gene (DR3)*, a member of the apoptosis inducing FAS family, was shown to have significant hypermethylation of CpG islands in RA patients, which may enhance resistance to apoptosis in RA synovial fibroblasts [57, 69].

The gender bias (3:1/F:M) of RA suggests an X-chromosome role in the development of RA. An increased presence of a skewed X-chromosome inactivation pattern has been shown in the peripheral blood of a RA patient. Instead of a random X-chromosome inactivation, 80% or more of the cells exhibited inactivation of the same X-chromosome. CD40 ligand (CD40L) is primarily expressed at the surface of activated $CD4^+$ T helper cells and plays a crucial role in the immune response. A promoter hypomethylation and increased expression of CD40L has been detected in $CD4^+$ T helper cells of female RA patients. Hypomethylation of genes on the inactive X chromosome may contribute to female-biased RA patients [70, 71]. The exposure to *IL-1* decreases the mRNA levels of *DNMT1* and *DNMT3a* in RASFs and also inhibits global methylation of RA fibroblast-like synoviocytes (FLS), suggesting the contribution to differential DNA methylation in RASFs through altered expression of *DNMTs*.

RA synovial tissue is characterized by an imbalance between HAT and HDAC activity. Cartilage destruction that is mediated by matrix metalloproteinases (MMPs) and enzymes from the ADAMTS family, are regulated by chromatin modifications and histone acetylations. HDAC inhibitors inhibit cartilage degradation by blocking the induction of certain MMPs by proinflammatory cytokines. Hyper acetylation of synovial cell histones, induce cyclin dependent kinase inhibitors (p16 and p21) expression via decreased *tumor necrosis factor-alpha (TNF-α)* expression [72, 73]. Additionally the hyper acetylation of histones down regulate HIF-1a (hypoxia inducible factor) and vascular endothelial growth factor (VEGF) to block angiogenesis in synovial cells [74].

DNA methylation has been shown to regulate the microRNAs (miRNAs) expressions in RA. Demethylation with 5-azacytidine (5-azaC) increases the expression of *miR*-203 in RA [75]. The increased levels of *miR-203* result in elevated secretion of MMP-1 and IL-6 via the NF-jB pathway, and thereby, contribute to the activated phenotype of RASFs. MeCP2 is a chromatin-binding protein that preferentially binds to methylated CpGs and is highly abundant in the synovium of RA rats. It has also been shown that increasing global methylation levels are regulated by overexpression of *MeCP2* in a RA rat model [76]. The promoter of *miR-124a* hypermethylation implicated in the proliferation of FLS [77]. The recent genome-wide methylation studies have reported hypomethylation and hypermethylation of multiple genes in tissue-derived FLS from patients with RA [78-82].

Micro RNAs have also been implicated in RA [63]. Primary human samples have increased expression of *miR-155, miR-146*, and *miR-223* and mouse arthritis models have demonstrated the functional relevance of *miR-155* and *miR-182* for regulating B and T cell function during disease [83-86]. *MiR-146* and *miR-155,* are shown to be induced by proinflammatory stimuli such as *IL-1, TNFα*, and *Toll-like receptors (TLRs)*. Another miRNA implicated in RA is *miR-124*, which targets *cyclin-dependent kinase 2 (CDK-2)*. In the basal state, *CDK2* represses cell proliferation and arrests the cell cycle at the G1 phase, but in pathologic conditions such as RA, its level decreases. *MiR-124* also targets *monocyte chemo attractant protein 1 (MCP-1)*, which is responsible for mononuclear phagocytes into the joint. Thus this miRNA increases cell proliferation and MCP-1 production in RA [87]. *MiR-203* is overexpressed in RASF, enhancing the secretion of MMP-1 and interleukin-6 [75]. Tumor necrosis factor-α induces *microRNA-18a* and activates RASF through a feedback loop in NF-κB signaling [88].

2.3. Systemic Lupus Erythematous (SLE)

Systemic Lupus Erythematous (SLE) is a systemic autoimmune disease with multiple organ involvement, which is characterized by autoantibody response to nuclear or cytoplasmic antigens. It is defined with the production of increased amount of autoantibodies, which potentially drive immune complex related inflammation in various tissues and organs [89]. Several genetic and epigenetic factors contribute to the development of SLE. Epigenetic disruption results in overexpression of certain genes of the key immune cells, such as T cells, has been associated with the pathophysiology of SLE. Self-antigens resulting in inflammation mediated multiple organ damage, displays global H3 and H4 hypoacetylation in the $CD4^+$ T cells of patients [90, 91].

Like RA, SLE is more common in females with peak disease prevalence between menarche and menopause. Estrogen increases the risk of SLE in women via increasing the type 1 interferon (IFN) production and survival of auto reactive B-lymphocytes [92, 93]. Epigenetic modifications might explain the female prevalence for SLE. Impaired DNA methylation presented on the inactive X-chromosomes of SLE patients was suggested to contribute to the female predominance of SLE [94].

The roles of DNA methylation in SLE first described in the 1960s and since then it become a very hot research topic [95]. Several studies have shown that there is a global hypomethylation of promoter regions, which contain the genes that are overexpressed in the SLE like *ITGAL, CD40LG, PRF1, CD70, IFGNR2, MMP14, LCN2,* and in the ribosomal RNA gene promoter (18S and 28S) [94, 96, 97]. The DNA hypomethylation may also affect the chromatin structure of T-cells and leads the overexpression of these genes. The genes overexpression will cause cell hyperactivity and perpetuation of inflammatory response [98]. Genome-wide methylation studies have provided new insights into the role of DNA methylation in the pathogenesis of SLE. The regulatory regions for interferon-responsive genes in naive $CD4^+$ T cells from patients with SLE are hypomethylated and 'poised' for expression; transcription does not occur until T-cell activation [99].

Histone modifications in SLE have been studied in animal models and in human diseases. These studies showed that, during apoptosis, histones could be modified to make them immunogenic. The antibodies are directed against components of the cell nucleus that are exposed at the cell surface during apoptosis [100, 101]. The nucleosomes, the primary pioneer antigen in SLE, are released in patients with SLE as a result of a corrupted apoptosis or an insufficient clearance of apoptotic residue. During apoptosis, the nucleosome is modified and creates more immunogenic epitopes leading to the formation of autoantibodies against unmodified chromatin components [102]. Histone modifications such as histone 3 lysine 4 trimethylation (H3K4me3), histone 3 lysine 8 (H4K8ac) triacetylation, histone 3 lysine 27 trimethylation (H3K27me3), and histone 2B lysine 12 acetylation (H2BK12ac) cause an increase in apoptotic nucleosomes. These apoptotic nucleosomes generate auto-immunogenicity that can cause activation of antigen presenting cells and autoantibody production with a subsequent inflammatory response [103]. There are other studies that have shown acetylation patterns of histone H3 and H4 in active SLE $CD4^+$ T cells [104]. Monocytes, which are important in SLE renal disease, have been shown to have an altered acetylation pattern of histone H4 thus increasing the expression of interferon (IFN) genes that play a key role in SLE pathogenesis [89, 105].

Epigenetic reprogramming has potential preventive and therapeutic purposes for inflammation and autoimmunity. Two methylation inhibitors, procainamide and hydralazine, were found to induce a lupus-like disease after long-term administration in wild-type mice, and the disease disappeared after the withdrawal of these drugs. DNA demethylation has been found in SLE CD4$^+$ T cells, but not in CD8$^+$ T cells or peripheral blood mononuclear cells (PBMCs) [106]. The recent researches have focused on specific cell types as well as certain SLE relevant genes. Among the other inhibitors procainamide, is a competitive inhibitor of DNMT1 and hydralazine inhibits T and B cell signal regulated kinase pathways that play an important role in the regulation of methylation [107, 108]. These two therapeutic mechanisms produce a reduction in DNMTs that will enhance the genetic expression of adhesion molecules on lupus-drug-induced lymphocytes [109].

Several studies suggest a role for miRNAs in the pathogenesis of SLE. MicroRNAs control the differentiation and function of B cells, which are considered key elements in the pathogenesis of systemic lupus erythematosus (SLE) [110]. For example the under expression of miR-1246 through the AKT-P53 signaling pathway, leads to the expression of *EBF1*, and promote the activation of B cells [111].

Peripheral blood mononuclear cells from patients with SLE showed differential expressions in some miRNAs (increased *miR-189, miR-61, miR-78, miR-21, miR-142-3p, miR-342, miR-299-3p, miR-198*, and *miR-298* and decreased *miR- 196a, miR-17-5p, miR-409-3p, miR-141, miR-383, miR-112*, and *miR-184*) [112]. The expression of *miR-21* which targets an important autoimmune gene *RASGRP1,* is increased in CD4$^+$ T cell of SLE patients and mediates the RAS–MAPK pathway, down regulating *DNMT1* expression [113]. Similarly, increased levels of *miR-21* found contributed to the B cell hyper responsiveness and aberrant T cell response in SLE by targeting the *PTEN* and *PDCD4* respectively [114-116]. Besides *miR-21, miR-126* was also reported to modulate DNA methylation in SLE CD4$^+$ T-cells by directly targeting DNMT1 [106].

MiR-146a is a well-recognized negative regulator in autoimmunity. *MiR-146a* was found under expressed in SLE [117] and this was possibly due to a germline genetic variant in the *miR-146a* promoter [118]. This genetic variant (rs57095329) in the promoter region of *miR-146a* confers association with its expression level. The SLE risk-associated G allele of the variant is linked to a reduced miR-146a expression in PBMCs, possibly by affecting the protein binding affinity and activity of the promoter. A reverse correlation of *miR-146a* levels with the expression of interferon inducible genes and SLE disease activity was reported, indicating a critical role of *miR-146a* in excessive type I interferon production and signaling activity in SLE [117]. *MiR- 146a* has been shown to negatively modulate type-I IFN production in macrophages through targeting TLR signaling genes *TRAF6, IRAK1* and *IRAK2* [119]. *MiR-146a* is often bundled with TLR7/9 stimulation, which plays a central role in SLE pathogenesis. Apart from PBMCs, TLR7/9 stimulation in plasmacytoid dendritic cells (pDCs) also induces *miR-146a* expression [120].

SLE patients have multiple clinical features caused by inflammation. *MiR-23b* was found under expressed in the affected tissues of patients with SLE. It has been reported that miR-23b targets *TAB2, TAB3* and *IKK-α* to inhibit *IL17, TNF* or *IL1β* signaling [121]. Many miRNAs contribute to the reduced IL-2 production in lupus patients. A significant reduction in *miR-31* expression in SLE T cells is positively correlated with the lowered IL-2 production [122].

2.4. Multiple Sclerosis (MS)

Multiple Sclerosis (MS) is another disease characterized by myelin destruction followed by a progressive degree of neurodegeneration. Inflammation is an important mechanism of CNS damage in MS, and proper transcriptional control of the immune responses is mediated in part by epigenetic regulation.

Peptidyl arginine deiminase type II (PADI2) has been found hypomethylated in MS patients [123]. *PADI2* plays a key role in the citrullination process of myelin basic protein (MBP) that promotes protein auto cleavage and increases the probability of creating new epitopes and also modulates the immune response. In MS, an increase has been found in demethylase enzyme activity, which will cause hypomethylation of the *PADI2* promoter region [123, 124]. Hypomethylation increases *PADI2* expression and MBP citrullination process via the production of immunodominant peptides, which will increase the autocleavage of MBP, causing irreversible changes in its biological properties, and will produce proteolytic digestion, myelin instability, and a chronic inflammation response [125]. *SHP-1* gene is the negative regulator of pro-inflammatory signals. Increased promoter methylation in leukocytes in more than one-third of the MS subjects leading to decreased *SHP-1* expression and increased leukocyte-mediated inflammation [126]. $CD4^+$ T-cells of relapsing-remitting multiple sclerosis patients showed *FOXP3* and *IL-17A* demethylation. *FOXP3* demethylation inhibits T_h1 and T_h2 cell differentiation and promotes T_h17 cell lineage commitment, whereas the *IL-17A* leads to increased differentiation towards the T_h17 cell lineage. Epigenetic control of the Th1/Th2/Th17 balance through to determine disease status [127, 128]. Genome-wide methylome analysis confirmed a global DNA hypermethylation of $CD8^+$ T cells for MS patients [129].

The oligodendrocyte identity is modulated by posttranslational modifications of histones. In patients with MS, there is a shift toward histone acetylation in the white matter. Hyperacetylation of H3 in the promoter region of inhibitory genes produce high levels of transcriptional inhibitors of oligodendrocyte differentiation such as *TCF7L2*, *ID2*, and *SOX2* [130]. H3 hyperacetylation is accompanied by an increase in number and activation state of microglia and astrocytes. While the expression of histone deacetylases showed *HDAC8* and *HDAC11* up-regulation, only a subgroup of patients showed overexpression of a number of genes including *TCF7L2, SOX2* and *IDS* [131].

Many studies have focused on miRNA involvement in MS pathogenesis. Among those, the *miR-326* plays a critical role in the pathogenesis of MS via upregulation of the Th-17 cell differentiation by targeting Ets-1, which acts as a negative regulator of Th-17 differentiation. *MiR*-326 was found significantly upregulated in patients with relapsing-remitting MS, which produced more severe symptoms [132]. Other miRNAs involved in MS are *miR-34a* and *miR-155*, are also upregulated in active MS lesions and contribute to MS pathogenesis by targeting CD47, which is a "don't eat me" signal Macrophages with low levels of CD47 are released via inhibitory control signals, leads to increased phagocytosis of myelin. *miR-1*55 also promotes development of inflammatory Th1 and Th17 cells [133].

Differentially expressed *miR-17-5p, miR-497, miR-193*, and *miR-126* have been identified in different lymphocyte subsets including $CD4^+$ T cells, $CD8^+$ T cells, B cells, and $CD4^+$ $CD25^+$ Treg cells from patients with MS. Direct involvement and contribution of deregulated miRNAs in MS still needs additional investigation [133, 134]. It is noteworthy that all miRNAs are involved in the pathogenesis of the disease. There are miRNAs that can serve as prognostic markers such as the expression of *miR-18*b and *miR-599* is related to relapse and miR-96 is involved in remission [135]. *MiR-124*, which is expressed in microglia, is brain specific and it might reduce the activation of myelin-specific T cells with a marked suppression of the disease, which would make it a key regulator of microglia quiescence and a good prognostic factor for MS [136].

2.5. Diabetes

The environmental factors and nutrition play an important role in the pathogenesis of diabetes. However, the knowledge on molecular mechanisms by which these factors trigger β-cell dysfunction and diabetes are still limited. Recent discoveries give an impression that epigenetic changes in response to environmental stimuli may play an important role in the development of diabetes. Chronic exposure to high concentrations of glucose, leading to oxidative stress and inflammation, may induce changes in the regulation of gene expression on insulin secretion and increased apoptosis. The combination of inflammation, mitochondrial dysfunction and ROS production provides an attractive model for the pathogenesis of diabetes and may explain the widespread alterations of epigenetic regulation of gene expression during the glycemic control [137, 138].

Type 1 Diabetes (T1D) is a T-cell-mediated autoimmune disease that develops in genetically susceptible individuals and affects their endocrine pancreas. There are some mechanisms by which epigenetics may play an important role in T1D by modulating lymphocyte maturation, cytokine gene expression and by differentiation of subtype Th cells ruled by epigenetic controls. In this autoimmune disease, in contrast to SLE and RA, there is a global hypermethylation activity caused by altered metabolism of homocysteine [139]. Glucose and insulin levels are determinants of methylation, they alter homocysteine metabolism by increasing cell homocysteine production [140, 141]. When there is an increase in the levels of homocysteine, methionine in cells will be catalyzed by DNMTs in S-adenosylmethionin, which will enhance DNMT activity, and lead to increased global DNA methylation. Elevation of maternal homocysteine during pregnancy as a result of a low protein diet can produce an altered methionine metabolism that will cause a decrease in islet mass and vascularity in the fetus with a subsequent glucose intolerance in adult life [142]. The reduced Foxp3 expression was found to be associated with Foxp3 promoter hypermethylation in adult autoimmune diabetes patients [143].

There are limited studies associated with histone modifications and T1D. Patients with T1D show a subset of genes with an increase in histone 3 lysine 9 dimethylation (H3K9me2) in lymphocytes. This subset of genes includes the *CLTA4,* which is a T1D susceptibility gene and has increased methylation of *H3K9* in its promoter region. Other genes that have altered H3K9me2 are *transforming growth factor-beta (TGF-β), NF-κB, p38 (mitogen-activated protein kinase), toll-like receptors (TLRs*), and *IL-6*. The transcription factor *NF-κB* is also upregulated by H3K4 methyltransferase thus causing an increase in inflammatory gene expression in diabetic mice. All these genes are associated with autoimmune and inflammation related pathways [144, 145]. Histone modifications are also among the mechanisms that cause cardiovascular complications in T1D patients. Chemical modification of H3K4 and H3K9 has been found to be related to the gene expression conferred by hyperglycemia. Transient hyperglycemia promotes gene activating epigenetic changes and signaling events critical in the development and progression of vascular complications. These epigenetic changes are H3K4 and H3K9 methylation in genes associated with vascular inflammation [146, 147].

There are some hypotheses that the function of regulatory T cells (Tregs) is influenced by changes in the expression of specific miRNAs in T1D. *MiR-510* expression was found to be increased and *miR-342* and *miR-191* expressions were found to be decreased in Tregs of diabetic patients. There are other studies, which demonstrate that miRNAs may be the cause of cytokine mediated beta-cell cytotoxicity. This cytotoxicity is achieved when IL-1B and TNF-α induce the expression of *miR-21, miR-34a,* and *miR-146a* in pancreatic islets thus producing beta-cell failure by increasing proinflammatory cytokines [148]. *MiR-101a* and *miR-30b* are key players and contribute to cytokine-mediated β-cell dysfunction in the development and progression of type 1 diabetes. IL-1β induces *miR-101a* and *miR-30b* expression and participates in β-cell dysfunction, including decreased insulin content, gene expression, and increased β-cell death. *MiR-101a* and *miR-30b* reduce pro-insulin expression and insulin content by directly targeting the transcriptional factor Neurod1. In addition, these miRNAs associated with diminished expression level of the antiapoptotic protein Bcl2 [149]. A significant increase of *miR-21* has been observed in the plasma and urine of pediatric T1D. In a type 1 diabetes mouse model, high glucose induced *miR-21* de-repress TORC1-pathway and leads to renal pathologies [150]. In a type 2 diabetes model *miR-21* down regulates phosphorylated SMAD-7, leading to up-regulation of TGF-β and nuclear factor-β signaling pathways, increasing renal fibrosis and inflammation [151].

2.6. Bacterial Infections

Recent studies revealed that bacteria could affect the chromatin structure and transcriptional program of host cells by influencing diverse epigenetic events (i.e., histone modifications, DNA methylation, chromatin-associated complexes, noncoding RNAs, and RNA splicing factors). Bacterial-induced epigenetic deregulations may affect host cell function either to promote host defense or to allow pathogen persistence. Their effects might generate specific, long-lasting imprints on host cells, leading to a memory of infection that influences immunity. The selective activation or silencing of specific genes not only depends on transcription factors, but also on their cross talk with epigenetic modulators, which regulate DNA accessibility by controlling the chromatin structure.

The importance of DNA methylation events associated with bacterial infections is attracting attention. The best documented example is H. pylori infection leading aberrant DNA methylation in the human gastric mucosa, strikingly at promoters of genes found to be methylated in gastric cancer cells [152]. H. pylori associated hypermethylation is seen at the *CDH1, USF1, USF2, WWOX, MLH1* genes [153]. H. pylori mediated inflammation triggered lymphocyte and macrophage infiltration, which appears to have a key role in induction of methylation [154]. Among signals resulting from chronic inflammation, elevated levels of interleukine 1b (IL1b) and nitric oxide (NO) are proposed to contribute to influence the recruitment of DNMTs.

In the intestine, following chronic inflammation, some *PRC2* target genes are subject to aberrant DNA methylation [155]. Bacteria-induced DNA methylation can also affect genes involved in cell proliferation. In human uroepithelial cells, infection with uropathogenic

Escherichia coli (UPEC) results in the up regulation of DNA methyltransferase (DNMT) activity and DNMT1 expression, and induces CpG methylation and down-regulation of *CDKN2A*, a G1-cell-cycle inhibitor regulator [156]. This may increase uroepithelial cell proliferation and pathogen persistence, by infection stimulated host cell apoptosis.

Other organs can also be targeted by bacteria-mediated epigenetic changes, including placenta. Maternal infection with Campylobacter rectus induces hypermethylation of the imprinted *IGF2* gene promoter in murine placental tissue [157] suggesting that bacterial infections during pregnancy might epigenetically affect genes that play an important role in fetal development.

Mechanisms controlling gene expression at the chromatin level show high levels of complexity. Various bacterial products can affect them in many ways, through activation of signaling cascades or directly in the nucleus. The most of the reported chromatin modifications induced by bacteria are histone modification (acetylation/[deacetylation and phosphorylation/dephosphorylation) events generated through activation of host cell signaling cascades. Among the host signaling pathways that certain bacteria activate, MAPKs, NF-kB, and PI3K pathways are known to activate the kinases. B. vulgatus induces TGF-1β anti inflammatory pathway, and H3 de-acetylation via HDAC recruitment at proinflammatory gene promoters [158]. Bacterial toxins also dampen the host innate immune responses by inhibiting H3 phosphorylation/acetylation. The lethal toxin (LT) from Bacillus anthracis, the agent of anthrax, cleaves and inactivates MAPKKs leading to disruption in MAPK signaling [159].

H. pylori is also capable of modifying histones via the modulation of the MAPK pathway. The peptidyl prolyl cis, trans isomerase HP0175 secreted by H. pylori binds the innate immunity receptor TLR4 and activates MAPKs in monocytes [160]. Another study showed that exposure of H. pylori to gastric epithelial cells promotes release of HDAC1 from the promoter of the cell-cycle regulator gene $p21^{WAF}$, hyperacetylation of H4, and increased expression of $p21^{WAF}$. Chromatin alterations might contribute to the effects of H. pylori on cell cycle progression, cellular proliferation, and cell death [161].

Pore-forming toxins use another signaling pathway to change histone marks. He-La cells, L. monocytogenes listeriolysin O (LLO) promotes histone deacetylation and dephosphorylation in a subset of immune genes such as *CXCL2, MKP2* and *IFIT3* which cause repression [162]. Other pore-forming toxins, such as PFO of Clostridium perfringens, PLY of Streptococcus pneumonia, and Aerolysin from Aeromonas hydrophila have similar effects, leading to histone modification [163].

Bacteria can also produce metabolites acting as inhibitors of chromatin-modifying enzymes. One such product is butyric acid, a short-chain fatty acid acting as a potent inhibitor of HDACs [164]. The adverse effect of Porphyromanas gingivalis in reactivating latent viruses, such as human immunodeficiency virus (HIV) and Epstein – Barr syndrome (EBV), result from production of butyrate by this bacterium [165]. It is proposed that viral genes maintained silent by HDAC-containing complexes are reactivated following inhibition of HDACs by butyric acid. Butyric acid also exerts anti-inflammatory effects on the host, via epigenetic up-regulation of anti-inflammatory genes. These observations give the idea of the usage of butyrate-producing probiotic bacteria as immunosuppressors [166].

Table 1. Selected epigenetic deregulations and inflammatory conditions

Disease	DNA Methylation	Histone Modification	Micro RNA
IBD	*ER, MYOD, GSPG2, p16, CDH1, GDNF, E-CAHERIN, DAPK, MDR1, SERPINA5, TNFRSF1A, AATK, GABRA5, MAPK10, STAT5A hypermethylation* [34-38]	*COL1, FOXP3* [32, 43]	*miR-125, miR-155, miR-192, miR-150, miR-21, miR-196* [49-52]
RA	*FOXP3, SFRP1, SFRP4, EBF3, IRX1, DR3 hypermethylation* [57, 68, 69] *L1 repeats, IL-6, IL-1, CXCL-12 CD40LG, IL1R2, CHI3L1, CASP1, STAT3, MAP3K5, MEFV*, and *WISP3 hypomethylation* [12, 64-67, 70, 71]	*EXH2, SITR1, ADAMTS, p16, p21, VEGF, HIF-1a* [72-74]	*miR-21, miR-30a-3p, miR-146a, miR-155, miR-223, miR-182, miR-203, miR-124a, miR-18a* [4, 14, 83-87]
SLE	*IL-10, IL-13, IFIT2, IRF5, STAT1, CD5, HRES1, IFNGR2, 18S and 28S hypomethylation* [93-96]	*IFN* genes [88, 102-104]	*Let-7a, miR-21, miR-29, miR-146, miR-182, miR-17-92, miR-155, miR-1246, miR126, miR-23b, miR-31* [110, 115-119, 120, 121]
MS	*PADI2-hypomethylation* [122, 123] *FOXP3* and *IL-17A hypomethylation [126]SHP1-hypermethylation* [125] Global hypermethylation in CD4$^+$ T cell [128]	*TCF7L2, ID2, IDS* and *SOX2* [125, 126]	*miR-21, miR-29b, miR-115-miR-326-, miR-124, miR34a, miR-17-5p, miR-497, miR-193*, and *miR-126, miR-18*b, *miR-599*, miR-96 [127-131]
Diabetes	*FOXP3- hypermethylation [142] Global hypermethylation [138]*	*CLTA4 TGF-β, NF-κB, p38, TLRs*, and *IL-6. NF-κB* [138, 139]	*miR-29, miR-510 miR-146a, miR-342, miR-191, miR-21, miR-34a, miR-101a* and *miR-30b* [142-144]
Bacterial infection	H. pylori induced hypermethylation seen at the *CDH1, USF1, USF2, WWOX, MLH1*[147] E. coli (UPEC) induced hypermethylation of *CDKN2A* [150]	B. vulgatus induces TGF-1β pathway [152]B. anthracis inactivates MAPKKs [153] H. pylori induces cell-cycle regulator gene $p21^{WAF}$ [155]	*MiR-155, miR-146, miR-125, let-7 and miR-21 (168)*

IBD; inflammatory bowel disease, RA; rheumatoid arthritis, SLE; systemic lupus erythematous, MS; multiple sclerosis.

miRNAs are important regulators of immune responses that are induced in response to pathogenic bacteria, such as H. pylori, Salmonella typhimurium, L. monocytogenes, and Mycobacterium bovis BCG [167, 168]. Whether a bacterial stimulus induces expression of miRNA or endogenous siRNA acts at the chromatin level in the nucleus is unknown. *MiR-155, miR-146, miR-125, let-7* and *miR-21* which are involved in immune responses and protection against harmful effects of inflammation, play important roles in the response of host cells to bacteria [169]. These miRNAs are involved in immune responses as well as in the protection against harmful effects of inflammation. miRNAs are involved in the regulation of proliferation, differentiation and apoptosis pathways induced by pathogenic bacteria in their host cells. A role for miRNAs that control bacteria populations and their production of metabolites leading to inflammatory conditions is beginning to emerge. It has got important implications both within the gut and in peripheral tissues.

3. Epigenetics, Inflammation and Cancer

Cancer has traditionally been described as a disease that is driven by the accumulation of genetic variations [170]. In addition to genetic alterations, it is well known now that epigenetic changes also play an important role in the development and progression of cancers [15, 171]. Self-sufficient proliferation, insensitivity to anti-proliferative signals, evasion of apoptosis, limitless replicative capacity, the maintenance of vascularization, invasion and metastasis are the hallmarks of cancer [170, 172]. In fact, the genetic background of cancer is relatively straightforward: mutation of tumor suppressors and/or oncogenes causes either loss or gain of function and abnormal expression. However, the epigenetic pathway to cancer is more complex and comprised by chromatin structure including DNA methylation, histone variants and modifications, nucleosome remodeling as well as small non-coding regulatory elements. During the carcinogenesis, the epigenome goes through multiple alterations, including a genome-wide loss of DNA methylation, frequent increases in promoter methylation of CpG islands, changes in nucleosome occupancy and modification profiles [173, 174]. Outcome of next generation sequencing studies has been the discovery of many inactivating mutations in genes that have potential to disrupt DNA methylation, histone modification and gene expression. The current studies have indicated that genetic and epigenetic mechanisms are not two separate states in cancer; they interwine and take advantage of each other during tumorigenesis. Alterations in epigenetic mechanisms can lead to genetic mutations and genetic mutations in epigenetic regulators lead to an altered epigenome [174].

Inflammation is part of our natural healing process and there are two types of inflammation; chronic and acute. Acute inflammation plays a beneficial role against infection and injury. However, inadequate resolution of inflammation and uncontrolled inflammatory reactions can arouse a state of chronic inflammation and may play role in cancer development and progression [5, 175]. The link between cancer and inflammation has long been recognized. Accumulating evidence from preclinical and clinical studies suggested that persistent inflammation functions as a driving force in the journey to cancer [176]. It was estimated earlier that out of 2.2 million cancer cases diagnosed in the world on average more than 15% cases could be rooted to infection [177]. For instance, cancer of stomach, liver, gallbladder, prostate, and pancreas are usually related to gastric inflammation [176]. There is about 1.2 million new patients diagnosed with colorectal cancer and patients suffering from inflammatory bowel disorders, such as ulcerative colitis and Chron's disease, have an increased risk of developing colorectal cancer [176, 178]. Beside that management of colitis with anti-inflammatory therapy reduces this risk [179, 180].

The association between inflammation and cancer has two paradigm; inflammation cause cancer (extrinsic mechanism), cancer cells produce toxins and alert the immune system, when cancer has started, inflammation makes the tumor grow, creates a blood supply then spreads the cancer (intrinsic mechanism) [181]. Chronic hepatitis based hepatocellular carcinoma and colon cancer caused by inflammatory bowel diseases are two examples for extrinsic mechanisms [178, 182]. The intratumoral inflammatory microenvironment acts as a favorable ground for premalignant cells to attain malignant properties by allowing tumor cells to escape from host immune attack, inducing epigenetic changes, enforcing epithelial to mesenchymal transition, and providing unlimited growth potential, thereby creating a vicious cancer-inflammation-cancer axis [176]. For instance, the mutation of RAS family proteins trigger RAS-RAF signaling and results in aggressive tumor formation [183, 184].

Several biochemical processes that altered during chronic inflammation have been implicated in carcinogenesis. These include shifting cellular redox balance toward oxidative stress, induction of genomic instability, increase DNA breakage, stimulation of cell proliferation, metastasis and angiogenesis, deregulation of cellular epigenetic control of gene expression. One of the mechanisms by which consistence inflammation can initiate tumorigenesis is the generation of oxidants and subsequent DNA damage leading to activation of oncogenes and/or silencing of tumor suppressors [185]. A wide variety of inflammatory and innate immune cells are often gathered at the site of infection. As a response to proinflammatory stimuli, activated inflammatory/immune cells generate reactive oxygen species (ROS) and reactive nitrogen species (RNS), which cause DNA damage. DNA damage can lead to activation of oncogenes [186]. Chronic exposure to ultraviolet (UV) B radiation is known to trigger inflammatory tissue damage and skin cancer through the activation of *ras* oncogene and loss of function of *p53* tumor suppressor gene [187]. NO, another reactive species, plays a role in inflammation related carcinogenesis by modification of DNA and inactivation of DNA repair enzymes [186]. 8-Oxo-7,8-dihydro-2'-deoxyguanosine (8-oxo-dG), a major biochemical hallmark of oxidative and mutagenic DNA damage, has been found to be produced in association with H.pylori induced gastric and TNF-α-induced pulmonary carcinogenesis. In general, inflammation driven activation of various protein kinases that include Janus-activated kinase (JAK), Akt, and mitogen-activated protein (MAP) kinases unbounded transmits growth signals, allowing cells to acquire a malignant phenotype [181, 188, 189].

Beside the mutations in the genomic DNA, epigenetic alterations can change gene expression via DNA methylation, histone modifications and non-coding regulatory elements. Traditionally, DNA methylation studies mostly have focused on CpG islands. DNA methylation, the addition of a methyl group to the 5-position of cytosine base in the DNA, represents a critical epigenetic control of gene expression [174, 190, 191]. The CpG hypermethylation of E-cadherin gene in intestinal metaplasia in H. pylori infected patients suggests DNA hypermetylation as a preneoplastic event in gastric cancer [192, 193]. Moreover, H. pylori infection has been shown to cause DNA involvement of epigenetic alterations in inflammation-associated cancers. Gene silencing via promoter hypermetylation in tumor suppressor genes p16, *RUNX3*, *MLH1* and *HPP1* are seen in ulcerative colitis and Barretts esophagus, which are closely associated with gastric carcinogenesis [193, 194]. Tissue specific expressed and imprinted genes can show hypomethylation, which may contribute to cancer cell phenotype. Recent genome wide methylation analyses have allowed to systematically discovering the key features of tumor methylomes. These include not just hypermethylated CpG islands, but also large (several 100 kb to several Mb) partially methylated domains that are localized at intergenic regions [195]. Another novel feature of methylomes is DNA methylation valleys (DMV) that are strongly hypomethylated in most tissues, and become hypermethylated in several cancers [196]. Abu-Remaileh et al. has shown the dynamic hypermethylation of DMVs and identify it as an early event in inflammation induced intestinal tumorigenesis. Aberrant methylation of DMVs may be novel epigenetic programs that links intestinal inflammation to colon cancer [197].

Epigenetic mechanisms may modulate the expression of pro-inflammatory cytokine TNF-α, interleukins, tumor suppressor genes, oncogenes and autocrine and paracrine activation of the transcription factor NF-κB [3, 198]. In areas of tissue inflammation, activated neutrophils and eosinophils release HOCl and HOBr, which react with DNA to produce 5-chlorocytosine and 5-bromocytosine, respectively [176, 199]. Neither methyl-binding proteins nor DNA methyltransferase-1 (DNMT-1) can distinguish between these inflammation-damaged 5-halocytosines and 5-methylcytosine. Thus, the formation and persistence of 5-halocytosine residues in the DNA of cells at the site of inflammation may lead to inappropriate de novo DNA methylation and represent another important link between inflammation and cancer development [176, 199]. Treatment of human multiple myeloma KAS 6/1 cells with a proinflammatory cytokine IL-6 was resulted an increased expression of *DNMT-1* and hypermethylation of the *p53* promoter. Zebularine is the DNMT inhibiter and restores the normal p53 function by demethylating *p53* promoter [200].

On the other hand, hypomethylation of epidermal growth factor (*EGFR*) gene in IL-6-transfected malignant cholangiocytes is involved in increasing *EGFR* expression, there by promoting growth of cholangiocarcinoma cells [201]. Moreover, silencing of cytokine signaling by epigenetic modification revealed resistance to apoptosis in cholangiocarcinoma cells via sustained inflammatory signaling mediated by IL-6/signal transducer and activators of transcription (STAT-3) and subsequent expression of myeloid cell lymphoma-1 (Mcl-1) [176, 202].

Gene expression changes can be modulated by the modification of histones and their variants. Indeed, aberrant histone modifications are already known hallmarks of different cancers types [27]. Histon related enzymes HDACs and HATs have been linked to chronic inflammatory responses and cancer [203, 204]. The acethylation of lysine residues on the N-terminus of histones by HATs activates gene transcription; on the other hand removal of an acetyl group from lysine residues in histone tails by HDACs leads to transcriptional suppression of genes. Hypermethylation itself may also lead to histone deacetylation and chromatin condensation, thereby instigating a transcriptionally silenced state [205, 206]. Transcriptional regulation of inflammatory cytokines and related pathways such as NF-κB by proinflammatory stimuli depends on histone regulations. Therefore, the inflammation induced alterations in histones and linking upregulation of COX-2 and NF-κB suggest that inflammation can change the cellular epigenetic machinery, thereby indirectly induce the genetic instability of cancer cells [176]. Beside intranuclear functions, histones can also be released into the extracellular space by damaged and activated cells, presenting toxic or pro-inflammatory activity *in vivo* and *in vitro* [26, 27]. When extracellular histones released, they selectively bind to Toll-like receptors to produce pro-inflammatory cytokines (TNF-α and IL-6), which in turn accelerates inflammatory response and tissue damage [207].

Several miRNAs can behave as oncogenes or tumor suppressors in cancer. Cytokine regulated miRNAs can function as a critical link between inflammation and cancer. UC patients revealed the upregulation of several miRNAs, including *miR-21* and *miR-155* according to healty controls [208]. Upregulation of *miR-let-7a* may play a role in inflammation-associated cholangiosarcoma. Stable overexpression of IL-6 induces *miR-let-7a*, which contributes phosphorylation of STAT3 in malignant cholangiocytes [181, 209]. As an alternative way, IL-6 can enhance the growth of human cholangiocarcinoma cells by downregulating *miR-370* [210]. Therefore, the role of miRNA in inflammation and cancer mostly depends on mechanisms by which tumor cells and tumor microenvironment regulate miRNA biogenesis and miRNA-mediated epigenetic control of proinflammatory gene expression [181].

CONCLUSION

Interactions of cells in the innate and adaptive immune systems, and inflammatory mediators orchestrate aspects of the acute and chronic inflammation that underlie the diseases of many organs. Understanding of molecular mechanisms in those processes will not only provide important insights into the aetiology of chronic inflammatory diseases, but also provide new targets for the development of novel anti-inflammatory drugs based on triggering pathways.

Epigenetic regulation of gene expression is a novel approach to treat diseases. Epigenetic reprogramming has potential preventive and therapeutic purposes in inflammation and autoimmunity. Inhibition of histone deacetylase enzymes (HDAC) initially used in the treatment of cancer, but recent reports suggesting inhibitors of HDAC (HDACi) could be utilized in treating inflammatory diseases as well. Chromatin regulatory protein families have also provided alternative potential therapeutic targets for new anti-cancer and anti-inflammatory drugs. Many of the inhibitors that have been described to date have targeted chromatin modulators. However, their non-selectivity and consequent side effects are still limiting factors in their broader clinical use. A number of effective small molecular inhibitors of individual human methyltransferases and lysine demethylases have been developed. Inhibition of chromatin reader domains is showing great promise as a selective pharmacological strategy.

As miRNAs have been functionally connected to the development of chronic inflammation, alterations to miRNA levels could be a diagnostic way to detect the presence of chronic inflammatory states in patients. miRNAs are currently being used as diagnostics for at least some types of diseases, including some forms of chronic inflammation such as colitis and IBD. The identification of miRNAs in blood serum as well as other biological fluids, samples obtained through non-invasive methods, has opened the door for diagnosis of various diseases. Specific secreted miRNAs in the serum can be used to diagnose chronic inflammatory diseases such as IBD.

The widely use of new generation whole genome technologies and meta-analysis, lead to understand epigenetic mechanism and their roles in disease development. Enlightenment of genetic and epigenetic modifications in inflammation will guide to create new therapeutic approach in practice. However, functional studies are still emerging to validate molecular targets for smart drugs development.

REFERENCES

[1] Libby P. Inflammatory mechanisms: the molecular basis of inflammation and disease. *Nutr Rev*, 2007, 65(12 Pt 2), S140-6.

[2] Eming, SA; Krieg, T; Davidson, JM. Inflammation in wound repair: molecular and cellular mechanisms. *J Invest Dermatol*, 2007, 127(3), 514-25.

[3] Hoesel, B; Schmid, JA. The complexity of NF-κB signaling in inflammation and cancer. *Mol Cancer*, 2013, 12(86).

[4] Disis, ML. Immune regulation of cancer. *J Clin Oncol*, 2010, 28(29), 4531-8.

[5] Balkwill, F; Mantovani, A. Inflammation and cancer: back to Virchow? *Lancet*, 2001, 357(9255), 539-45.

[6] Wertz, IE; O'Rourke, KM; Zhou, H; Eby, M; Aravind, L; Seshagiri, S; et al. De-ubiquitination and ubiquitin ligase domains of A20 downregulate NF-kappaB signalling. *Nature*, 2004, 430(7000), 694-9.

[7] Trompouki, E; Hatzivassiliou, E; Tsichritzis, T; Farmer, H; Ashworth, A; Mosialos, G. CYLD is a deubiquitinating enzyme that negatively regulates NF-kappaB activation by TNFR family members. *Nature*, 2003, 424(6950), 793-6.

[8] Holliday, R; Pugh, JE. DNA modification mechanisms and gene activity during development. *Science*, 1975, 187(4173), 226-32.

[9] Jones, PA; Liang, G. Rethinking how DNA methylation patterns are maintained. *Nat Rev Genet*, 2009, 10(11), 805-11.

[10] Jones, PA. Functions of DNA methylation: islands, start sites, gene bodies and beyond. *Nat Rev Genet*, 2012, 13(7), 484-92.

[11] Simonsson, S; Gurdon, J. DNA demethylation is necessary for the epigenetic reprogramming of somatic cell nuclei. *Nat Cell Biol*, 2004, 6(10), 984-90.

[12] Lai, AY; Mav, D; Shah, R; Grimm, SA; Phadke, D; Hatzi, K; et al. DNA methylation profiling in human B cells reveals immune regulatory elements and epigenetic plasticity at Alu elements during B-cell activation. *Genome Res*, 2013, 23(12), 2030-41.

[13] Zhang, Z; Zhang, R. Epigenetics in autoimmune diseases: Pathogenesis and prospects for therapy. *Autoimmun Rev*, 2015, 14(10), 854-63.

[14] Raisch, J; Darfeuille-Michaud, A; Nguyen, HT. Role of microRNAs in the immune system, inflammation and cancer. *World J Gastroenterol*, 2013, 19(20), 2985-96.

[15] Baylin, SB; Jones, PA. A decade of exploring the cancer epigenome - biological and translational implications. *Nat Rev Cancer*, 2011, 11(10), 726-34.

[16] Schübeler, D. Function and information content of DNA methylation. *Nature*, 2015, 517(7534), 321-6.

[17] Campos, EI; Reinberg, D. Histones: annotating chromatin. *Annu Rev Genet*, 2009, 43(559-99.

[18] Luger, K; Richmond, TJ. The histone tails of the nucleosome. *Curr Opin Genet Dev*, 1998, 8(2), 140-6.

[19] Singh, R; Mortazavi, A; Telu, KH; Nagarajan, P; Lucas, DM; Thomas-Ahner, JM; et al. Increasing the complexity of chromatin: functionally distinct roles for replication-dependent histone H2A isoforms in cell proliferation and carcinogenesis. *Nucleic Acids Res*, 2013, 41(20), 9284-95.

[20] Morillon, A; Karabetsou, N; Nair, A; Mellor, J. Dynamic lysine methylation on histone H3 defines the regulatory phase of gene transcription. *Mol Cell*, 2005, 18(6), 723-34.

[21] Eberharter, A; Becker, PB. Histone acetylation: a switch between repressive and permissive chromatin. Second in review series on chromatin dynamics. *EMBO Rep*, 2002, 3(3), 224-9.

[22] Cantley, MD; Haynes, DR. Epigenetic regulation of inflammation: progressing from broad acting histone deacetylase (HDAC) inhibitors to targeting specific HDACs. *Inflammopharmacology*, 2013, 21(4), 301-7.

[23] Jiang, C; Pugh, BF. Nucleosome positioning and gene regulation: advances through genomics. *Nat Rev Genet*, 2009, 10(3), 161-72.

[24] Martens, JA; Winston, F. Recent advances in understanding chromatin remodeling by Swi/Snf complexes. *Curr Opin Genet Dev*, 2003, 13(2), 136-42.

[25] Wang, KC; Yang, YW; Liu, B; Sanyal, A; Corces-Zimmerman, R; Chen, Y; et al. A long noncoding RNA maintains active chromatin to coordinate homeotic gene expression. *Nature*, 2011, 472(7341), 120-4.

[26] Allam, R; Kumar, SV; Darisipudi, MN; Anders, HJ. Extracellular histones in tissue injury and inflammation. *J Mol Med* (Berl), 2014, 92(5), 465-72.

[27] Chen, R; Kang, R; Fan, XG; Tang, D. Release and activity of histone in diseases. *Cell Death Dis*, 2014, 5(e1370.

[28] Grammatikakis, I; Panda, AC; Abdelmohsen, K; Gorospe, M. Long noncoding RNAs(lncRNAs) and the molecular hallmarks of aging. *Aging (Albany NY)*, 2014, 6(12), 992-1009.

[29] Marques-Rocha, JL; Samblas, M; Milagro, FI; Bressan, J; Martínez, JA; Marti, A. Noncoding RNAs, cytokines, and inflammation-related diseases. *FASEB J*, 2015, 29(9), 3595-611.

[30] Alexander, M; O'Connell, RM. Noncoding RNAs and chronic inflammation: Micro-managing the fire within. *Bioessays*, 2015, 37(9), 1005-15.

[31] Manichanh, C; Borruel, N; Casellas, F; Guarner, F. The gut microbiota in IBD. *Nat Rev Gastroenterol Hepatol*, 2012, 9(10), 599-608.

[32] Däbritz, J; Menheniott, TR. Linking immunity, epigenetics, and cancer in inflammatory bowel disease. *Inflamm Bowel Dis*, 2014, 20(9), 1638-54.

[33] Glória, L; Cravo, M; Pinto, A; de Sousa, LS; Chaves, P; Leitão, CN; et al. DNA hypomethylation and proliferative activity are increased in the rectal mucosa of patients with long-standing ulcerative colitis. *Cancer*, 1996, 78(11), 2300-6.

[34] Schuebel, KE; Chen, W; Cope, L; Glöckner, SC; Suzuki, H; Yi, JM; et al. Comparing the DNA hypermethylome with gene mutations in human colorectal cancer. *PLoS Genet*, 2007, 3(9), 1709-23.

[35] Lin, Z; Hegarty, JP; Cappel, JA; Yu, W; Chen, X; Faber, P; et al. Identification of disease-associated DNA methylation in intestinal tissues from patients with inflammatory bowel disease. *Clin Genet*, 2011, 80(1), 59-67.

[36] Kellermayer, M. [To the Editors: The Hungarian Healthcare Society's assessment and recommendations for medical specialty education in Hungary]. *Orv Hetil*, 2012, 153(4), 158.

[37] Tahara, T; Shibata, T; Nakamura, M; Yamashita, H; Yoshioka, D; Okubo, M; et al. Effect of MDR1 gene promoter methylation in patients with ulcerative colitis. *Int J Mol Med*, 2009, 23(4), 521-7.

[38] Häsler, R; Feng, Z; Bäckdahl, L; Spehlmann, ME; Franke, A; Teschendorff, A; et al. A functional methylome map of ulcerative colitis. *Genome Res*, 2012, 22(11), 2130-7.

[39] Karatzas, PS; Mantzaris, GJ; Safioleas, M; Gazouli, M. DNA methylation profile of genes involved in inflammation and autoimmunity in inflammatory bowel disease. *Medicine (Baltimore)*, 2014, 93(28), e309.

[40] Stylianou, E. Epigenetics: the fine-tuner in inflammatory bowel disease? *Curr Opin Gastroenterol*, 2013, 29(4), 370-7.

[41] Nimmo, ER; Prendergast, JG; Aldhous, MC; Kennedy, NA; Henderson, P; Drummond, HE; et al. Genome-wide methylation profiling in Crohn's disease identifies altered epigenetic regulation of key host defense mechanisms including the Th17 pathway. *Inflamm Bowel Dis*, 2012, 18(5), 889-99.

[42] Tsaprouni, LG; Ito, K; Powell, JJ; Adcock, IM; Punchard, N. Differential patterns of histone acetylation in inflammatory bowel diseases. *J Inflamm (Lond)*, 2011, 8(1), 1.

[43] Tao, R; de Zoeten, EF; Ozkaynak, E; Chen, C; Wang, L; Porrett, PM; et al. Deacetylase inhibition promotes the generation and function of regulatory T cells. *Nat Med*, 2007, 13(11), 1299-307.

[44] Glauben, R; Siegmund, B. Inhibition of histone deacetylases in inflammatory bowel diseases. *Mol Med*, 2011, 17(5-6), 426-33.

[45] McKenna, LB; Schug, J; Vourekas, A; McKenna, JB; Bramswig, NC; Friedman, JR; et al. MicroRNAs control intestinal epithelial differentiation, architecture, and barrier function. *Gastroenterology*, 2010, 139(5), 1654-64, 1664.e1.

[46] Ventham, NT; Kennedy, NA; Nimmo, ER; Satsangi, J. Beyond gene discovery in inflammatory bowel disease: the emerging role of epigenetics. *Gastroenterology*, 2013, 145(2), 293-308.

[47] Pekow, JR; Kwon, JH. MicroRNAs in inflammatory bowel disease. *Inflamm Bowel Dis*, 2012, 18(1), 187-93.

[48] Chapman, CG; Pekow, J. The emerging role of miRNAs in inflammatory bowel disease: a review. *Therap Adv Gastroenterol*, 2015, 8(1), 4-22.

[49] Wu, F; Zikusoka, M; Trindade, A; Dassopoulos, T; Harris, ML; Bayless, TM; et al. MicroRNAs are differentially expressed in ulcerative colitis and alter expression of macrophage inflammatory peptide-2 alpha. *Gastroenterology*, 2008, 135(5), 1624-1635 e24.

[50] Bian, Z; Li, L; Cui, J; Zhang, H; Liu, Y; Zhang, CY; et al. Role of miR-150-targeting c-Myb in colonic epithelial disruption during dextran sulphate sodium-induced murine experimental colitis and human ulcerative colitis. *J Pathol*, 2011, 225(4), 544-53.

[51] Fasseu, M; Treton, X; Guichard, C; Pedruzzi, E; Cazals-Hatem, D; Richard, C; et al. Identification of restricted subsets of mature microRNA abnormally expressed in inactive colonic mucosa of patients with inflammatory bowel disease. *PLoS One*, 2010, 5(10).

[52] Brest, P; Lapaquette, P; Souidi, M; Lebrigand, K; Cesaro, A; Vouret-Craviari, V; et al. A synonymous variant in IRGM alters a binding site for miR-196 and causes deregulation of IRGM-dependent xenophagy in Crohn's disease. *Nat Genet*, 2011, 43(3), 242-5.

[53] Wu, F; Guo, NJ; Tian, H; Marohn, M; Gearhart, S; Bayless, TM; et al. Peripheral blood microRNAs distinguish active ulcerative colitis and Crohn's disease. *Inflamm Bowel Dis*, 2011, 17(1), 241-50.

[54] Petronis, A; Petroniene, R. Epigenetics of inflammatory bowel disease. *Gut*, 2000, 47(2), 302-6.

[55] Jonsson, B; Kobelt, G; Smolen, J. The burden of rheumatoid arthritis and access to treatment: uptake of new therapies. *Eur J Health Econ*, 2008, 8 Suppl 2(S61-86.

[56] Padyukov, L; Seielstad, M; Ong, RT; Ding, B; Ronnelid, J; Seddighzadeh, M; et al. A genome-wide association study suggests contrasting associations in ACPA-positive versus ACPA-negative rheumatoid arthritis. *Ann Rheum Dis*, 2011, 70(2), 259-65.

[57] Klein, K; Ospelt, C; Gay, S. Epigenetic contributions in the development of rheumatoid arthritis. *Arthritis Res Ther*, 2012, 14(6), 227.

[58] Karouzakis, E; Gay, RE; Michel, BA; Gay, S; Neidhart, M. DNA hypomethylation in rheumatoid arthritis synovial fibroblasts. *Arthritis Rheum*, 2009, 60(12), 3613-22.

[59] Karouzakis, E; Gay, RE; Gay, S; Neidhart, M. Increased recycling of polyamines is associated with global DNA hypomethylation in rheumatoid arthritis synovial fibroblasts. *Arthritis Rheum*, 2012, 64(6), 1809-17.

[60] Lafyatis, R; Remmers, EF; Roberts, AB; Yocum, DE; Sporn, MB; Wilder, RL. Anchorage-independent growth of synoviocytes from arthritic and normal joints. Stimulation by exogenous platelet-derived growth factor and inhibition by transforming growth factor-beta and retinoids. *J Clin Invest*, 1989, 83(4), 1267-76.

[61] Kinne, RW; Boehm, S; Iftner, T; Aigner, T; Vornehm, S; Weseloh, G; et al. Synovial fibroblast-like cells strongly express jun-B and C-fos proto-oncogenes in rheumatoid- and osteoarthritis. *Scand J Rheumatol Suppl*, 1995, 101(121-5.

[62] Fu, LH; Ma, CL; Cong, B; Li, SJ; Chen, HY; Zhang, JG. Hypomethylation of proximal CpG motif of interleukin-10 promoter regulates its expression in human rheumatoid arthritis. *Acta Pharmacol Sin*, 2011, 32(11), 1373-80.

[63] Song, YJ; Li, G; He, JH; Guo, Y; Yang, L. Bioinformatics-Based Identification of MicroRNA-Regulated and Rheumatoid Arthritis-Associated Genes. *PLoS One*, 2015, 10(9), e0137551.

[64] Kuchen, S; Seemayer, CA; Rethage, J; von Knoch, R; Kuenzler, P; Beat, AM; et al. The L1 retroelement-related p40 protein induces p38delta MAP kinase. *Autoimmunity*, 2004, 37(1), 57-65.

[65] Garaud, S; Le Dantec, C; Jousse-Joulin, S; Hanrotel-Saliou, C; Saraux, A; Mageed, RA; et al. IL-6 modulates CD5 expression in B cells from patients with lupus by regulating DNA methylation. *J Immunol*, 2009, 182(9), 5623-32.

[66] Karouzakis, E; Rengel, Y; Jungel, A; Kolling, C; Gay, RE; Michel, BA; et al. DNA methylation regulates the expression of CXCL12 in rheumatoid arthritis synovial fibroblasts. *Genes Immun*, 2011, 12(8), 643-52.

[67] Bertoncelj, V; Pelipenko, J; Kristl, J; Jeras, M; Cukjati, M; Kocbek, P. Development and bioevaluation of nanofibers with blood-derived growth factors for dermal wound healing. *Eur J Pharm Biopharm*, 2014, 88(1), 64-74.

[68] Park, SH; Kim, SK; Choe, JY; Moon, Y; An, S; Park, MJ; et al. Hypermethylation of EBF3 and IRX1 genes in synovial fibroblasts of patients with rheumatoid arthritis. *Mol Cells*, 2013, 35(4), 298-304.

[69] Takami, N; Osawa, K; Miura, Y; Komai, K; Taniguchi, M; Shiraishi, M; et al. Hypermethylated promoter region of DR3, the death receptor 3 gene, in rheumatoid arthritis synovial cells. *Arthritis Rheum*, 2006, 54(3), 779-87.

[70] Chabchoub, G; Uz, E; Maalej, A; Mustafa, CA; Rebai, A; Mnif, M; et al. Analysis of skewed X-chromosome inactivation in females with rheumatoid arthritis and autoimmune thyroid diseases. *Arthritis Res Ther*, 2009, 11(4), R106.

[71] Liao, J; Liang, G; Xie, S; Zhao, H; Zuo, X; Li, F; et al. CD40L demethylation in CD4(+) T cells from women with rheumatoid arthritis. *Clin Immunol*, 2012, 145(1), 13-8.

[72] Huber, LC; Stanczyk, J; Jüngel, A; Gay, S. Epigenetics in inflammatory rheumatic diseases. *Arthritis Rheum*, 2007, 56(11), 3523-31.

[73] Nishida, K; Komiyama, T; Miyazawa, S; Shen, ZN; Furumatsu, T; Doi, H; et al. Histone deacetylase inhibitor suppression of autoantibody-mediated arthritis in mice via regulation of p16INK4a and p21(WAF1/Cip1) expression. *Arthritis Rheum*, 2004, 50(10), 3365-76.

[74] Manabe, H; Nasu, Y; Komiyama, T; Furumatsu, T; Kitamura, A; Miyazawa, S; et al. Inhibition of histone deacetylase down-regulates the expression of hypoxia-induced vascular endothelial growth factor by rheumatoid synovial fibroblasts. *Inflamm Res*, 2008, 57(1), 4-10.

[75] Stanczyk, J; Ospelt, C; Karouzakis, E; Filer, A; Raza, K; Kolling, C; et al. Altered expression of microRNA-203 in rheumatoid arthritis synovial fibroblasts and its role in fibroblast activation. *Arthritis Rheum*, 2011, 63(2), 373-81.

[76] Miao, CG; Yang, YY; He, X; Li, J. New advances of DNA methylation and histone modifications in rheumatoid arthritis, with special emphasis on MeCP2. *Cell Signal*, 2013, 25(4), 875-82.

[77] Zhou, Q; Long, L; Shi, G; Zhang, J; Wu, T; Zhou, B. Research of the methylation status of miR-124a gene promoter among rheumatoid arthritis patients. *Clin Dev Immunol*, 2013, 2013(524204.

[78] Nakano, K; Whitaker, JW; Boyle, DL; Wang, W; Firestein, GS. DNA methylome signature in rheumatoid arthritis. *Ann Rheum Dis*, 2013, 72(1), 110-7.

[79] Wilkinson, J. Study reveals a DNA methylome signature in rheumatoid arthritis. *Epigenomics*, 2012, 4(5), 481.

[80] Ai, R; Whitaker, JW; Boyle, DL; Tak, PP; Gerlag, DM; Wang, W; et al. DNA Methylome Signature in Synoviocytes From Patients With Early Rheumatoid Arthritis Compared to Synoviocytes From Patients With Longstanding Rheumatoid Arthritis. *Arthritis Rheumatol*, 2015, 67(7), 1978-80.

[81] de la Rica, L; Urquiza, JM; Gomez-Cabrero, D; Islam, AB; Lopez-Bigas, N; Tegner, J; et al. Identification of novel markers in rheumatoid arthritis through integrated analysis of DNA methylation and microRNA expression. *J Autoimmun*, 2013, 41(6-16.

[82] Ballestar, E. Genetics: Insights into RA pathogenesis from DNA methylome analysis. *Nat Rev Rheumatol*, 2015, 11(7), 386-8.

[83] Kurowska-Stolarska, M; Alivernini, S; Ballantine, LE; Asquith, DL; Millar, NL; Gilchrist, DS; et al. MicroRNA-155 as a proinflammatory regulator in clinical and experimental arthritis. *Proc Natl Acad Sci U S A*, 2011, 108(27), 11193-8.

[84] Fulci, V; Scappucci, G; Sebastiani, GD; Giannitti, C; Franceschini, D; Meloni, F; et al. miR-223 is overexpressed in T-lymphocytes of patients affected by rheumatoid arthritis. *Hum Immunol*, 2010, 71(2), 206-11.

[85] Nakasa, T; Miyaki, S; Okubo, A; Hashimoto, M; Nishida, K; Ochi, M; et al. Expression of microRNA-146 in rheumatoid arthritis synovial tissue. *Arthritis Rheum*, 2008, 58(5), 1284-92.

[86] Stittrich, AB; Haftmann, C; Sgouroudis, E; Kühl, AA; Hegazy, AN; Panse, I; et al. The microRNA miR-182 is induced by IL-2 and promotes clonal expansion of activated helper T lymphocytes. *Nat Immunol*, 2010, 11(11), 1057-62.

[87] Nakamachi, Y; Kawano, S; Takenokuchi, M; Nishimura, K; Sakai, Y; Chin, T; et al. MicroRNA-124a is a key regulator of proliferation and monocyte chemoattractant protein 1 secretion in fibroblast-like synoviocytes from patients with rheumatoid arthritis. *Arthritis Rheum*, 2009, 60(5), 1294-304.

[88] Trenkmann, M; Brock, M; Gay, RE; Michel, BA; Gay, S; Huber, LC. Tumor necrosis factor alpha-induced microRNA-18a activates rheumatoid arthritis synovial fibroblasts through a feedback loop in NF-kappaB signaling. *Arthritis Rheum*, 2013, 65(4), 916-27.

[89] Miceli-Richard, C. Epigenetics and lupus. *Joint Bone Spine*, 2015, 82(2), 90-3.

[90] Mak, A; Kow, NY. The pathology of T cells in systemic lupus erythematosus. *J Immunol Res*, 2014, 2014(419029.

[91] Richardson, BC; Patel, DR. Epigenetics in 2013. DNA methylation and miRNA: key roles in systemic autoimmunity. *Nat Rev Rheumatol*, 2014, 10(2), 72-4.

[92] Hughes, GC; Choubey, D. Modulation of autoimmune rheumatic diseases by oestrogen and progesterone. *Nat Rev Rheumatol*, 2014, 10(12), 740-51.

[93] Schwartzman-Morris, J; Putterman, C. Gender differences in the pathogenesis and outcome of lupus and of lupus nephritis. *Clin Dev Immunol*, 2012, 2012(604892.

[94] Lu, Q; Wu, A; Tesmer, L; Ray, D; Yousif, N; Richardson, B. Demethylation of CD40LG on the inactive X in T cells from women with lupus. *J Immunol*, 2007, 179(9), 6352-8.

[95] Cannat, A; Seligmann, M. Induction by isoniazid and hydrallazine of antinuclear factors in mice. *Clin Exp Immunol*, 1968, 3(1), 99-105.

[96] Oelke, K; Lu, Q; Richardson, D; Wu, A; Deng, C; Hanash, S; et al. Overexpression of CD70 and overstimulation of IgG synthesis by lupus T cells and T cells treated with DNA methylation inhibitors. *Arthritis Rheum*, 2004, 50(6), 1850-60.

[97] Javierre, BM; Fernandez, AF; Richter, J; Al-Shahrour, F; Martin-Subero, JI; Rodriguez-Ubreva, J; et al. Changes in the pattern of DNA methylation associate with twin discordance in systemic lupus erythematosus. *Genome Res*, 2010, 20(2), 170-9.

[98] Wu, H; Zhao, M; Chang, C; Lu, Q. The real culprit in systemic lupus erythematosus: abnormal epigenetic regulation. *Int J Mol Sci*, 2015, 16(5), 11013-33.

[99] Coit, P; Jeffries, M; Altorok, N; Dozmorov, MG; Koelsch, KA; Wren, JD; et al. Genome-wide DNA methylation study suggests epigenetic accessibility and transcriptional poising of interferon-regulated genes in naïve CD4+ T cells from lupus patients. *J Autoimmun*, 2013, 43(78-84.

[100] Schett, G; Smole, J; Zimmermann, C; Hiesberger, H; Hoefler, E; Fournel, S; et al. The autoimmune response to chromatin antigens in systemic lupus erythematosus: autoantibodies against histone H1 are a highly specific marker for SLE associated with increased disease activity. *Lupus*, 2002, 11(11), 704-15.

[101] Bruns, A; Bläss, S; Hausdorf, G; Burmester, GR; Hiepe, F. Nucleosomes are major T and B cell autoantigens in systemic lupus erythematosus. *Arthritis Rheum*, 2000, 43(10), 2307-15.

[102] van Bavel, CC; Dieker, JW; Tamboer, WP; van der Vlag, J; Berden, JH. Lupus-derived monoclonal autoantibodies against apoptotic chromatin recognize acetylated conformational epitopes. *Mol Immunol*, 2010, 48(1-3), 248-56.

[103] Hu, N; Qiu, X; Luo, Y; Yuan, J; Li, Y; Lei, W; et al. Abnormal histone modification patterns in lupus CD4+ T cells. *J Rheumatol*, 2008, 35(5), 804-10.
[104] van Bavel, CC; Dieker, JW; Kroeze, Y; Tamboer, WP; Voll, R; Muller, S; et al. Apoptosis-induced histone H3 methylation is targeted by autoantibodies in systemic lupus erythematosus. *Ann Rheum Dis*, 2011, 70(1), 201-7.
[105] Zhang, Z; Song, L; Maurer, K; Petri, MA; Sullivan, KE. Global H4 acetylation analysis by ChIP-chip in systemic lupus erythematosus monocytes. *Genes Immun*, 2010, 11(2), 124-33.
[106] Zhao, M; Liu, S; Luo, S; Wu, H; Tang, M; Cheng, W; et al. DNA methylation and mRNA and microRNA expression of SLE CD4+ T cells correlate with disease phenotype. *J Autoimmun*, 2014, 54(127-36.
[107] Lee, BH; Yegnasubramanian, S; Lin, X; Nelson, WG. Procainamide is a specific inhibitor of DNA methyltransferase 1. *J Biol Chem*, 2005, 280(49), 40749-56.
[108] Deng, C; Lu, Q; Zhang, Z; Rao, T; Attwood, J; Yung, R; et al. Hydralazine may induce autoimmunity by inhibiting extracellular signal-regulated kinase pathway signaling. *Arthritis Rheum*, 2003, 48(3), 746-56.
[109] Cornacchia, E; Golbus, J; Maybaum, J; Strahler, J; Hanash, S; Richardson, B. Hydralazine and procainamide inhibit T cell DNA methylation and induce autoreactivity. *J Immunol*, 1988, 140(7), 2197-200.
[110] Duroux-Richard, I; Cuenca, J; Ponsolles, C; Pineiro, AB; Gonzalez, F; Roubert, C; et al. MicroRNA Profiling of B Cell Subsets from Systemic Lupus Erythematosus Patients Reveals Promising Novel Biomarkers. *Int J Mol Sci*, 2015, 16(8), 16953-65.
[111] Luo, S; Liu, Y; Liang, G; Zhao, M; Wu, H; Liang, Y; et al. The role of microRNA-1246 in the regulation of B cell activation and the pathogenesis of systemic lupus erythematosus. *Clin Epigenetics*, 2015, 7(1), 24.
[112] Dai, Y; Huang, YS; Tang, M; Lv, TY; Hu, CX; Tan, YH; et al. Microarray analysis of microRNA expression in peripheral blood cells of systemic lupus erythematosus patients. *Lupus*, 2007, 16(12), 939-46.
[113] Aletaha, D; Landewe, R; Karonitsch, T; Bathon, J; Boers, M; Bombardier, C; et al. Reporting disease activity in clinical trials of patients with rheumatoid arthritis: EULAR/ACR collaborative recommendations. *Arthritis Rheum*, 2008, 59(10), 1371-7.
[114] Pan, W; Zhu, S; Yuan, M; Cui, H; Wang, L; Luo, X; et al. MicroRNA-21 and microRNA-148a contribute to DNA hypomethylation in lupus CD4+ T cells by directly and indirectly targeting DNA methyltransferase 1. *J Immunol*, 2010, 184(12), 6773-81.
[115] Wu, XN; Ye, YX; Niu, JW; Li, Y; Li, X; You, X; et al. Defective PTEN regulation contributes to B cell hyperresponsiveness in systemic lupus erythematosus. *Sci Transl Med*, 2014, 6(246), 246ra99.
[116] Stagakis, E; Bertsias, G; Verginis, P; Nakou, M; Hatziapostolou, M; Kritikos, H; et al. Identification of novel microRNA signatures linked to human lupus disease activity and pathogenesis: miR-21 regulates aberrant T cell responses through regulation of PDCD4 expression. *Ann Rheum Dis*, 2011, 70(8), 1496-506.
[117] Tang, Y; Luo, X; Cui, H; Ni, X; Yuan, M; Guo, Y; et al. MicroRNA-146A contributes to abnormal activation of the type I interferon pathway in human lupus by targeting the key signaling proteins. *Arthritis Rheum*, 2009, 60(4), 1065-75.

[118] Luo, X; Yang, W; Ye, DQ; Cui, H; Zhang, Y; Hirankarn, N; et al. A functional variant in microRNA-146a promoter modulates its expression and confers disease risk for systemic lupus erythematosus. *PLoS Genet*, 2011, 7(6), e1002128.

[119] A B, R K, N R, Meza, V; Morelatto, R; Piccinni, D. [Malignant melanoma of the oral cavity]. *Rev Fac Cien Med Univ Nac Cordoba*, 2008, 65(2), 70-3.

[120] Karrich, JJ; Jachimowski, LC; Libouban, M; Iyer, A; Brandwijk, K; Taanman-Kueter, EW; et al. MicroRNA-146a regulates survival and maturation of human plasmacytoid dendritic cells. *Blood*, 2013, 122(17), 3001-9.

[121] Zhu, S; Pan, W; Song, X; Liu, Y; Shao, X; Tang, Y; et al. The microRNA miR-23b suppresses IL-17-associated autoimmune inflammation by targeting TAB2, TAB3 and IKK-alpha. *Nat Med*, 2012, 18(7), 1077-86.

[122] Fan, W; Liang, D; Tang, Y; Qu, B; Cui, H; Luo, X; et al. Identification of microRNA-31 as a novel regulator contributing to impaired interleukin-2 production in T cells from patients with systemic lupus erythematosus. *Arthritis Rheum*, 2012, 64(11), 3715-25.

[123] Mastronardi, FG; Noor, A; Wood, DD; Paton, T; Moscarello, MA. Peptidyl argininedeiminase 2 CpG island in multiple sclerosis white matter is hypomethylated. *J Neurosci Res*, 2007, 85(9), 2006-16.

[124] Mastronardi, FG; Wood, DD; Mei, J; Raijmakers, R; Tseveleki, V; Dosch, HM; et al. Increased citrullination of histone H3 in multiple sclerosis brain and animal models of demyelination: a role for tumor necrosis factor-induced peptidylarginine deiminase 4 translocation. *J Neurosci*, 2006, 26(44), 11387-96.

[125] Calabrese, R; Zampieri, M; Mechelli, R; Annibali, V; Guastafierro, T; Ciccarone, F; et al. Methylation-dependent PAD2 upregulation in multiple sclerosis peripheral blood. *Mult Scler*, 2012, 18(3), 299-304.

[126] Kumagai, C; Kalman, B; Middleton, FA; Vyshkina, T; Massa, PT. Increased promoter methylation of the immune regulatory gene SHP-1 in leukocytes of multiple sclerosis subjects. *J Neuroimmunol*, 2012, 246(1-2), 51-7.

[127] Jadidi-Niaragh, F; Mirshafiey, A. Th17 cell, the new player of neuroinflammatory process in multiple sclerosis. *Scand J Immunol*, 2011, 74(1), 1-13.

[128] Janson, PC; Linton, LB; Bergman, EA; Marits, P; Eberhardson, M; Piehl, F; et al. Profiling of CD4+ T cells with epigenetic immune lineage analysis. *J Immunol*, 2011, 186(1), 92-102.

[129] Bos, SD; Page, CM; Andreassen, BK; Elboudwarej, E; Gustavsen, MW; Briggs, F; et al. Genome-wide DNA methylation profiles indicate CD8+ T cell hypermethylation in multiple sclerosis. *PLoS One*, 2015, 10(3), e0117403.

[130] Pedre, X; Mastronardi, F; Bruck, W; Lopez-Rodas, G; Kuhlmann, T; Casaccia, P. Changed histone acetylation patterns in normal-appearing white matter and early multiple sclerosis lesions. *J Neurosci*, 2011, 31(9), 3435-45.

[131] Pedre, X; Mastronardi, F; Bruck, W; López-Rodas, G; Kuhlmann, T; Casaccia, P. Changed histone acetylation patterns in normal-appearing white matter and early multiple sclerosis lesions. *J Neurosci*, 2011, 31(9), 3435-45.

[132] Du, C; Liu, C; Kang, J; Zhao, G; Ye, Z; Huang, S; et al. MicroRNA miR-326 regulates TH-17 differentiation and is associated with the pathogenesis of multiple sclerosis. *Nat Immunol*, 2009, 10(12), 1252-9.

[133] Junker, A; Krumbholz, M; Eisele, S; Mohan, H; Augstein, F; Bittner, R; et al. MicroRNA profiling of multiple sclerosis lesions identifies modulators of the regulatory protein CD47. *Brain*, 2009, 132(Pt 12), 3342-52.

[134] De Santis, G; Ferracin, M; Biondani, A; Caniatti, L; Rosaria Tola, M; Castellazzi, M; et al. Altered miRNA expression in T regulatory cells in course of multiple sclerosis. *J Neuroimmunol*, 2010, 226(1-2), 165-71.

[135] Otaegui, D; Baranzini, SE; Armañanzas, R; Calvo, B; Muñoz-Culla, M; Khankhanian, P; et al. Differential micro RNA expression in PBMC from multiple sclerosis patients. *PLoS One*, 2009, 4(7), e6309.

[136] Ponomarev, ED; Veremeyko, T; Barteneva, N; Krichevsky, AM; Weiner, HL. MicroRNA-124 promotes microglia quiescence and suppresses EAE by deactivating macrophages via the C/EBP-α-PU.1 pathway. *Nat Med*, 2011, 17(1), 64-70.

[137] Ling, C; Groop, L. Epigenetics: a molecular link between environmental factors and type 2 diabetes. *Diabetes*, 2009, 58(12), 2718-25.

[138] Gilbert, ER; Liu, D. Epigenetics: the missing link to understanding β-cell dysfunction in the pathogenesis of type 2 diabetes. *Epigenetics*, 2012, 7(8), 841-52.

[139] Meda, F; Folci, M; Baccarelli, A; Selmi, C. The epigenetics of autoimmunity. *Cell Mol Immunol*, 2011, 8(3), 226-36.

[140] Schalinske, KL. Interrelationship between diabetes and homocysteine metabolism: hormonal regulation of cystathionine beta-synthase. *Nutr Rev*, 2003, 61(4), 136-8.

[141] McCarty, MF. Insulin secretion as a potential determinant of homocysteine levels. *Med Hypotheses*, 2000, 55(5), 454-5.

[142] Arany, E; Strutt, B; Romanus, P; Remacle, C; Reusens, B; Hill, DJ. Taurine supplement in early life altered islet morphology, decreased insulitis and delayed the onset of diabetes in non-obese diabetic mice. *Diabetologia*, 2004, 47(10), 1831-7.

[143] Wang, Z; Zheng, Y; Hou, C; Yang, L; Li, X; Lin, J; et al. DNA methylation impairs TLR9 induced Foxp3 expression by attenuating IRF-7 binding activity in fulminant type 1 diabetes. *J Autoimmun*, 2013, 41(50-9.

[144] Miao, F; Smith, DD; Zhang, L; Min, A; Feng, W; Natarajan, R. Lymphocytes from patients with type 1 diabetes display a distinct profile of chromatin histone H3 lysine 9 dimethylation: an epigenetic study in diabetes. *Diabetes*, 2008, 57(12), 3189-98.

[145] Li, Y; Reddy, MA; Miao, F; Shanmugam, N; Yee, JK; Hawkins, D; et al. Role of the histone H3 lysine 4 methyltransferase, SET7/9, in the regulation of NF-kappaB-dependent inflammatory genes. Relevance to diabetes and inflammation. *J Biol Chem*, 2008, 283(39), 26771-81.

[146] Reddy, MA; Natarajan, R. Epigenetic mechanisms in diabetic vascular complications. *Cardiovasc Res*, 2011, 90(3), 421-9.

[147] Reddy, MA; Natarajan, R. Role of epigenetic mechanisms in the vascular complications of diabetes. *Subcell Biochem*, 2013, 61(435-54.

[148] Roggli, E; Britan, A; Gattesco, S; Lin-Marq, N; Abderrahmani, A; Meda, P; et al. Involvement of microRNAs in the cytotoxic effects exerted by proinflammatory cytokines on pancreatic beta-cells. *Diabetes*, 2010, 59(4), 978-86.

[149] Zheng, Y; Wang, Z; Tu, Y; Shen, H; Dai, Z; Lin, J; et al. miR-101a and miR-30b contribute to inflammatory cytokine-mediated beta-cell dysfunction. *Lab Invest*, 2015.

[150] Dey, N; Das, F; Mariappan, MM; Mandal, CC; Ghosh-Choudhury, N; Kasinath, BS; et al. MicroRNA-21 orchestrates high glucose-induced signals to TOR complex 1, resulting in renal cell pathology in diabetes. *J Biol Chem*, 2011, 286(29), 25586-603.

[151] Zhong, X; Chung, AC; Chen, HY; Dong, Y; Meng, XM; Li, R; et al. miR-21 is a key therapeutic target for renal injury in a mouse model of type 2 diabetes. *Diabetologia*, 2013, 56(3), 663-74.

[152] Maekita, T; Nakazawa, K; Mihara, M; Nakajima, T; Yanaoka, K; Iguchi, M; et al. High levels of aberrant DNA methylation in Helicobacter pylori-infected gastric mucosae and its possible association with gastric cancer risk. *Clin Cancer Res*, 2006, 12(3 Pt 1), 989-95.

[153] Bierne, H; Hamon, M; Cossart, P. Epigenetics and bacterial infections. *Cold Spring Harb Perspect Med*, 2012, 2(12), a010272.

[154] Hur, K; Niwa, T; Toyoda, T; Tsukamoto, T; Tatematsu, M; Yang, HK; et al. Insufficient role of cell proliferation in aberrant DNA methylation induction and involvement of specific types of inflammation. *Carcinogenesis*, 2011, 32(1), 35-41.

[155] Hahn, MA; Hahn, T; Lee, DH; Esworthy, RS; Kim, BW; Riggs, AD; et al. Methylation of polycomb target genes in intestinal cancer is mediated by inflammation. *Cancer Res*, 2008, 68(24), 10280-9.

[156] Tolg, C; Sabha, N; Cortese, R; Panchal, T; Ahsan, A; Soliman, A; et al. Uropathogenic E. coli infection provokes epigenetic downregulation of CDKN2A (p16INK4A) in uroepithelial cells. *Lab Invest*, 2011, 91(6), 825-36.

[157] Bobetsis, YA; Barros, SP; Lin, DM; Weidman, JR; Dolinoy, DC; Jirtle, RL; et al. Bacterial infection promotes DNA hypermethylation. *J Dent Res*, 2007, 86(2), 169-74.

[158] Haller, D; Holt, L; Kim, SC; Schwabe, RF; Sartor, RB; Jobin, C. Transforming growth factor-beta 1 inhibits non-pathogenic Gram negative bacteria-induced NF-kappa B recruitment to the interleukin-6 gene promoter in intestinal epithelial cells through modulation of histone acetylation. *J Biol Chem*, 2003, 278(26), 23851-60.

[159] Bardwell, AJ; Abdollahi, M; Bardwell, L. Anthrax lethal factor-cleavage products of MAPK (mitogen-activated protein kinase) kinases exhibit reduced binding to their cognate MAPKs. *Biochem J*, 2004, 378(Pt 2), 569-77.

[160] Pathak, SK; Basu, S; Bhattacharyya, A; Pathak, S; Banerjee, A; Basu, J; et al. TLR4-dependent NF-kappaB activation and mitogen- and stress-activated protein kinase 1-triggered phosphorylation events are central to Helicobacter pylori peptidyl prolyl cis-, trans-isomerase (HP0175)-mediated induction of IL-6 release from macrophages. *J Immunol*, 2006, 177(11), 7950-8.

[161] Xia, G; Schneider-Stock, R; Diestel, A; Habold, C; Krueger, S; Roessner, A; et al. Helicobacter pylori regulates p21(WAF1) by histone H4 acetylation. *Biochem Biophys Res Commun*, 2008, 369(2), 526-31.

[162] Hamon, MA; Batsché, E; Régnault, B; Tham, TN; Seveau, S; Muchardt, C; et al. Histone modifications induced by a family of bacterial toxins. *Proc Natl Acad Sci U S A*, 2007, 104(33), 13467-72.

[163] Hamon, MA; Cossart, P. K+ efflux is required for histone H3 dephosphorylation by Listeria monocytogenes listeriolysin O and other pore-forming toxins. *Infect Immun*, 2011, 79(7), 2839-46.

[164] Riggs, MG; Whittaker, RG; Neumann, JR; Ingram, VM. n-Butyrate causes histone modification in HeLa and Friend erythroleukaemia cells. *Nature*, 1977, 268(5619), 462-4.

[165] Imai, K; Ochiai, K. Role of histone modification on transcriptional regulation and HIV-1 gene expression: possible mechanisms of periodontal diseases in AIDS progression. *J Oral Sci*, 2011, 53(1), 1-13.

[166] Licciardi, PV; Wong, SS; Tang, ML; Karagiannis, TC. Epigenome targeting by probiotic metabolites. *Gut Pathog*, 2010, 2(1), 24.

[167] O'Connell, RM; Rao, DS; Baltimore, D. microRNA regulation of inflammatory responses. *Annu Rev Immunol*, 2012, 30(295-312.

[168] Fassi Fehri, L; Koch, M; Belogolova, E; Khalil, H; Bolz, C; Kalali, B; et al. Helicobacter pylori induces miR-155 in T cells in a cAMP-Foxp3-dependent manner. *PLoS One*, 2010, 5(3), e9500.

[169] Zeng, FR; Tang, LJ; He, Y; Garcia, RC. An update on the role of miRNA-155 in pathogenic microbial infections. *Microbes Infect*, 2015, 17(9), 613-21.

[170] Hanahan, D; Weinberg, RA. Hallmarks of cancer: the next generation. *Cell*, 2011, 144(5), 646-74.

[171] Beck, S; Bernstein, BE; Campbell, RM; Costello, JF; Dhanak, D; Ecker, JR; et al. A blueprint for an international cancer epigenome consortium. A report from the AACR Cancer Epigenome Task Force. *Cancer Res*, 2012, 72(24), 6319-24.

[172] Colotta, F; Allavena, P; Sica, A; Garlanda, C; Mantovani, A. Cancer-related inflammation, the seventh hallmark of cancer: links to genetic instability. *Carcinogenesis*, 2009, 30(7), 1073-81.

[173] Sharma, SV; Lee, DY; Li, B; Quinlan, MP; Takahashi, F; Maheswaran, S; et al. A chromatin-mediated reversible drug-tolerant state in cancer cell subpopulations. *Cell*, 2010, 141(1), 69-80.

[174] You, JS; Jones, PA. Cancer genetics and epigenetics: two sides of the same coin? *Cancer Cell*, 2012, 22(1), 9-20.

[175] Philip, M; Rowley, DA; Schreiber, H. Inflammation as a tumor promoter in cancer induction. *Semin Cancer Biol*, 2004, 14(6), 433-9.

[176] Kundu, JK; Surh, YJ. Inflammation: gearing the journey to cancer. *Mutat Res*, 2008, 659(1-2), 15-30.

[177] Pal, S; Bhattacharjee, A; Ali, A; Mandal, NC; Mandal, SC; Pal, M. Chronic inflammation and cancer: potential chemoprevention through nuclear factor kappa B and p53 mutual antagonism. *J Inflamm (Lond)*, 2014, 11(23.

[178] Söderlund, S; Brandt, L; Lapidus, A; Karlén, P; Broström, O; Löfberg, R; et al. Decreasing time-trends of colorectal cancer in a large cohort of patients with inflammatory bowel disease. *Gastroenterology*, 2009, 136(5), 1561-7, quiz 1818-9.

[179] Eaden, J; Abrams, K; Ekbom, A; Jackson, E; Mayberry, J. Colorectal cancer prevention in ulcerative colitis: a case-control study. *Aliment Pharmacol Ther*, 2000, 14(2), 145-53.

[180] Eaden, JA; Mayberry, JF. Colorectal cancer complicating ulcerative colitis: a review. *Am J Gastroenterol*, 2000, 95(10), 2710-9.

[181] Kundu, JK; Surh, YJ. Emerging avenues linking inflammation and cancer. *Free Radic Biol Med*, 2012, 52(9), 2013-37.

[182] Werling, K; Szepesi, A; Szentirmay, Z; Schaff, Z; Tulassay, Z; Szalay, F. Effect of hepatitis C virus on hepatocyte proliferation and DNA ploidy in patients with chronic hepatitis C. *Z Gastroenterol*, 2000, 38(7), 553-4, 556-8.

[183] Rajalingam, K; Schreck, R; Rapp, UR; Albert, S. Ras oncogenes and their downstream targets. *Biochim Biophys Acta*, 2007, 1773(8), 1177-95.

[184] Strano, S; Dell'Orso, S; Di Agostino, S; Fontemaggi, G; Sacchi, A; Blandino, G. Mutant p53: an oncogenic transcription factor. *Oncogene*, 2007, 26(15), 2212-9.

[185] Federico, A; Morgillo, F; Tuccillo, C; Ciardiello, F; Loguercio, C. Chronic inflammation and oxidative stress in human carcinogenesis. Int *J Cancer*, 2007, 121(11), 2381-6.

[186] Marnett, LJ. Oxyradicals and DNA damage. *Carcinogenesis*, 2000, 21(3), 361-70.

[187] Halliday, GM. Inflammation, gene mutation and photoimmunosuppression in response to UVR-induced oxidative damage contributes to photocarcinogenesis. *Mutat Res*, 2005, 571(1-2), 107-20.

[188] Xu, H; Chaturvedi, R; Cheng, Y; Bussiere, FI; Asim, M; Yao, MD; et al. Spermine oxidation induced by Helicobacter pylori results in apoptosis and DNA damage: implications for gastric carcinogenesis. *Cancer Res*, 2004, 64(23), 8521-5.

[189] Babbar, N; Casero, RA. Tumor necrosis factor-alpha increases reactive oxygen species by inducing spermine oxidase in human lung epithelial cells: a potential mechanism for inflammation-induced carcinogenesis. *Cancer Res*, 2006, 66(23), 11125-30.

[190] Bergman, Y; Cedar, H. DNA methylation dynamics in health and disease. *Nat Struct Mol Biol*, 2013, 20(3), 274-81.

[191] Esteller, M. Epigenetics in cancer. *N Engl J Med*, 2008, 358(11), 1148-59.

[192] Chan, AO; Lam, SK; Wong, BC; Kwong, YL; Rashid, A. Gene methylation in non-neoplastic mucosa of gastric cancer: age or Helicobacter pylori related? *Am J Pathol*, 2003, 163(1), 370-1, author reply 371-3.

[193] Chan, AO; Lam, SK; Wong, BC; Wong, WM; Yuen, MF; Yeung, YH; et al. Promoter methylation of E-cadherin gene in gastric mucosa associated with Helicobacter pylori infection and in gastric cancer. *Gut*, 2003, 52(4), 502-6.

[194] Kawasaki, H; Abe, H. in cancer and inflammation. *Personalized Medicine Universe*, 2012, 1(1), 7-12.

[195] Hansen, KD; Timp, W; Bravo, HC; Sabunciyan, S; Langmead, B; McDonald, OG; et al. Increased methylation variation in epigenetic domains across cancer types. *Nat Genet*, 2011, 43(8), 768-75.

[196] Xie, W; Schultz, MD; Lister, R; Hou, Z; Rajagopal, N; Ray, P; et al. Epigenomic analysis of multilineage differentiation of human embryonic stem cells. *Cell*, 2013, 153(5), 1134-48.

[197] Abu-Remaileh, M; Bender, S; Raddatz, G; Ansari, I; Cohen, D; Gutekunst, J; et al. Chronic inflammation induces a novel epigenetic program that is conserved in intestinal adenomas and in colorectal cancer. *Cancer Res*, 2015, 75(10), 2120-30.

[198] Lawrence, T. The nuclear factor NF-kappaB pathway in inflammation. *Cold Spring Harb Perspect Biol*, 2009, 1(6), a001651.

[199] Valinluck, V; Sowers, LC. Inflammation-mediated cytosine damage: a mechanistic link between inflammation and the epigenetic alterations in human cancers. *Cancer Res*, 2007, 67(12), 5583-6.

[200] Hodge, DR; Peng, B; Cherry, JC; Hurt, EM; Fox, SD; Kelley, JA; et al. Interleukin 6 supports the maintenance of p53 tumor suppressor gene promoter methylation. *Cancer Res*, 2005, 65(11), 4673-82.

[201] Wehbe, H; Henson, R; Meng, F; Mize-Berge, J; Patel, T. Interleukin-6 contributes to growth in cholangiocarcinoma cells by aberrant promoter methylation and gene expression. *Cancer Res*, 2006, 66(21), 10517-24.

[202] Isomoto, H; Mott, JL; Kobayashi, S; Werneburg, NW; Bronk, SF; Haan, S, et al. Sustained IL-6/STAT-3 signaling in cholangiocarcinoma cells due to SOCS-3 epigenetic silencing. *Gastroenterology*, 2007, 132(1), 384-96.

[203] Gray, SG; The, BT. Histone acetylation/deacetylation and cancer: an "open" and "shut" case? *Curr Mol Med*, 2001, 1(4), 401-29.

[204] Huang, L. Targeting histone deacetylases for the treatment of cancer and inflammatory diseases. *J Cell Physiol*, 2006, 209(3), 611-6.

[205] Santos-Rosa, H; Caldas, C. Chromatin modifier enzymes, the histone code and cancer. *Eur J Cancer*, 2005, 41(16), 2381-402.

[206] Yang, SH; Sharrocks, AD. SUMO promotes HDAC-mediated transcriptional repression. *Mol Cell*, 2004, 13(4), 611-7.

[207] Xu, J; Zhang, X; Monestier, M; Esmon, NL; Esmon, CT. Extracellular histones are mediators of death through TLR2 and TLR4 in mouse fatal liver injury. *J Immunol*, 2011, 187(5), 2626-31.

[208] Takagi, T; Naito, Y; Mizushima, K; Hirata, I; Yagi, N; Tomatsuri, N; et al. Increased expression of microRNA in the inflamed colonic mucosa of patients with active ulcerative colitis. *J Gastroenterol Hepatol*, 2010, 25 Suppl 1(S129-33.

[209] Meng, F; Henson, R; Wehbe-Janek, H; Smith, H; Ueno, Y; Patel, T. The MicroRNA let-7a modulates interleukin-6-dependent STAT-3 survival signaling in malignant human cholangiocytes. *J Biol Chem*, 2007, 282(11), 8256-64.

[210] Meng, F; Wehbe-Janek, H; Henson, R; Smith, H; Patel, T. Epigenetic regulation of microRNA-370 by interleukin-6 in malignant human cholangiocytes. *Oncogene*, 2008, 27(3), 378-86.

In: Epigenetics: Mechanisms and Clinical Perspectives
Editor: Yildiz Dincer
ISBN: 978-1-63484-490-1

Chapter 8

A New Molecular Approach to Obesity: Epigenetic Perspective

Burcu Bayoglu*[*], PhD, and Mujgan Cengiz, PhD
Istanbul University Cerrahpasa Medical Faculty,
Department of Medical Biology, Istanbul, Turkey

Abstract

Obesity is a public health problem leading to morbidity and mortality throughout the world. It arises from the interactions between genetics and environmental factors. In recent years, susceptibility to obesity has been also linked to epigenetic factors. Epigenetics, is the study of heritable changes in gene expression which do not involve in the underlying DNA sequence. The epigenetic mechanisms include DNA methylation, covalent histone modifications, chromatin folding, miRNAs, and polycomb group repressive complexes. Both dietary factors and individual behaviors affect obesity development via epigenetic mechanisms. Epigenetic mechanisms are also linked to programmed changes in gene expression as a result of early environmental exposures during pregnancy which alter offspring growth and development. There is evidence that nutrient and environmental exposures during pregnancy may affect fetal/newborn development and result in offspring obesity or metabolic syndrome which is a cluster of metabolic abnormalities. Obesity related genes, epiobesigenes, display methylation patterns playing important roles in the development of obesity which are potential future epigenetic biomarkers of obesity. The susceptibility genes have been reported as FGF2, PTEN, CDKN1A, and ESR1, functional in adipogenesis; SOCS1 and SOCS3, functional in inflammation, and COX7A1, LPL, CAV1, and IGFBP3 which are functional in fat metabolism and insulin signaling. It is important to prevent this era's epidemic, obesity, since it leads to chronic diseases including hypertension, atherosclerosis, insulin resistance, and moreover metabolic syndrome. It may be easier to prevent the progress of the disease by revealing the epigenetic mechanisms especially methylation profiles of the susceptibility genes.

[*] E-mail: burcubayoglu@yahoo.com.

Keywords: obesity, adipogenesis, epigenetics, epiobesigenic genes, DNA methylation, histon modifications

Abbreviations

AcH3	histone acetylation;
ADIPOQ	adiponectin;
Avy	agouti;
BMI	body mass index;
CASP8	caspase 8;
CAV-1	caveolin-1;
CDKN1A	cyclin-dependent kinase inhibitor 1A;
CEBPA	CCAAT/enhancer binding protein (C/EBP) a;
CG	Cytosine-Guanine;
COX7A1	cytochrome c oxidase subunit VIIa polypeptide 1;
DNA	deoxyribonucleic acid;
DNMTs	DNA methyltransferases;
ESR1	Oestrogen receptor-alpha; FABP4, fatty acid-binding protein 4, adipocyte;
FASN,	fatty acid synthase;
FGF2	fibroblast growth factor-2;
FTO	fat mass and obesity associated;
GLUT4	insulin-responsive glucose transporter 4;
H2K4me3	trimethylation of histone H3 at lysine 4;
H3K4	histone lysine methylation;
H3K9	histone H3 lysine 9;
H3K9me2	dimethylation of histone H3 at lysine 9;
HDAC	histone deacetylase;
HIF1A	hypoxia inducible factor 1;
HMT	histone methylase;
IFNG	Interferon γ;
IGF2	insulin-like growth factor 2;
IGFBP3	insulin-like growth factor-binding protein 3;
INS	insulin;
INSR	insulin receptor;
H3-K27	lysine 27 of histone H3;
LEP	leptin;
LPL	lipoprotein lipase;
MC4R	melanocortin 4 receptor;
miRNA	microRNA;
ncRNAs	non-coding RNAs;
NR0B2	nuclear receptor small heterodimer partner fibroblast growth factor-2;
NPY	neuropeptide Y;
NR3C1	glucocorticoid receptor; nts, nucleotides;

PBMC	peripheral blood mononuclear cell;
PcG	polycomb group;
PcG	polycomb group proteins;
PEPCK	phosphoenolpyruvate carboxykinase;
PIK3CG	phosphoinositide-3-kinase, catalytic, gamma pp;
PLA2G4A	plasma secretory phospholipase A(2) type IIA;
POMC	proopiomelanocortin;
PPARA	peroxisome proliferator-activated receptor a;
PPARγ	peroxisome proliferator-activated receptor gamma;
PPARGC1A	peroxisome proliferator-activated receptor gamma-coactivator-1 alpha;
PTEN	phosphatase and tensin homologue;
PWS	Prader Willi Sydrome;
PRC2	polycomb repressive complex 2;
RARA,	retinoic acid receptor-alpha;
SOCS1	suppressor of cytokine signalling 1;
SOCS3	suppressor of cytokine signalling 3;
SOD3	extracellular superoxide dismutase;
SSTR2	somatostatin receptor 2;
HSD11B2	11 beta-hydroxysteroid dehydrogenase 2;
T2D	type 2 diabetes;
TNF	tumor necrosis factor;
UCP1	uncoupling protein 1.

1. Obesity and Its Molecular Pathogenesis

Obesity is a major health problem leading to morbidity and mortality throughout the world [1]. Obesity has a multifactorial etiology which involves interactions between genetic background, hormones and environmental factors [2]. However, the etiology of obesity is not well understood. Obesity also plays a crucial role for developing metabolic syndrome which includes insulin resistance, hypertension, elevated triglycerides and atherosclerosis [3]. Metabolic syndrome is a growing cause of morbidity and mortality worldwide. It is characterized by the cluster of metabolic disturbances including obesity, hyperlipidemia, hypertension, and elevated fasting blood sugar. The risk for metabolic syndrome has largely been attributed to adult lifestyle factors such as poor nutrition, lack of exercise, and smoking, however; there is now strong evidence suggesting that predisposition to metabolic syndrome development begins in *utero* [4]. Obesity is related with both type 2 diabetes (T2D) and metabolic syndrome development. Other than individual life style choices, there is evidence that developing obesity is in part due to genetic disposition and epigenetic alterations [5].

Body mass index (BMI) is the commonly used measure of weight status. BMI is a calculation of the ratio of someone's height and weight ($BMI=kg/m^2$).

Research have shown that, BMI is a good estimate of fatness and correlates with chronic diseases including heart disease, diabetes, cancer and overall mortality. For adult men and women, a BMI between 18.5 and 24.9 is considered healthy. BMI between 25.0 and 29.9 is defined as overweight and a BMI of 30 or higher is considered obese [6].

When energy intake exceeds energy expenditure, the excess energy is often stored as body fat and cause obesity. In recent years, it is suggested that the availability of energy-dense food, environmental changes create an obesogenic environment. Since obesity is multifactorial, an individual can become obese in the absence of an obesogenic environment or on the contrary, an individual does not become obese in an obesogenic environment. Various factors causing obesity can be endocrine disrupters, which are the chemicals that mimic the effect of hormones in the body, and also reproductive factors including intrauterine effects, epigenetics and maternal age [7]. An obese phenotype has occurred among people in a short period of time, just one or two generations.

This shows that environmental or epigenetic factors play a major role rather than genetic mechanisms in the etiology of obesity. Although, high fat diet with decreased physical activity seems to be the strong contributor of the prevalence of obesity, some studies support that obesity may have its origin in *utero* [8-12]. Obesity and the metabolic abnormalities related with it, are developmentally programmed since, although normal postnatal nutrition is carried out, obesity exist in offspring because of the suboptimal nutrition in *utero* [13]. Similarly to type 2 diabetes, hypertension, atherosclerosis and other metabolic disorders, predisposition to obesity, and weight loss have been associated with epigenetic pattern alterations. Nutritional factors and also different non-nutritional risk factors accompanying obesity are involved in the epigenetic modifications, and they affect adipogenesis, insulin sensitivity, especially hyperglycemia, inflammation, endocrine disruptors, hypoxia, and oxidative stress [14], (Figure 1).

In recent years, the epigenetic regulation of the gene expression in obesity seems to be a potential considerable contributor. Epigenetics is defined as the study of heritable gene expression changes that occur in the absence of a change in the DNA sequence itself [15]. In other words, epigenetics is the study of gene expression's molecular control that is not caused by DNA sequence.

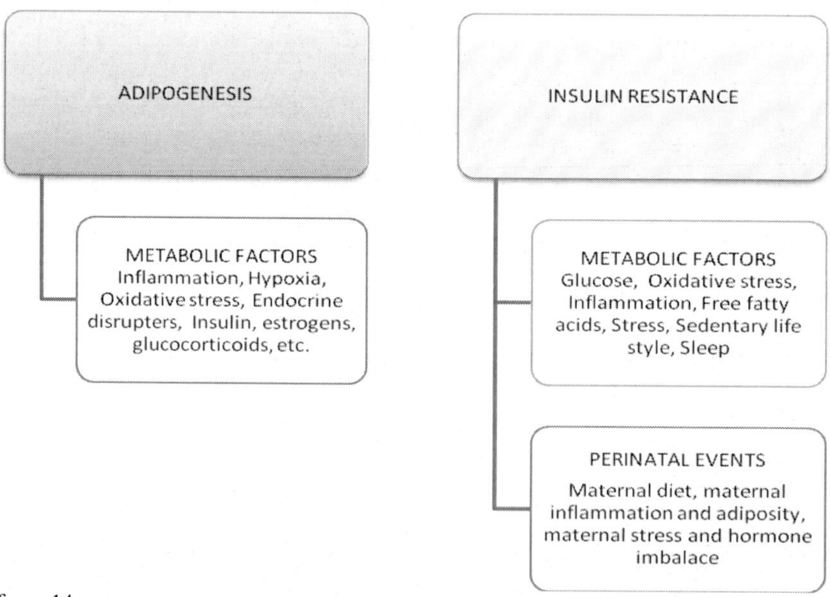

Modified from 14.

Figure 1. Metabolic and perinatal factors related to adipogenesis, insulin resistance.

Epigenetic modifications include methylation, covalent modifications of histones, DNA packaging around nucleosomes, chromatin folding and chromatin attachment to the nuclear matrix [16].

Epigenetics may help to understand genetic fetal programming, monozygotic twin differences, and chronic disease onset in adults, which interact with dietary intake and nutritional processes [14]. For instance, DNA sequence methylation usually reduces gene expression. Epigenetic factors may be passed on mitotically, through cell division, or meiotically, which is called transgenerational inheritance.

Thus, epigenetic programming of parental or maternal gametes, the fetus, or early postnatal development may be influenced by environmental factors [7] (Figure 2).

2. Epigenetics: Basic Mechanisms

DNA methylation, histone-tail acetylation, poly-ADP-ribosylation and ATP-dependent chromatin-remodeling are all included in chromatin-remodeling mechanisms [16]. DNA methylation is the covalent attachment of a methyl group onto the Cytosine residues' C5 position on CpG islands. CpG islands, are the genomic regions containing high number of Cytosine-Guanine (CG) dinucleotides. In the methylation of CpG islands, cytosine is converted to 5-methylcytosine and since CpG islands are mostly located on the promoter region, methylation process is often associated with gene silencing [17] (Figure 3). DNA methyltransferases (DNMTs) catalyze the addition of the methyl group to the DNA. Three major enzymes function in maintaining DNA methylation patterns.

1- DNA methyltransferases 3A (DNMT3A)
2- DNA methyltransferases 3B (DNMT3B)
3- DNA methyltransferases 1 (DNMT1)

DNMT3A and 3B are *de novo* methyltransferases, DNMT1 controls the methylation patterns they are copied in each cell division [18]. DNMTs interact with histone deacetylases, histone methyltransferases and methyl-cytosine-binding proteins to establish and maintain DNA methylation patterns in a complex regulatory pathway in the cell. In histone modifications, methylation of histone H3 lysine 9 (H3K9) means condensed and inactive chromatin whereas acetylation is related with active gene expression [17]. DNA methylation typically occurs during DNA replication, so that methylation patterns maintain in cell division via hemimethylated DNA. However during replication and embryogenesis, *de novo* DNA methylation can also occur [18].

To date four current models have been discussed for how DNA methylation can mediate gene silencing [19]. In the first model, DNA methylation can prevent the binding of transcriptional activator to the target DNA sequence and this directly inhibits transactivation [20]. The second model explains that the DNMT protein may link to histone deacetylase (HDAC) and histone methylase (HMT) proteins which allows the coupling of enzymatic activities [21].

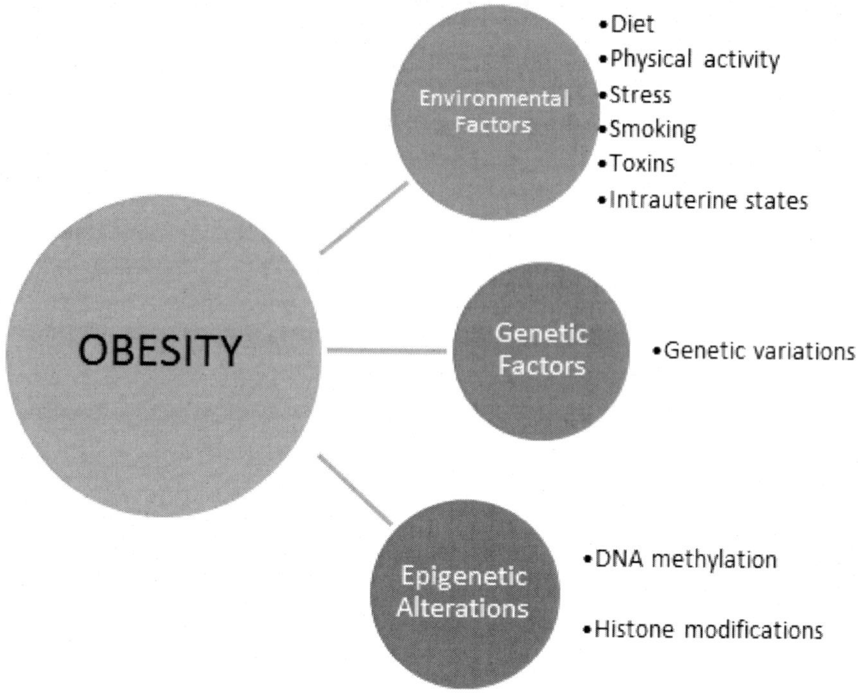

Figure 2. The interplay between environmental, genetic and epigenetic factors in the etiology of obesity.

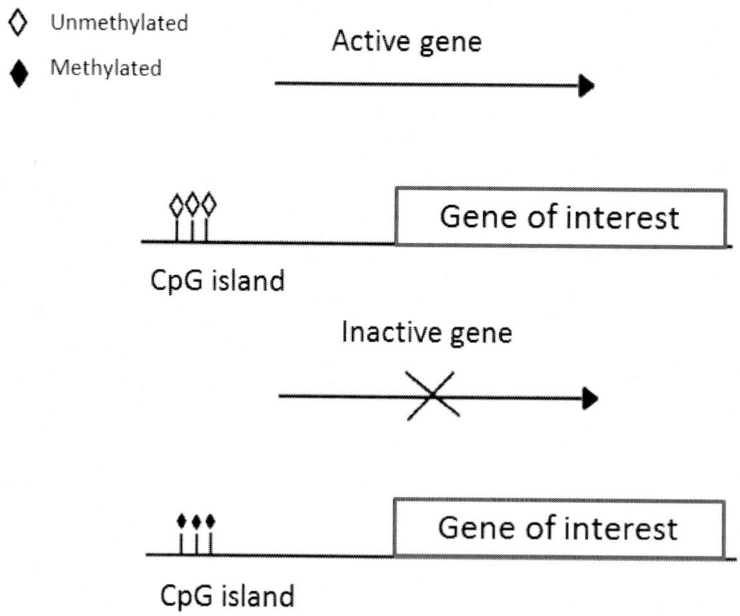

Figure 3. DNA methylation of CpG islands.

Next, DNA methylation within the gene exhibits an effect of repression on transcriptional elongation. Methylation occurs generally in the promoter region, however it can occur downstream the gene too [22]. Finally, methylated DNA can be recognized directly by methyl-CpG-binding proteins. And these proteins can recruit transcriptional repressors to silence the surrounding chromatin [23, 24].

There are also other mechanisms related with epigenetic pathways. One of them is chromatin packaging into open and closed states. The open state of the chromatin is called the euchromatin and the closed state is called the heterochromatin. DNA is packaged around histones and forms the chromatin. Histone tails are translationally modified with methylation/demethylation and acetylation/deacetylation mechanisms by histone modifying enzymes. Histone methylation may activate or repress transcription, depending on which lysine residue is methylated. For instance, trimethylation of histone H3 at lysine 4 (H2K4me3) means gene transcription is active, however dimethylation of histone H3 at lysine 9 (H3K9me2) means transcriptional silencing. Unlike methylation, acetylation activates, deacetylation represses gene expression [13, 25, 26].

An additional mechanism for the regulation of gene expression and silencing is polycomb group (PcG) complexes. PcG complexes are associated with remodeling of chromatins; for example they get involved in histone H3K27 methylation or H2AK119 ubiquitination. In this mechanism, chromatin is epigenetically repressed by these protein complexes. They are recruited to the promoter region of gene of interest and prevent transcription factors to bind the promoter of DNA sequences [27].

In the epigenetic regulation of gene expression, non-coding RNAs (ncRNAs) also take place. NcRNAs are transcribed from DNA, but they are not translated to proteins. Their main function is regulation in gene expression at the transcriptional and post-transcriptional states. NcRNAs are classified into short ncRNAs (<30 nts) and the long ncRNAs (>200 nts). Three major short ncRNAs participate in gene silencing; microRNAs (miRNAs), short inhibitory RNAs, and piwi-interacting RNAs [28, 29]. Long ncRNAs play a regulatory role during development [30]. They are expressed cell type-specific [31, 32]. Long ncRNAs are also found to be associated with adipogenesis [33, 34]. Recently, it is suggested that ncRNAs do not just regulate gene expression in translational level but also they take part in DNA methylation and histone modifications, and thus regulating the transcription of their target genes [35, 36]. Epigenetic gene silencing is also related with the microRNAs (miRNAs). miRNAs are small ncRNA genes composed of 22-nucleotides. They are specific gene regulators and control the gene expression [37]. miRNAs are transcriptionally regulated by DNA methylation and histone modifications. And so, they have significant roles in targeting key enzymes which control chromatin structure and histone modifications [17].

Epigenome is the term covering overall epigenetic state of a cell. During gametogenesis and between fertilization and blastocyst formation, epigenome may be susceptible to dysregulate [38], however during gestation, lactation and throughout life course DNA methylation patterns are maintained [39]. DNA methylation patterns change during aging process, thus a good working epigenetic homeostatic mechanism must be activated to prevent epigenetic abnormalities [40]. Otherwise, epigenetic lesions would accumulate to develop nutrition-related disorders like obesity [39]. Epigenomics is the study of genome-wide epigenetic modifications. Epigenetics is the study trying to understand relations between genes and environmental factors such as diet, inactivity, smoking, and etc. to generate a phenotype [41]. Epigenome is dynamic, since it is changed in response to nutrition, physical exercise, weight loss and aging [42-45]. In *utero* nutrition, environmental exposures, maternal or fetal stress may change the fetus's gene expression via epigenetic mechanisms, thus change the cells' and organs' structure and function leading to metabolic anomalies [13].

3. Obesity: From an Epigenetic Perspective

Obesity is a multifactorial disorder involving complex interactions among genetic factors, hormones, and some environmental factors including sedentary life style and inadequate dietary habits [2]. Obesity has been associated with lifestyle and accordingly, studies have investigated thermoenergetic, life style and balance between energy intake and expenditure. Nowadays, genetic studies and molecular pathways are being investigated in obesity pathogenesis. These studies include gene mutations and polymorphisms, candidate gene approach etc [46]. Also, in recent years epigenetic regulation of gene expression has been related with obesity pathogenesis as a potential significant contributor.

There are largely animal and limited human studies reporting the association between epigenetic mechanisms and obesity. Tissue gene expression can be changed via epigenetic regulation in accordance with environmental conditions throughout life. Nutritional and environmental factors may interact with the genotype and modulate epigenetic marks in somatic cells and affect the phenotype. This is shown in monozygous twins and in *agouti* mouse.

In monozygous twins, the offspring have divergent DNA methylation and histone acetylation patterns. Lymphocyte and muscle biopsy samples were taken from 3-74 years old Spanish monozygotic twins. They are analyzed for large-scaled and locus-specific DNA methylation and histone acetylation. Older twins showed differences in their epigenetic marks and gene expression profiles; however younger twins showed no differences, suggesting that there is environmental effect on gene expression regulation, despite a similar genotype [44].

Animal obesity model of *agouti* mouse with the *agouti* viable yellow (*Avy*) mutation is used for obesity studies and this model can be modulated through an epigenetic mechanism [47]. In *agouti* mouse, the diet affects methylation status and the animal's coat color [48, 49]. In *agouti* mice, when the *agouti* gene is normally methylated, the mice are thin and have brown color. However, when the *agouti* gene completely unmethylated, the mice are obese and diabetic and also they have yellow coat color. This mutation causes the ectopic expression of the agouti protein. As a result of this mutation, the agouti protein binds to the melanocortin 4 receptor in the hypothalamus antagonistically and leads to hypothalamic dysfunction and induces hyperphagic obesity. It is known that, methylation controls gene expression, therefore, obesity phenotype depends on methylation [47]. Body weight can be affected by epigenetic regulation with dietary factors. In *agouti* mice, maternal diet with methyl supplementation affects the methylation of the *Avy* allele and offspring obesity [49]. The epigenetic regulation of hedonic reward pathways and metabolic regulation of energy balance may be changed with maternal and post-weaning high fat diet in mice and results in altering methylation of the leptin promoter in rats [50-53]. Yellow fat mice and brown thin mice have the same genotype but they have different phenotypes as a result of epigenetic alterations. These methylation patterns can also be transmitted across generations [48, 49]. Interestingly, when yellow fat mice were fed with methyl-rich diet during pregnancy, they produced offspring with brown coat without obesity. This result showed the interaction between nutrition and epigenetic modifications influencing the phenotype [54].

Epigenetic modulation of imprinted genes is also well known, for example the impact of insulin-like growth factor 2 (*IGF2)* on body weight and obesity is well studied [55]. By several autocrine, paracrine and endocrine hormonal factors, adipose tissue growth can be induced. For the increase in adipose tissue, IGF2 expression is increased methylation of differentially methylated regions in the H19 imprinting control region. Therefore, studies on the IGF2/H19 locus showed that hypomethylated CpG islands in the 5′ region of the H19 gene are critical binding sites for the 11-zinc finger protein CTCF [56]. Activation of IGF2 gene transcription is regulated via maternal diets such as low dietary protein and folate through increased methylation of ICR/H19 DMR [57]. Also, another mechanism for elevated IGF2 expression is acetylation of histone 4 (H4) at lysines 16 and 8 [58]. In this mechanism, from polycomb group proteins (PcG), polycomb repressive complex 2 (PRC2) is recruited through the interaction of CTCF with Suz12 PRC2 subunit and leads to allele-specific methylation at lysine 27 of histone H3 (H3-K27) and represses the maternal IGF2 promoters [59].

Epigenetic alterations may also occur in nonimprinted genes involved in energy metabolism. Genes functioning in adipogenesis, glucose homeostasis, inflammation and/or insulin signaling are regulated by epigenetic mechanisms, for example hormone encoding genes like, leptin; nuclear receptors like adipogenic and lipogenic transcription factors, PPARγ and PPARα, respectively; gluconeogenic enzymes, phosphoenolpyruvate carboxykinase (PEPCK); and transmembrane proteins, uncoupling protein 1 [53, 60-67]. For example, leptin plasma levels can be modulated by diet with epigenetic mechanisms. It is also suggested that leptin methylation affects leptin gene expression in rats since high-fat diet in rats increased the leptin gene promoter methylation in retroperitoneal adipocytes and this was related with lower circulating leptin levels [53]. Also, adipocyte differentiation and the induction of adipogenic transcription factors via epigenetic mechanisms, such as methylation of PPARγ and C/EBPα, may drive adipogenesis. Adipocyte differentiation is also regulated by both histone lysine methylation (H3K4) and acetylation (AcH3). Increased histone trimethylation (H3K4me3) and acetylation (AcH3K9/K14) are also consistent with the PPARγ and C/EBPα upregulation during terminal adipocyte differentiation. When H3K4me3 and AcH3K9/K14 increases, adipogenic genes may be induced contributing to early adipocyte differentiation and obesity [68-70].

Hepatic PEPCK is the rate-limiting enzyme in the metabolic pathway of gluconeogenesis. High-calorie diet causes hypomethylation of the *PEPCK* gene in rats. And this was reported to be associated with the increased *PEPCK* mRNA levels, which shows increased liver gluconeogenesis [71]. Methylation alterations were found to be associated with changes in target gene expression and glucose and lipid metabolism [13].

Recently, DNA methylation and specific histone methylation (H3K4 and H3K9) and some miRNAs were found to be related with BMI [72-74]. In a study performed with genome wide methylation analysis, human obesity was reported to be associated with methylation changes in blood leukocyte DNA [72]. Another study related with epigenetic modifications was performed in overweight individuals with and without type 2 diabetes. Histone methylation of K4 and K9 in primary human adipocytes of overweight individuals emerged 40% lower levels of K4 dimethylation in overweight nondiabetic individulas. Also, K4 trimethylation was reported 40 % higher in adipocytes of overweight diabetic subjects when compared with normal-weight and overweight non-diabetic subjects. In the same study, obese and lean individuals showed varying DNA methylation levels in specific genes. Also, in cultured preadipocytes and mature adipocytes miRNA expression profiles showed different expressions [73]. Similary, in human subcutaneous adipose tissue biopsies, differential and dysregulated miRNA expression was observed in obese and lean individuals of genome wide miRNA profiling. In human adipose tissue, miRNA expression pattern was associated with obesity and glucose metabolism parameters [74].

In the development of obesity, animal studies were also performed to reveal the role of the epigenome. But this relation was more specific to histone modifications and miRNAs rather than DNA methylation [65, 75-78]. For example, the main adipogenic transcription factor PPAR and its target genes regulation have been associated with histone modifications (methylation and acetylation) and miRNAs, and leads to obesity, hyperphagia, and hyperlipidemia [75, 79-81]. In another animal study, maternal obesity, high-fat diet in mice were shown to increase the key transcription factor, Zfp423, which is responsible for adipogenic lineage commitment during fetal development. The repressive histone methylation, H3K27me3, was observed lower in Zfp423 promoter of fetal tissues from maternal obese mice [82].

Increased mRNA expression and CpG site/island DNA methylation changes of pro-adipogenic factors (Zfp423 and C/EBP-β) have been shown in 3 week old rat offspring from obese pregnancies [83]. On the other hand, mice with genetically increased or decreased levels of DNA methyltransferases, DNMT1 and DNMT3b, do not gain weight or have increased adiposity. In the same study, *de novo* DNMT3a was more than doubled in white adipose tissue of obese mice, and a transgenic mouse having threefold increased *DNMT3a* mRNA levels in adipose tissue, did not show increased body weight or adiposity [84]. In that study, it was argued that, perhaps it is histone deacetylases which was functional in tissue specific expression rather than DNA methyltransferases. However, specific histone methyltransferases play a significant role in adipogenesis; mice with mutations in the histone H3, lysine 4 methyltransferase *MLL3* show decreased adipogenesis and they are protected from high-fat diet induced obesity [85]. In contrast, mice with loss of function of the H3K9-specific demethylase JmjC domain with histone demethylase 2A are obese and hyperlipidemic [75, 76]. Thus, it has been shown that mutations in epigenetic-modifiers may lead to obesity [13].

3.1. Programmed Obesity and the Epigenome

During, gametogenesis and early embryogenesis, epigenetic modifications are not completely erased. And some memory of the epigenetic state transfers to the offspring [86, 87]. This is called the transgenerational epigenetic inheritance [48]. Also, mother's nutrition or perinatal lifestyle choices in *utero* or lactation periods, may change the fetus's developmental program and may lead to obesity development in the future [88]. However the role of the nutrition on epigenetic pattern modifications during adulthood and its transfer to the gametes is still unknown [89]. So the effects of the dietary and environmental factors on DNA methylation patterns can lead to the susceptibility to develop obesity and related metabolic diseases [39].

Environmental exposures during early life may induce offspring's epigenome changes. This may lead to increased risk of obesity and metabolic syndrome in later life. Studies showed that programmed obesity in humans occurs via DNA methylation changes. Histone modifications and chromatin structure changes has not been shown yet in programmed obesity. Exposure to maternal famine or obesity decreases the DNA methylation of the imprinted *IGF2* gene in the offspring. 60-year-old adults, prenatally exposed to famine, in the

Dutch Famine (1944–1945) cohort, exhibit hypomethylation of whole blood *IGF2* gene, and hypermethylation of two obesity-related non-imprinted genes, *TNF* and leptin, compared to their unexposed, same-sex siblings. This association showed early developmental period is crucial for establishing and maintaining epigenetic marks [90, 91]. Recently, paternal obesity was found to be associated with *IGF2* hypomethylation in umbilical cord blood leukocytes of newborns [92, 93]. Also, specific maternal features like gestational weight gain and gestational diabetes have been related with signature DNA methylation in cord blood and elevated placental leptin gene methylation, respectively [42, 94]. Also, in early pregnancy maternal carbohydrate intake was related with the umbilical cord methylation levels of retinoid X receptor-α which is a nuclear receptor gene and make heterodimers with adipogenic transcription factor PPARγ, and also related with adiposity in later childhood [95]. Animal studies show that both DNA methylation and histone modification changes are related with developmental programming of obesity in rodent models of maternal under or overnutrition; for example, high-fat diet, maternal obesity [41, 96-99]. Genes that regulate growth factors [100], adipogenesis [101], brain appetite and satiety/reward pathways [51, 102-104] and glucose homeostasis [105] have been detected to be regulated by epigenetic changes. Human and animal studies provide evidence of programmed obesity and metabolic syndrome resulting from early nutritional exposures and suggesting epigenetic modification of nonimprinted genes may be a major contributing factor [13].

3.2. Epiobesigenic Genes

In the onset, development and therapy of type 2 diabetes and obesity, epigenetic research became to have advances [16]. Some gene promoters are susceptible to epigenetic regulation and may have potential role in the development of obesity. These genes are called epiobesigenic genes [39]. Some human genes regulated by epigenetic mechanisms related with obesity are shown in Table 1.

In human obesity, many genes have been identified. These gene effects' may be modified by epigenetic mechanisms [144]. The most known example of obesity is the Prader Willi Sydrome (PWS) caused by an epigenetic mechanism in human. PWS occurs due to a lack of expression from parental chromosome 15q11-q13. This lack of expression arise from the epigenetic silencing on the maternal and paternal copy of the genes, instead of just the maternal copy, in 25% of cases. PWS is characterized with hyperphagia, infertility, and obesity [145], however obese phenotype is not passed on, due to infertility [7].

Leptin and proopiomelanocortin gene mutations also cause human obesity, and these genes have both CpG islands in which their expression can be controlled through methylation [114]. Some of these susceptibility genes are associated with various metabolic diseases like hypertension (*HSD11B2*) [134], atherosclerosis (*PPARG*) [61], diabetes (*PPARGC1*) and endotoxin tolerance [66]. Also, some of these genes may play a role in obesity since they affect adipogenesis process [61, 107-116, 134], inflammation [66, 114, 123-126, 128] and insulin signaling [66, 131-135]. When unmethylated CpG islands are hypermethylated, this means that transcriptional activation is repressed. The presence of CpG islands in the promoter region of a gene shows that, the gene might be partly regulated by CpG methylation [134]. Normal and pathologic gene expressions are being studied with methylation pattern mapping of CpG islands.

Table 1. Some human genes regulated by epigenetic mechanisms (Epiobesigenes) involved in obesity pathogenesis (Modified from 14, 39, 106)

Gene symbol	Gene name	Role in obesity	Epigenetic mechanism	Ref.
PPARGC1A	Peroxisome proliferator-activated receptor gamma-coactivator-1 alpha	Adipogenesis	Important in human islet insulin secretion	[107]
NR0B2	Nuclear receptor small heterodimer partner Fibroblast growth factor-2	Adipogenesis	Tumour suppressor methylation related	[108]
FGF2	Fibroblast growth factor-2	Adipogenesis	Homocysteine disrupts endothelial cells through altered promoter DNA methylation	[109]
PPARG	Peroxisome proliferator-activated receptor gamma	Adipogenesis	Changes in DNA methylation during cellular ageing and atherosclerosis	[61]
PTEN	Phosphatase and tensin homologue	Adipogenesis	Epigenetic role in colorectal cancer and gliomas	[110]
RARA	Retinoic acid receptor-alpha	Adipogenesis	Promoter hypermethylationis associated with prostate and mammary cancer	[111]
CDKN1A	Cyclin-dependent kinase inhibitor 1A (p21, Cip1)	Adipogenesis and cell cycle	Aberrant promoter methylations related to cancer	[112]
LEP	Leptin	Adipogenesis and inflammation, appetite regulation	Post-zygotic development, adipocyte maturation and cellular ageing, DNA methylation	[53, 61, 113, 114]
ESR1	Oestrogen receptor-alpha	Adipogenesis	Prognostic value of Oestrogen receptor hypermethylation	[115]
NR3C1	Glucocorticoid receptor	Adipogenesis, stress	Methylation status is sensitive to prenatal maternal mood, Histone acetylation	[116]
CEBPA	CCAAT/enhancer binding protein (C/EBP) a	Adipogenesis	Histone acetylation and methylation	[117]
PPARA	Peroxisome proliferator-activated receptor a	Adipogenesis	DNA methylation	[118]

Gene symbol	Gene name	Role in obesity	Epigenetic mechanism	Ref.
Avy	Agouti	Appetite control and adipogenesis	DNA methylation	[119, 120]
MC4R	Melanocortin 4 receptor	Appetite regulation	DNA methylation	[121]
NPY	Neuropeptide Y	Appetite regulation	DNA methylation	[122]
POMC	Proopiomelanocortin	Appetite regulation	DNA methylation and histone acetylation and methylation	[122]
TNF	Tumor necrosis factor	Inflammation and insulin resistance	Epigenetic silencing during endotoxin tolerance and myeloid differentiation, DNA methylation, Methylation status is sensitive to prenatal maternal mood	[66]
PLA2G4A	Plasma secretory phospholipase A(2) type IIA	Inflammation	Malignant prostate cells	[123]
SOD3	Extracellular superoxide dismutase	Inflammation	Development of foam cells	[124]
SOCS1	Suppressor of cytokine signalling 1	Inflammation	Severity of liver fibrosis and hepatocarcionma	[125]
SOCS3	Suppressor of cytokine signalling 3	Inflammation	Role in cellular growth and migration and melanomas	[126]
IFNG	Interferon γ	Inflammation	DNA methylation	[127]
CASP8	Caspase 8	Inflammation and apoptosis	Hypermethylation in neuroblastomas and medulloblastomas	[128]
COX7A1	Cytochrome c oxidase subunit VIIa polypeptide 1 (muscle)	Energy metabolism	Age influences DNA methylation	[129]
LPL	Lipoprotein lipase	Fat metabolism	Changes in DNA methylation during cellular ageing	[61]
FABP4	Fatty acid-binding protein 4, adipocyte	Fat metabolism	Changes in DNA methylation during cellular ageing	[61]
FASN	Fatty acid synthase	Lipid storage	DNA methylation	[130]
CAV1	Caveolin-1	Insulin signalling	Aberrant methylation is associated with hepatocellular carcinoma	[131]
PIK3CG	Phosphoinositide-3-kinase, catalytic, gamma pp	Insulin signalling	CpG hypermethylation is associated with progression of colorectal cancer	[132]

Table 1. (Continued)

Gene symbol	Gene name	Role in obesity	Epigenetic mechanism	Ref.
SSTR2	Somatostatin receptor 2	Insulin resistance	Tissue-specific related	[133]
HSD11B2	11 beta-hydroxysteroid dehydrogenase 2	Insulin resistance and adiposity	*In vivo* epigenetic repression and relation to hypertension	[134]
IGFBP3	Insulin-like growth factor-binding protein 3	Insulin resistance	Hypermethylation is associated with non-small cell lung cancer	[135]
IGF2	Insulin-like growth factor 2	Glucose homeostasis	DNA methylation and histone acetylation	[136]
ADIPOQ	Adiponectin	Glucose homeostasis	DNA methylation and histone acetylation	[137]
GLUT4	Insulin-responsive glucose transporter 4	Glucose homeostasis	DNA methylation and histone acetylation	[117]
INS	Insulin	Glucose homeostasis	DNA methylation and histone acetylation	[138, 139]
INSR	Insulin receptor	Adipogenesis, Insulin sensitivity disruption	DNA methylation	[140]
FTO	Fat mass and obesity associated	Body weight homeostasis	DNA methylation	[141]
HIF1A	Hypoxia inducible factor 1	Hypoxia	DNA methylation and histone acetylation and methylation	[142]
UCP1	Uncoupling protein 1	Thermogenesis	DNA methylation	[143]

Recently, primers are being designed for studying methylation patterns of genes regulated by methylation process. CpG islands of susceptible epiobesigenes are analyzed to emerge their epigenetic regulation [146]. With these analysis, it has been found that, *FGF-2, PTEN, CDKN1A,* and *ESR1* genes have enough CpG islands in their promoter region to be potential susceptibility genes for obesity and adipogenesis [147, 148-150]. In leptin regulation, *SOCS1* and *SOCS3* (suppressor of cytokine signaling 1 and 3) take place and cytokine signaling affected in obese state is attenuated [151-153]. Also, in epigenetic regulation of obesity, energy homeostasis genes, *COX7a, LPL,* and insulin-related genes like *CAV1* and *IGFBP3* can be potential targets [2, 154-156]. There are also other genes which do not have CpG risk sequences and have roles in body weight control [66].

For example, Glucocorticoid receptor and *TNF-α* genes are under epigenetic control but don't have remarkable CpG islands although they have significant roles in obesity pathogenesis [66, 157, 158].

CpG methylation in somatic cells may be changed by some environmental factors in postnatal life. These environmental factors may cause variation or reversibility in the DNA methylation patterns. One of these environmental factors is aging. Aging affects the DNA methylation patterns in tissue-specific way [159]. Methylation of such genes like hepatic glucokinase increases with age [160]. This shows that, DNA methylation can play a role in age-related susceptibility to hepatic insulin resistance and diabetes. Other factors influencing DNA methylation is inflammation [161], oxidative stress [162], and hypoxia [163]. In adipose tissue of obese subjects, these factors are all exacerbated [147].

In recent years, the relationship between obesity and epigenetic pathways are reported. Methylation of many genes is reversible since it can be altered by exercise or diet throughout life. Alternatively, methylation of several genes at birth are associated with obesity later in life since epigenetic regulation of inheritance may be more stable in human [164]. It has been shown that, hypocaloric diet may be associated with TNF-α promoter in PBMC cells [39]. Recently, studies have also been demonstrated the effects of different dietary compounds on epigenetic marks and, on gene expression regulation in disease states. Some of the nutritional factors influencing the epigenetic marks on some disease states are shown in Table 2 [14, 165].

Table 2. Some dietary factors influencing epigenetic marks and have beneficial effects on obesity (modified from 14, 165)

Dietary factor	Related Metabolic Disease	Epigenetic mechanism	Ref.
Methyl donors			
Methionine	Insulin resistance, obesity	Histone and DNA methylation	[166]
Folate	Insulin resistance, adiposity	DNA methylation	[166]
Vitamin B-12	Insulin resistance, obesity	DNA methylation	[166]
Serine, Glycine and Histidine	Amino acid metabolism	Histone and DNA methylation	[98]
Phytochemicals			
Curcumin	Inflammation, obesity	Histone acetylation, DNA methylation, and microRNA	[167]
Epigallocatechin 3-gallate	Obesity, insulin resistance	Histone acetylation and DNA methylation	[168]
Genistein	Obesity	Histone acetylation and DNA methylation	[169]
Sulforaphane	Adipocyte differentiation	Histone acetylation	[170]
Resveratrol	Obesity	Histone acetylation	[171, 172]
Fatty acids			
Eicosapentaenoic acid	n-3 Polyunsaturated fatty acid metabolism	DNA methylation	[173]
Docosahexaenoic acid	n-3 Polyunsaturated fatty acid metabolism	DNA methylation	[174]

Studies are going on to investigate the relationship between obesity, obesity related diseases and epigenetics. In a recent study, increased DNA methylation has been associated with T2D in a Swedish population [175]. Also, global DNA hypermethylation has been related with diabetic retinopathy in T2D [176].

In another study, *BCL11A* gene polymorphism has been established as a risk for T2D, and a recent study suggested a possible male gender-specific association between *BCL11A* gene methylation and T2D [177]. Besides, cholesterol and the different types of fatty acids in the diet have been shown to affect on genome-wide DNA methylation patterns [178]. Recently, a GWAS on 459 European origin individuals was performed to investigate the relationship between DNA methylation and BMI, and the analysis reported that increased BMI is associated with the elevation of methylation at the *HIF3A* locus in blood cells and in adipose tissue [179]. This study generated a support for the relationship between the epigenome and the development of obesity in genetically susceptible individuals.

CONCLUSION

In human obesity, many genes have been identified. The effects of these genes may be modified by epigenetic mechanisms. There is preliminary but compelling evidence from human and animal studies that changes in the epigenetic states may contribute to this era's epidemic, obesity. However, there are still questions remaining. Does the epigenetic analysis of a specific tissue mirror the epigenome of another tissue?, and are these epigenetic changes reversible? Further studies in epigenetics will enlighten the obesity etiology and related metabolic disease phenotypes and will provide therapeutic approaches to improve obesity prognoses.

REFERENCES

[1] James P. T., Leach R., Kalamara E., Shayeghi M. The worldwide obesity epidemic. *Obes. Res.*, 2001; 9: 228S–233S.

[2] Martinez J. A., Parra M. D., Santos J. L., Moreno-Aliaga M. J., Marti A., Martinez-Gonzalez M. A. Genotype-dependent response to energy-restricted diets in obese subjects: towards personalized nutrition. *Asia Pac. J. Clin. Nutr.*, 2008; 17(Suppl. 1): 119–122.

[3] Reilly M. P., Rader D. J. The metabolic syndrome: more than the sum of its parts? *Circulation,* 2003; 108: 1546–1551.

[4] Smith C. J., Ryckman K. K. Epigenetic and developmental influences on the risk of obesity, diabetes, and metabolic syndrome. *Diabetes Metab. Syndr. Obes.*, 2015 Jun. 29; 8: 295-302.

[5] Mau T., Yung R. Potential of epigenetic therapies in non-cancerous conditions. *Front. Genet.*, 2014 Dec. 19; 5: 438.

[6] http://www.hsph.harvard.edu/obesity-prevention-source/obesity-definition/ (Last accessed: 01.08.2015).

[7] Dhurandhar E. J., Keith S. W. The aetiology of obesity beyond eating more and exercising less. *Best Pract. Res. Clin. Gastroenterol.,* 2014 Aug.; 28(4): 533-44.

[8] de Rooij S. R., Painter R. C., Holleman F., Bossuyt P. M., Roseboom T. J. The metabolic syndrome in adults prenatally exposed to the Dutch famine. *Am. J. Clin. Nutr.*, 2007; 86: 1219–1224.

[9] Desai M., Gayle D., Babu J., Ross M. G. Programmed obesity in intrauterine growthrestricted newborns: modulation by newborn nutrition. *Am. J. Physiol. Regul. Integr. Comp. Physiol.*, 2005; 288: R91–R96.

[10] Desai M., Jellyman J. K., Han G., Beall M., Lane R. H., Ross M. G. Maternal obesity and high-fat diet program offspring metabolic syndrome. *Am. J. Obstet. Gynecol.*, 2014; 211: 237.e1–237.e13.
[11] Vickers M. H., Krechowec S. O., Breier B. H. Is later obesity programmed in utero? *Curr. Drug Targets*, 2007; 8: 923–934.
[12] Remacle C., Bieswal F., Reusens B. Programming of obesity and cardiovascular disease. *Int. J. Obes. Relat. Metab. Disord.*, 2004; 28: S46–S53.
[13] Desai M., Jellyman J. K., Ross M. G. Epigenomics, gestational programming and risk of metabolic syndrome. *Int. J. Obes.*, (Lond) 2015 Apr.; 39(4): 633-41.
[14] Martínez J. A., Milagro F. I., Claycombe K. J., Schalinske K. L. Epigenetics in adipose tissue, obesity, weight loss, and diabetes. *Adv. Nutr.,* 2014 Jan. 1; 5(1): 71-81.
[15] Junien C., Nathanielsz P. Report on the IASO Stock Conference 2006: early and lifelong environmental epigenomic programming of metabolic syndrome, obesity and type II diabetes. *Obes. Rev.,* 2007; 8: 487–502.
[16] Dolinoy D. C., Jirtle R. L. Environmental epigenomics in human health and disease. *Environ. Mol. Mutagen.*, 2008; 49: 4–8.
[17] Chuang J. C., Jones P. A. Epigenetics and microRNAs. *Pediatr. Res.,* 2007; 61: 24–29.
[18] Klose R. J., and Bird A. P. Genomic DNA methylation: the mark and its mediators. *Trends Biochem. Sci.,* 2006; 31: 89-97.
[19] Feng J., and Fan G. The role of DNA methylation in the central nervous system and neuropsychiatric disorders. *Int. Rev. Neurobiol.,* 2009; 89: 67-84.
[20] Watt F. and Molloy P. L. Cytosine methylation prevents binding to DNA of a HeLa cell transcription factor required for optimal expression of the adenovirus major late promoter. *Genes Dev.*, 1988; 2: 1136-1143.
[21] Fuks F., Burgers W. A., Brehm A., Hughes-Davies L., Kouzarides T. DNA methyltransferases Dnmt1 associates with histone deacetylase activity. *Nat. Genet.*, 2000; 24: 88-91.
[22] Lorincz M. C., Dickerson D. R., Schmitt M., Groudine M. Intra- genic DNA methylation alters chromatin structure and elongation efficiency in mammalian cells. *Nat. Struct. Mol. Biol.*, 2004; 11: 1068-1075.
[23] Nan X., Campoy F. J., Bird A. MeCP2 is a transcriptional repressor with abundant binding sites in genomic chromatin. *Cell,* 1997; 88: 471-481.
[24] Nan X., Ng H. H., Johnson C. A., Laherty C. D., Turner B. M., Eisenman R. N., et al. Transcriptional repression by the methyl-CpG-binding protein MeCP2 involves a histone deacetylase complex. *Nature,* 1998; 393: 386-389.
[25] Jenuwein T., Allis C. D. Translating the histone code. *Science,* 2001; 293: 1074–1080.
[26] Cosgrove M. S. Histone proteomics and the epigenetic regulation of nucleosome mobility. *Expert Rev. Proteomics*, 2007; 4: 465–478.
[27] Schwartz Y. B., Pirrotta V. Polycomb complexes and epigenetic states. *Curr. Opin. Cell Biol.,* 2008; 20: 266–273.
[28] Higgs P. G., Lehman N. The RNA World: molecular cooperation at the origins of life. *Nat. Rev. Genet.*, 2015; 16: 7–17.
[29] Taft R. J., Pang K. C., Mercer T. R., Dinger M., Mattick J. S. Non-coding RNAs: regulators of disease. *J. Pathol.*, 2010; 220: 126–139.

[30] Rinn J. L., Kertesz M., Wang J. K., Squazzo S. L., Xu X., Brugmann S. A. et al. Functional demarcation of active and silent chromatin domains in human HOX loci by noncoding RNAs. *Cell,* 2007; 129: 1311–1323.

[31] Mercer T. R., Dinger M. E., Sunkin S. M., Mehler M. F., Mattick J. S. Specific expression of long noncoding RNAs in the mouse brain. *Proc. Natl. Acad. Sci. USA,* 2008; 105: 716–721.

[32] Clemson C. M., Hutchinson J. N., Sara S. A., Ensminger A. W., Fox A. H., Chess A. et al. An architectural role for a nuclear noncoding RNA: NEAT1 RNA is essential for the structure of paraspeckles. *Mol. Cell,* 2009; 33: 717–726.

[33] Sun L., Goff L. A., Trapnell C., Alexander R., Lo K. A., Hacisuleyman E. et al. Long noncoding RNAs regulate adipogenesis. *Proc. Natl. Acad. Sci. USA,* 2013; 110: 3387–3392.

[34] Zhao X. Y., Li S., Wang G. X., Yu Q., Lin J. D. A long noncoding RNA transcriptional regulatory circuit drives thermogenic adipocyte differentiation. *Mol. Cell*, 2014; 55: 372–382.

[35] Chen K., Rajewsky N. The evolution of gene regulation by transcription factors and microRNAs. *Nat. Rev. Genet*., 2007; 8: 93–103.

[36] Ding X. C., Weiler J., Grosshans H. Regulating the regulators: mechanisms controlling the maturation of microRNAs. *Trends Biotechnol.*, 2009; 27: 27–36.

[37] Saetrom P., Snøve O. Jr., Rossi J. J. Epigenetics and microRNAs. *Pediatr. Res*., 2007 May; 61(5 Pt 2): 17-23.

[38] Nafee T. M., Farrell W. E., Carroll W. D., Fryer A. A., Ismail K. M. Epigenetic control of fetal gene expression. *BJOG*, 2008; 115: 158–168.

[39] Campión J., Milagro F. I., Goyenechea E., Martínez J. A. TNF-alpha promoter methylation as a predictive biomarker for weight-loss response. *Obesity*, (Silver Spring). 2009 Jun.; 17(6): 1293-7.

[40] Zhu J. DNA methylation and hepatocellular carcinoma. *J. Hepatobiliary Pancreat. Surg*., 2006; 13: 265–273.

[41] Campión J., Milagro F., Martínez J. A. Epigenetics and obesity. *Prog. Mol. Biol. Transl. Sci*., 2010; 94: 291–347.

[42] Bouchard L., Rabasa-Lhoret R., Faraj M., Lavoie M. E., Mill J., Pérusse L. et al. Differential epigenomic and transcriptomic responses in subcutaneous adipose tissue between low and high responders to caloric restriction. *Am. J. Clin. Nutr*., 2010; 91: 309–320.

[43] McGee S. L., Hargreaves M. Exercise and skeletal muscle glucose transporter 4 expression: molecular mechanisms. *Clin. Exp. Pharmacol. Physiol.*, 2006; 33: 395–399.

[44] Fraga M. F., Ballestar E., Paz M. F., Ropero S., Setien F., Ballestar M. L. et al. Epigenetic differences arise during the lifetime of monozygotic twins. *Proc. Natl. Acad. Sci. USA*, 2005; 102: 10604–10609.

[45] Moleres A., Campión J., Milagro F. I., Marcos A., Campoy C., Garagorri J. M. et al. Differential DNA methylation patterns between high and low responders to a weight loss intervention in overweight or obese adolescents: the EVASYON study. *FASEB J.*, 2013; 27: 2504–2512.

[46] Marti A., Martinez-Gonzalez M. A., Martinez J. A. Interaction between genes and lifestyle factors on obesity. *Proc. Nutr. Soc.,* 2008; 67: 1–8.

[47] Wolff G. L., Roberts D. W., Mountjoy K. G. Physiological consequences of ectopic agouti gene expression: the yellow obese mouse syndrome. *Physiol. Genomics*, 1999; 1: 151-63.

[48] Morgan H. D., Sutherland H. G., Martin D. I., Whitelaw E. Epigenetic inheritance at the agouti locus in the mouse. *Nat. Genet.*, 1999; 23: 314–318.

[49] Waterland R. A., Jirtle R. L. Transposable elements: targets for early nutritional effects on epigenetic gene regulation. *Mol. Cell Biol.*, 2003; 23: 5293–5300.

[50] Strakovsky R. S., Zhang X., Zhou D., Pan Y. X. Gestational high fat diet programs hepatic phosphoenolpyruvate carboxykinase gene expression and histone modification in neonatal offspring rats. *J. Physiol.*, 2011; 589: 2707-17.

[51] Vucetic Z., Kimmel J., Totoki K., Hollenbeck E., Reyes T. M. Maternal high-fat diet alters methylation and gene expression of dopamine and opioid-related genes. *Endocrinology,* 2010; 151: 4756-64.

[52] Vucetic Z., Kimmel J., Reyes T. M. Chronic high-fat diet drives postnatal epigenetic regulation of mu-opioid receptor in the brain. Neuropsychopharmacol e off *Publ. Am. Coll. Neuropsychopharmacol.,* 2011; 36: 1199-206.

[53] Milagro F. I., Campion J., Garcia-Diaz D. F., Goyenechea E, Paternain L., Martinez J. A. High fat diet-induced obesity modifies the methylation pattern of leptin promoter in rats. *J. Physiol. Biochem*., 2009; 65: 1-9.

[54] Waterland R. A., Travisano M., Tahiliani K. G. Diet-induced hypermethylation at agouti viable yellow is not inherited transgenerationally through the female. *FASEB J.*, 2007; 21: 3380–3385.

[55] Weinstein L. S., Xie T., Qasem A., Wang J., Chen M. The role of GNAS and other imprinted genes in the development of obesity. *Int. J. Obes.,* (Lond). 2010 Jan.; 34(1):6-17.

[56] Pant V., Mariano P., Kanduri C., Mattsson A., Lobanenkov V., Heuchel R., Ohlsson R. The nucleotides responsible for the direct physical contact between the chromatin insulator protein CTCF and the H19 imprinting control region manifest parent of origin-specific long-distance insulation and methylation-free domains. *Genes Dev.,* 2003; 17: 586–90.

[57] Gao H., Sathishkumar K. R., Yallampalli U., Balakrishnan M., Li X.,Wu G., Yallampalli C. Maternal protein restriction regulates IGF2 system in placental labyrinth. *Front. Biosci*., (Elite Ed) 2012; 4: 1434–50.

[58] Grandjean V., O'Neill L., Sado T., Turner B., Ferguson-Smith A. Relationship between DNA methylation, histone H4 acetylation and gene expression in the mouse imprinted Igf2–H19 domain. *FEBS Lett.,* 2001; 488: 165–9.

[59] Li T., Hu J. F., Qiu X., Ling J., Chen H., Wang S., Hou A., Vu T. H., Hoffman A. R. CTCF regulates allelic expression of Igf2 by orchestrating a promoter- polycomb repressive complex 2 intrachromosomal loop. *Mol. Cell Biol*., 2008; 28: 6473–82.

[60] Yang B. T., Dayeh T. A., Kirkpatrick C. L., Taneera J., Kumar R., Groop L. et al. Insulin promoter DNA methylation correlates negatively with insulin gene expression and positively with HbA(1c) levels in human pancreatic islets. *Diabetologia,* 2011; 54: 360–367.

[61] Noer A., Boquest A. C., Collas P. Dynamics of adipogenic promoter DNA methylation during clonal culture of human adipose stem cells to senescence. *BMC Cell Biol.,* 2007; 8: 18.

[62] Stepanow S., Reichwald K., Huse K., Gausmann U., Nebel A., Rosenstiel P. et al. Allele-specific, age-dependent and BMI-associated DNA methylation of human MCHR1. *PLoS One,* 2011; 6: 17711.

[63] Yokomori N., Tawata M., Onaya T. DNA demethylation during the differentiation of 3T3-L1 cells affects the expression of the mouse GLUT4 gene. *Diabetes,* 1999; 48: 685–690.

[64] Fujiki K., Kano F., Shiota K., Murata M. Expression of the peroxisome proliferator activated receptor gamma gene is repressed by DNA methylation in visceral adipose tissue of mouse models of diabetes. *BMC Biol.*, 2009; 7: 38.

[65] Musri M. M., Corominola H., Casamitjana R., Gomis R., Párrizas M. Histone H3 lysine 4 dimethylation signals the transcriptional competence of the adiponectin promoter in preadipocytes. *J. Biol. Chem.*, 2006; 281: 17180–17188.

[66] Sullivan K. E., Reddy A. B., Dietzmann K., Suriano A. R., Kocieda V. P., Stewart M. et al. Epigenetic regulation of tumor necrosis factor alpha. *Mol. Cell Biol.,* 2007; 27: 5147–5160.

[67] Kilpeläinen T. O., Zillikens M. C., Stančákova A., Finucane F. M., Ried J. S., Langenberg C. et al. Genetic variation near IRS1 associates with reduced adiposity and an impaired metabolic profile. *Nat. Genet.*, 2011; 43: 753–760.

[68] Musri M. M., Párrizas M. Epigenetic regulation of adipogenesis. *Curr. Opin. Clin. Nutr. Metab. Care,* 2012; 15: 342–349.

[69] Ge K. Epigenetic regulation of adipogenesis by histone methylation. *Biochim. Biophys. Acta*, 2012; 1819: 727–732.

[70] Li H. X., Xiao L., Wang C., Gao J. L., Zhai Y. G. Review: Epigenetic regulation of adipocyte differentiation and adipogenesis. *J. Zhejiang Univ. Sci. B,* 2010; 11: 784–791.

[71] Burdge G. C., Hoile S. P., Uller T., Thomas N. A., Gluckman P. D., Hanson M. A. et al. Progressive, transgenerational changes in offspring phenotype and epigenotype following nutritional transition. *PLoS One,* 2011; 6: e28282.

[72] Wang X., Zhu H., Snieder H., Su S, Munn D., Harshfield G. et al. Obesity related methylation changes in DNA of peripheral blood leukocytes. *BMC Med.*, 2010; 8: 87.

[73] Jufvas A., Sjödin S., Lundqvist K., Amin R., Vener A. V., Strålfors P. Global differences in specific histone H3 methylation are associated with overweight and type 2 diabetes. *Clin. Epigenetics*, 2013; 5: 15.

[74] Ortega F. J., Moreno-Navarrete J. M., Pardo G., Sabater M., Hummel M., Ferrer A. et al. MiRNA expression profile of human subcutaneous adipose and during adipocyte differentiation. *PLoS One,* 2010; 5: e9022.

[75] Tateishi K., Okada Y., Kallin E. M., Zhang Y. Role of Jhdm2a in regulating metabolic gene expression and obesity resistance. *Nature,* 2009; 458: 757–761.

[76] Inagaki T., Tachibana M., Magoori K., Kudo H., Tanaka T., Okamura M. et al. Obesity and metabolic syndrome in histone demethylase JHDM2a-deficient mice. *Genes Cells*, 2009; 14: 991–1001.

[77] Lu Y. L., Jing W., Feng L. S., Zhang L., Xu J. F., You T. J. et al. Effects of hypoxic exercise training on microRNA expression and lipid metabolism in obese rat livers. *J. Zhejiang Univ. Sci. B,* 2014; 15: 820–829.

[78] Son Y. H., Ka S., Kim A. Y., Kim J. B. Regulation of Adipocyte Differentiation via MicroRNAs. *Endocrinol. Metab.*, (Seoul) 2014; 29: 122–135.

[79] Vinnikov I. A., Hajdukiewicz K., Reymann J., Beneke J., Czajkowski R., Roth L. C. et al. Hypothalamic miR-103 protects from hyperphagic obesity in mice. *J. Neurosci.*, 2014; 34: 10659–10674.

[80] Funato H., Oda S., Yokofujita J., Igarashi H., Kuroda M. Fasting and high-fat diet alter histone deacetylase expression in the medial hypothalamus. *PLoS One*, 2011; 6: e18950.

[81] McGregor R. A., Choi M. S. microRNAs in the regulation of adipogenesis and obesity. *Curr. Mol. Med.*, 2011; 11: 304–316.

[82] Yang Q. Y., Liang J. F., Rogers C. J., Zhao J. X., Zhu M. J., Du M. Maternal obesity induces epigenetic modifications to facilitate Zfp423 expression and enhance adipogenic differentiation in fetal mice. *Diabetes,* 2013; 62: 3727–3735.

[83] Borengasser S. J., Zhong Y., Kang P., Lindsey F., Ronis M. J., Badger T. M. et al. Maternal obesity enhances white adipose tissue differentiation and alters genome-scale DNA methylation in male rat offspring. *Endocrinology*, 2013; 154: 4113–4125.

[84] Kamei Y., Suganami T., Ehara T., Kanai S., Hayashi K., Yamamoto Y. et al. Increased expression of DNA methyltransferase 3a in obese adipose tissue: studies with transgenic mice. *Obesity,* (Silver Spring) 2010; 18: 314–321.

[85] Lee J., Saha P. K., Yang Q. H., Lee S., Park J. Y., Suh Y. et al. Targeted inactivation of MLL3 histone H3-Lys-4 methyltransferase activity in the mouse reveals vital roles for MLL3 in adipogenesis. *Proc. Natl. Acad. Sci. USA,* 2008; 105: 19229–19234.

[86] Morgan D. K., Whitelaw E. The case for transgenerational epigenetic inheritance in humans. *Mamm Genome*, 2008; 9: R182.

[87] Rakyan V. K., Chong S., Champ M. E., Cuthbert P. C., Morgan H. D., Luu K. V., Whitelaw E. Transgenerational inheritance of epigenetic states at the murine Axin(Fu) allele occurs after maternal and paternal transmission. *Proc. Natl. Acad. Sci. U S A*, 2003; 100: 2538–2543.

[88] Vickers M. H. Developmental programming and adult obesity: the role of leptin. *Curr. Opin. Endocrinol. Diabetes Obes.*, 2007; 14: 17–22.

[89] Cobiac L. Epigenomics and nutrition. *Forum Nutr.*, 2007; 60: 31–41.

[90] Heijmans B. T., Tobi E. W., Stein A. D., Putter H., Blauw G. J., Susser E. S. et al. Persistent epigenetic differences associated with prenatal exposure to famine in humans. *Proc. Natl. Acad. Sci. USA*, 2008; 105: 17046–17049.

[91] Tobi E. W., Lumey L. H., Talens R. P., Kremer D., Putter H., Stein A. D. et al. DNA methylation differences after exposure to prenatal famine are common and timing- and sex-specific. *Hum. Mol. Genet.*, 2009; 18: 4046–4053.

[92] Soubry A., Murphy S. K., Wang F., Huang Z., Vidal A. C., Fuemmeler B. F., Kurtzberg J., Murtha A., Jirtle R. L., Schildkraut J. M., Hoyo C. Newborns of obese parents have altered DNA methylation patterns at imprinted genes. *Int. J. Obes.*, (Lond). 2015 Apr.; 39(4):650-7.

[93] Soubry A., Schildkraut J. M., Murtha A., Wang F., Huang Z., Bernal A. et al. Paternal obesity is associated with IGF2 hypomethylation in newborns: results from a Newborn Epigenetics Study (NEST) cohort. *BMC Med.,* 2013; 11: 29.

[94] Morales E., Groom A., Lawlor D. A., Relton C. L. DNA methylation signatures in cord blood associated with maternal gestational weight gain: results from the ALSPAC cohort. *BMC Res. Notes*, 2014; 7: 278.

[95] Godfrey K. M., Sheppard A., Gluckman P. D., Lillycrop K. A., Burdge G. C., McLean C. et al. Epigenetic gene promoter methylation at birth is associated with child's later adiposity. *Diabetes,* 2011; 60: 1528–1534.
[96] Lavebratt C., Almgren M., Ekström T. J. Epigenetic regulation in obesity. *Int. J. Obes.*, (Lond) 2012; 36: 757–765.
[97] Lillycrop K. A., Burdge G. C. Epigenetic changes in early life and future risk of obesity. *Int. J. Obes.*, (Lond) 2011; 35: 72–83.
[98] Wang J., Wu Z., Li D., Li N., Dindot S. V., Satterfield M. C., Bazer F. W., Wu G. Nutrition, epigenetics, and metabolic syndrome. *Antioxid. Redox. Signal,* 2012; 17: 282–301.
[99] Ganu R. S., Harris R. A., Collins K., Aagaard K. M. Maternal diet: a modulator for epigenomic regulation during development in nonhuman primates and humans. *Int. J. Obes. Suppl.,* 2012; 2: S14–S18.
[100] Tosh D. N., Fu Q., Callaway C. W., McKnight R. A., McMillen I. C., Ross M. G. et al. Epigenetics of programmed obesity: alteration in IUGR rat hepatic IGF1 mRNA expression and histone structure in rapid vs delayed postnatal catch-up growth. *Am. J. Physiol. Gastrointest. Liver Physiol.*, 2010; 299: G1023–G1029.
[101] Desai M., Jellyman J. K., Han G., Beall M. H., Ross M. G. Mechanism of programmed obesity in intrauterine growth restricted newborns: epigenetic mediated early induction of adipocyte differentiation contributes to enhanced adipogenesis. *J. Dev. Orig. Health Dis.*, 2013: 4(Suppl. 2): S138.
[102] Cho C. E., Sánchez-Hernández D., Reza-López S. A., Huot P. S., Kim Y. I., Anderson G. H. High folate gestational and post-weaning diets alter hypothalamic feeding pathways by DNA methylation in Wistar rat offspring. *Epigenetics,* 2013; 8: 710–719.
[103] Stevens A., Begum G., Cook A., Connor K., Rumball C., Oliver M. et al. Epigenetic changes in the hypothalamic proopiomelanocortin and glucocorticoid receptor genes in the ovine fetus after periconceptional undernutrition. *Endocrinology,* 2010; 151: 3652–3664.
[104] Plagemann A., Harder T., Brunn M., Harder A., Roepke K., Wittrock-Staar M. et al. Hypothalamic proopiomelanocortin promoter methylation becomes altered by early overfeeding: an epigenetic model of obesity and the metabolic syndrome. *J. Physiol.,* 2009; 587: 4963–4976.
[105] Raychaudhuri N., Raychaudhuri S., Thamotharan M., Devaskar S. U. Histone code modifications repress glucose transporter 4 expression in the intrauterine growth-restricted offspring. *J. Biol. Chem.*, 2008; 283: 13611–13626.
[106] Drummond E. M., Gibney E. R. Epigenetic regulation in obesity. *Curr. Opin. Clin. Nutr. Metab. Care,* 2013 Jul.; 16(4): 392-7.
[107] Ling C., Del Guerra S., Lupi R., Ronn T., Granhall C., Luthman H., Masiello P., Marchetti P., Groop L., Del Prato S. Epigenetic regulation of PPARGC1A in human type 2 diabetic islets and effect on insulin secretion. *Diabetologia,* 2008; 51: 615–622.
[108] He N., Park K., Zhang Y., Huang J., Lu S., Wang L. Epigenetic inhibition of nuclear receptor small heterodimer partner is associated with and regulates hepatocellular carcinoma growth. *Gastroenterology,* 2008; 134: 793–802.
[109] Chang P. Y., Lu S. C., Lee C. M., Chen Y. J., Dugan T. A., Huang W. H., Chang S. F., Liao W. S., Chen C. H., Lee Y. T. Homocysteine inhibits arterial endothelial cell

growth through transcriptional downregulation of fibroblast growth factor-2 involving G protein and DNA methylation. *Circ. Res.*, 2008; 102: 933–941.

[110] Goel A., Arnold C. N., Niedzwiecki D., Carethers J. M., Dowell J. M., Wasserman L., Compton C., Mayer R. J., Bertagnolli M. M., Boland C. R. Frequent inactivation of PTEN by promoter hypermethylation in microsatellite instability-high sporadic colorectal cancers. *Cancer Res.*, 2004; 64: 3014–3021.

[111] Roupret M., Hupertan V., Catto J. W., Yates D. R., Rehman I., Proctor L. M., Phillips J., Meuth M., Cussenot O., Hamdy F. C. Promoter hypermethylation in circulating blood cells identifies prostate cancer progression. *Int. J. Cancer*, 2008; 122: 952–956.

[112] Majid S., Kikuno N., Nelles J., Noonan E., Tanaka Y., Kawamoto K., Hirata H., Li L. C., Zhao H., Okino S. T., Place R. F., Pookot D., Dahiya R. Genistein induces the p21WAF1/CIP1 and p16INK4a tumor suppressor genes in prostate cancer cells by epigenetic mechanisms involving active chromatin modification. *Cancer Res.*, 2008; 68: 2736–2744.

[113] Stoger R. In vivo methylation patterns of the leptin promoter in human and mouse. *Epigenetics*, 2006; 1: 155–162.

[114] Melzner I., Scott V., Dorsch K., Fischer P., Wabitsch M., Bruderlein S., Hasel C., Moller P. Leptin gene expression in human preadipocytes is switched on by maturation-induced demethylation of distinct CpGs in its proximal promoter. *J. Biol. Chem.*, 2002; 277: 45420–45427.

[115] Lai J. C., Cheng Y. W., Chiou H. L., Wu M. F., Chen C. Y., Lee H. Gender difference in estrogen receptor-alpha promoter hypermethylation and its prognostic value in non-small cell lung cancer. *Int. J. Cancer*, 2005; 117: 974–980.

[116] Oberlander T. F., Weinberg J., Papsdorf M., Grunau R., Misri S., Devlin A. M. Prenatal exposure to maternal depression, neonatal methylation of human glucocorticoid receptor gene (NR3C1) and infant cortisol stress responses. *Epigenetics*, 2008; 3: 97–106.

[117] Zheng S., Rollet M., Pan Y. X. Maternal protein restriction during pregnancy induces CCAAT/enhancer-binding protein (C/EBPb) expression through the regulation of histone modification at its promoter region in female offspring rat skeletal muscle. *Epigenetics*, 2011;6: 161–70.

[118] Lillycrop K. A., Phillips E. S., Torrens C., Hanson M. A., Jackson A. A., Burdge G. C. Feeding pregnant rats a protein-restricted diet persistently alters the methylation of specific cytosines in the hepatic PPAR alpha promoter of the offspring. *Br. J. Nutr.*, 2008; 100: 278–82.

[119] Ebrahim S. Epigenetics: the next big thing. *Int. J. Epidemiol.*, 2012; 41: 1–3.

[120] Delaney C., Garg S. K., Fernandes C., et al. Maternal Diet Supplemented with Methyl-Donors Protects against Atherosclerosis in F1 ApoE(_/_) Mice. *Plos One*, 2013; 8:e56253.

[121] Widiker S., Karst S., Wagener A., Brockmann G. A. High-fat diet leads to a decreased methylation of the Mc4r gene in the obese BFMI and the lean B6 mouse lines. *J. Appl. Genet.*, 2010;51:193–7.

[122] Crujeiras A. B., Campion J., Díaz-Lagares A., Milagro F. I., Goyenechea E., Abete I., Casanueva F. F., Martínez J. A. Association of weight regain with specific methylation levels in the NPY and POMC promoters in leukocytes of obese men: a translational study. *Regul. Pept.*, 2013;186:1–6.

[123] Menschikowski M., Hagelgans A., Gussakovsky E., Kostka H., Paley E. L., Siegert G. Differential expression of secretory phospholipases A2 in normal and malignant prostate cell lines: regulation by cytokines, cell signaling pathways, and epigenetic mechanisms. *Neoplasia,* 2008; 10: 279–286.

[124] Jiang Y., Jiang J., Xiong J., Cao J., Li N., Li G., Wang S. Homocysteine-induced extracellular superoxide dismutase and its epigenetic mechanisms in monocytes. *J. Exp. Biol.,* 2008; 211: 911–920.

[125] Okochi O., Hibi K., Sakai M., Inoue S., Takeda S., Kaneko T., Nakao A. Methylation-mediated silencing of SOCS-1 gene in hepatocellular carcinoma derived from cirrhosis. *Clin. Cancer Res.,* 2003; 9: 5295–5298.

[126] Tokita T., Maesawa C., Kimura T., Kotani K., Takahashi K., Akasaka T., Masuda T. Methylation status of the SOCS3 gene in human malignant melanomas. *Int. J. Oncol.,* 2007; 30: 689–694.

[127] Bollati V., Baccarelli A., Sartori S., Tarantini L., Motta V., Rota F., Costa G. Epigenetic effects of shiftwork on blood DNA methylation. *Chronobiol. Int.,* 2010; 27:1093–104.

[128] Lazcoz P., Munoz J., Nistal M., Pestana A., Encio I., Castresana J. S. Frequent promoter hypermethylation of RASSF1A and CASP8 in neuroblastoma. *BMC Cancer,* 2006; 6: 254.

[129] Ronn T., Poulsen P., Hansson O., Holmkvist J., Almgren P., Nilsson P., Tuomi T., Isomaa B., Groop L., Vaag A., Ling C. Age influences DNA methylation and gene expression of COX7A1 in human skeletal muscle. *Diabetologia,* 2008; 51: 1159–1168.

[130] Lomba A., Martinez J. A., Garcia-Diaz D. F., Paternain L., Marti A., Campion J., Milagro F. I. Weight gain induced by an isocaloric pair-fed high fat diet: a nutriepigenetic study on FAS and NDUFB6 gene promoters. *Mol. Genet. Metab.,* 2010;101:273–8.

[131] Hirasawa Y., Arai M., Imazeki F., Tada M., Mikata R., Fukai K., Miyazaki M., Ochiai T., Saisho H., Yokosuka O. Methylation status of genes upregulated by demethylating agent 5-aza-2′- deoxycytidine in hepatocellular carcinoma. *Oncology,* 2006; 71: 77–85.

[132] Semba S., Itoh N., Ito M., Youssef E. M., Harada M., Moriya T., Kimura W., Yamakawa M. Down-regulation of PIK3CG, a catalytic subunit of phosphatidylinositol 3-OH kinase, by CpG hypermethylation in human colorectal carcinoma. *Clin. Cancer Res.,* 2002; 8: 3824–3831.

[133] Torrisani J., Hanoun N., Laurell H., Lopez F.,Maoret J. J., Souque A., Susini C., Cordelier P., Buscail L. Identification of an upstream promoter of the human somatostatin receptor, hSSTR2, which is controlled by epigenetic modifications. *Endocrinology,* 2008; 149: 3137–3147.

[134] Alikhani-Koopaei R., Fouladkou F., Frey F. J., Frey B. M. Epigenetic regulation of 11 beta-hydroxysteroid dehydrogenase type 2 expression. *J. Clin. Invest.,* 2004; 114: 1146–1157.

[135] Chang Y. S., Wang L., Liu D., Mao L., Hong W. K., Khuri F. R., Lee H. Y. Correlation between insulin-like growth factor-binding protein 3 promoter methylation and prognosis of patients with stage I non-small cell lung cancer. *Clin. Cancer Res.,* 2002; 8: 3669–3675.

[136] Gong L., Pan Y. X., Chen H. Gestational low protein diet in the rat mediates Igf2 gene expression in male offspring via altered hepatic DNA methylation. *Epigenetics,* 2010; 5:619–626.

[137] Bouchard L., Hivert M. F., Guay S. P., St-Pierre J., Perron P., Brisson D. Placental adiponectin gene DNA methylation levels are associated with mothers' blood glucose concentration. *Diabetes,* 2012; 61: 1272–80.

[138] Kuroda A., Rauch T. A., Todorov I., et al. Insulin gene expression is regulated by dna methylation. *Plos One,* 2009; 4: e6953.

[139] Mosley A. L., Ozcan S. Glucose regulates insulin gene transcription by hyperacetylation of histone H4. *J. Biol. Chem.*, 2003; 278: 19660–19666.

[140] Plagemann A., Roepke K., Harder T., et al. Epigenetic malprogramming of the insulin receptor promoter due to developmental overfeeding. *J. Perinat. Med.,* 2010; 38: 393–400.

[141] Toperoff G., Aran D., Kark J. D., Rosenberg M., Dubnikov T., Nissan B.,Wainstein J., Friedlander Y., Levy-Lahad E., Glaser B., et al. Genome-wide survey reveals predisposing diabetes type 2-related DNA methylation variations in human peripheral blood. *Hum. Mol. Genet.,* 2012;21:371–83.

[142] Brigati C., Banelli B., di Vinci A., Casciano I., Allemanni G., Forlani A., Borzì L., Romani M. Inflammation, HIF-1, and the epigenetics that follows. *Mediators Inflamm.*, 2010; 2010: 263914.

[143] Shore A., Karamitri A., Kemp P., Speakman J. R., Lomax M. A. Role of Ucp1 enhancer methylation and chromatin remodelling in the control of Ucp1 expression in murine adipose tissue. *Diabetologia,* 2010;53: 1164–73.

[144] Rankinen T., Zuberi A., Chagnon Y. C., Weisnagel S. J., Argyropoulos G., Walts B., et al. The human obesity gene map: the 2005 update. *Obes. e Silver Spring,* 2006; 14: 529-644.

[145] Goldstone A. P. Prader-Willi syndrome: advances in genetics, pathophysiology and treatment. *Trends Endocrinol. Metab. e Tem.,* 2004; 15: 12-20.

[146] Li L. C., Dahiya R. MethPrimer: designing primers for methylation PCRs. *Bioinformatics,* 2002; 18: 1427–1431.

[147] Campion J., Milagro F. I., Fernandez D., Martinez J. A. Diferential gene expression and adiposity reduction induced by ascorbic acid supplementation in a cafeteria model of obesity. *J. Physiol. Biochem.,* 2006; 62: 71–80.

[148] Patel N. G., Kumar S., Eggo M. C. Essential role of fibroblast growth factor signaling in preadipoctye differentiation. *J. Clin. Endocrinol. Metab.*, 2005; 90: 1226–1232.

[149] Sasaoka T., Wada T., Tsuneki H. Lipid phosphatases as a possible therapeutic target in cases of type 2 diabetes and obesity. *Pharmacol. Ther.*, 2006; 112: 799–809.

[150] Shi H., Maier S., Nimmrich I., Yan P. S., Caldwell C. W., Olek A., Huang T. H. Oligonucleotide-based microarray for DNA methylation analysis: principles and applications. *J. Cell Biochem.*, 2003; 88: 138–143.

[151] Alexander W. S., Hilton D. J. The role of suppressors of cytokine signaling (SOCS) proteins in regulation of the immune response. *Annu. Rev. Immunol.*, 2004; 22: 503–529.

[152] Howard J. K., Flier J. S. Attenuation of leptin and insulin signaling by SOCS proteins. *Trends Endocrinol. Metab.*, 2006; 17: 365–371.

[153] Guilherme A., Virbasius J. V., Puri V., Czech M. P. Adipocyte dysfunctions linking obesity to insulin resistance and type 2 diabetes. *Nat. Rev. Mol. Cell Biol.*, 2008; 9: 367–377.

[154] Huttemann M., Muhlenbein N., Schmidt T. R., Grossman L. I., Kadenbach B. Isolation and sequence of the human cytochrome c oxidase subunit VIIaL gene. *Biochim. Biophys. Acta*, 2000; 1492: 252–258.

[155] Lopez I. P., Milagro F. I., Marti A., Moreno-Aliaga M. J., Martinez J. A., De Miguel C. Gene expression changes in rat white adipose tissue after a high-fat diet determined by differential display. *Biochem. Biophys. Res. Commun.*, 2004; 318: 234–239.

[156] Pellegrini M., Marzocca G. Obesity, insulin-like growth factors and colorectal cancer. *Panminerva Med.,* 2006; 48: 143–144.

[157] Lillycrop K. A., Slater-Jefferies J. L., Hanson M. A., Godfrey K. M., Jackson A. A., Burdge G. C. Induction of altered epigenetic regulation of the hepatic glucocorticoid receptor in the offspring of rats fed a protein-restricted diet during pregnancy suggests that reduced DNA methyltransferase-1 expression is involved in impaired DNA methylation and changes in histone modifications. *Br. J. Nutr.*, 2007; 97: 1064–1073.

[158] Achard V., Boullu-Ciocca S., Desbriere R., Grino M. Perinatal programming of central obesity and the metabolic syndrome: role of glucocorticoids. *Metab. Syndr. Relat. Disord.,* 2006; 4: 129–137.

[159] Richardson B. Impact of aging on DNA methylation. *Ageing Res. Rev.,* 2003; 2: 245–261.

[160] Jiang M. H., Fei J., Lan M. S., Lu Z. P., Liu M., Fan W. W., Gao X., Lu D. R. Hypermethylation of hepatic Gck promoter in ageing rats contributes to diabetogenic potential. *Diabetologia,* 2008; 51: 1525–1533.

[161] Stenvinkel P., Karimi M., Johansson S., Axelsson J., Suliman M., Lindholm B., Heimburger O., Barany P., Alvestrand A., Nordfors L., Qureshi A. R., Ekstrom T. J., Schalling M. Impact of inflammation on epigenetic DNA methylation – a novel risk factor for cardiovascular disease? *J. Intern. Med.,* 2007; 261: 488–499.

[162] Franco R., Schoneveld O., Georgakilas A. G., Panayiotidis M. I. Oxidative stress, DNA methylation and carcinogenesis. *Cancer Lett.,* 2008; 266: 6–11.

[163] Shahrzad S., Bertrand K., Minhas K., Coomber B. L. Induction of DNA hypomethylation by tumor hypoxia. *Epigenetics*, 2007; 2: 119–125.

[164] van Dijk S. J., Molloy P. L., Varinli H., Morrison J. L., Muhlhausler B. S. Epigenetics and human obesity. *Int. J. Obes. e Lond.*, 2014.

[165] Milagro F. I., Mansego M. L., De Miguel C., Martínez J. A. Dietary factors, epigenetic modifications and obesity outcomes: progresses and perspectives. *Mol. Aspects Med.*, 2013 Jul.-Aug.; 34(4): 782-812.

[166] Sinclair, K. D., Allegrucci, C., Singh, R., Gardner, D. S., Sebastian, S., Bispham, J., Thurston, A., Huntley, J. F., Rees, W. D., Maloney, C. A., Lea, R. G., Craigon, J., McEvoy, T. G., Young, L. E., 2007. DNA methylation, insulin resistance, and blood pressure in offspring determined by maternal periconceptional B vitamin and methionine status. *Proc. Nat. Acad. Sci. U.S.A.*, 104, 19351–19356.

[167] Shao W., Yu Z., Chiang Y., Yang Y., Chai T., Foltz W., Lu H., Fantus I. G., Jin T. Curcumin prevents high fat diet induced insulin resistance and obesity via attenuating lipogenesis in liver and inflammatory pathway in adipocytes. *PLoS ONE,* 2012;7:e28784.

[168] Yun J. M., Jialal I., Devaraj S. Effects of epigallocatechin gallate on regulatory T cell number and function in obese v. lean volunteers. *Br. J. Nutr.*, 2010;103:1771–7.

[169] Dolinoy D. C., Weidman J. R., Waterland R. A., Jirtle R. L. Maternal genistein alters coat color and protects Avy mouse offspring from obesity by modifying the fetal epigenome. *Environ. Health Perspect.,* 2006; 114: 567–72.

[170] Choi K. M., Lee Y. S., Sin D. M., Lee S., Lee M. K., Lee Y. M., Hong J. T., Yun Y. P., Yoo H. S. Sulforaphane inhibits mitotic clonal expansion during adipogenesis through cell cycle arrest. *Obesity,* (Silver Spring). 2012;20: 1365–71.

[171] Bujanda L., Hijona E., Larzabal M., Beraza M., Aldazabal P., García-Urkia N., Sarasqueta C., Cosme A., Irastorza B., González A., et al. Resveratrol inhibits nonalcoholic fatty liver disease in rats. *BMC Gastroenterol.*, 2008;8:40.

[172] Baile C. A., Yang J. Y., Rayalam S., Hartzell D. L., Lai C. Y., Andersen C., Della-Fera M. A. Effect of resveratrol on fat mobilization. *Ann. NY Acad. Sci.,* 2011; 1215: 40–47.

[173] Ceccarelli V., Racanicchi S., Martelli M. P., Nocentini G., Fettucciari K., Riccardi C., Marconi P., Di Nardo P., Grignani F., Binaglia L., Vecchini A. Eicosapentaenoic acid demethylates a single CpG that mediates expression of tumor suppressor CCAAT/enhancer-binding protein delta in U937 leukemia cells. *J. Biol. Chem.*, 2011; 286: 27092–27102.

[174] Kulkarni A., Dangat K., Kale A., Sable P., Chavan-Gautam P., Joshi S. Effects of altered maternal folic acid, vitamin B12 and docosahexaenoic acid on placental global DNA methylation patterns in Wistar rats. *PLoS One,* 2011; 6: e17706.

[175] Gu T., Gu H. F., Hilding A., Sjöholm L. K., Ostenson C. G., Ekström T. J., Brismar K. Increased DNA methylation levels of the insulin-like growth factor binding protein 1 gene are associated with type 2 diabetes in Swedish men. *Clin. Epigenetics*, 2013 Nov. 19; 5(1): 21.

[176] Maghbooli Z., Hossein-nezhad A., Larijani B., Amini M., Keshtkar A. Global DNA methylation as a possible biomarker for diabetic retinopathy. *Diabetes Metab. Res. Rev.,* 2015 Feb.; 31(2): 183-9.

[177] Tang L., Wang L., Ye H., Xu X., Hong Q., Wang H., Xu L., Bu S., Zhang L., Cheng J., Liu P., Ye M., Mai Y., Duan S. BCL11A gene DNA methylation contributes to the risk of type 2 diabetes in males. *Exp. Ther. Med.*, 2014 Aug.; 8(2): 459-463.

[178] Voisin S., Almén M. S., Moschonis G., Chrousos G. P., Manios Y., Schiöth H. B. Dietary fat quality impacts genome-wide DNA methylation patterns in a cross-sectional study of Greek preadolescents. *Eur. J. Hum. Genet.*, 2015 May; 23(5): 654-62.

[179] Dick K. J., Nelson C. P., Tsaprouni L., Sandling J. K., Aïssi D., Wahl S., Meduri E., Morange P. E., Gagnon F., Grallert H., Waldenberger M., Peters A., Erdmann J., Hengstenberg C., Cambien F., Goodall A. H., Ouwehand W. H., Schunkert H., Thompson J. R., Spector T. D., Gieger C., Trégouët D. A., Deloukas P., Samani N. J. DNA methylation and body-mass index: a genome-wide analysis. *Lancet,* 2014 Jun. 7; 383 (9933): 1990-8.

INDEX

B

C

D

E

F

G

H

I

J

K

L

M

N

O

P

R

S

T

U

V

W

X

Y

Z